AF324297

METAMATERIALS AND NANOPHOTONICS

Principles, Techniques and Applications

METAMATERIALS AND NANOPHOTONICS

Principles, Techniques and Applications

Kosmas L. Tsakmakidis
National and Kapodistrian University of Athens, Greece

Konstantinos G. Baskourelos
National and Kapodistrian University of Athens, Greece

Marek S. Wartak
Wilfrid Laurier University, Canada

World Scientific

NEW JERSEY · LONDON · SINGAPORE · BEIJING · SHANGHAI · HONG KONG · TAIPEI · CHENNAI · TOKYO

Published by

World Scientific Publishing Co. Pte. Ltd.

5 Toh Tuck Link, Singapore 596224

USA office: 27 Warren Street, Suite 401-402, Hackensack, NJ 07601

UK office: 57 Shelton Street, Covent Garden, London WC2H 9HE

Library of Congress Control Number: 2022942047

British Library Cataloguing-in-Publication Data
A catalogue record for this book is available from the British Library.

METAMATERIALS AND NANOPHOTONICS
Principles, Techniques and Applications

ISBN 978-981-126-186-2 (hardcover)
ISBN 978-981-126-187-9 (ebook for institutions)
ISBN 978-981-126-188-6 (ebook for individuals)

For any available supplementary material, please visit
https://www.worldscientific.com/worldscibooks/10.1142/13010#t=suppl

Desk Editor: Joseph Ang

Typeset by Stallion Press
Email: enquiries@stallionpress.com

Kosmas Tsakmakidis dedicates this book to his late advisor,
Professor Theodoros D. Tsiboukis.

Konstantinos Baskourelos dedicates this book to the memory of his father,
who was his first teacher in mathematics.

Marek Wartak dedicates this book to P. Rusek—friend, physicist and
mountaineer—who lost his life in the French Alps.

Contents

Introduction

An important scientific breakthrough in modern electromagnetics and optics has been the conception and practical implementation of man-made materials exhibiting negative electric permittivity, magnetic permeability and refractive index, known also as 'double-negative' or 'left-handed' or, simply, 'negative-refractive-index' 'metamaterials'. Their potential applicability in diverse realms of science (such as telecommunications, radars and defence, nanolithography with light, microelectronics, medical imaging, and so forth) has prompted an overwhelming excitement within the wider scientific community. This book is an attempt to capture and detail in a pedagogical manner the essential physics behind the science of metamaterials and the closely related field of nanophotonics.

The book begins, in **Chapter 1**, with an introduction to dielectric waveguide theory, which we shall find useful when we later examine, in Chapter 6, the slowing and stopping of light in negative-index metamaterial waveguides. Chapter 1 introduces the relations that describe the vectorial field propagation in waveguides and summarises several semi-analytic methods that are sometimes used in the investigation of such structures. Particular emphasis is paid upon the derivation and finite-difference discretisation of the vectorial wave equations, as well as on the development of fully vectorial (FV) mode-solvers, which nowadays constitute an indispensable tool for the analysis of three-dimensional waveguides.

Chapter 2 introduces the finite-difference time-domain (FDTD) method for the numerical solution of Maxwell's equations of electromagnetism. This method is widely used by the scientific community in assessing and optimising the performance of optical devices and is also extensively deployed in the analysis of metamaterials. The method is also used later on, in Chapter 3, to show that a negative-index slab can bring light to a double focus, but in Chapter 2 we are mainly concerned with the theoretical aspects behind the operation of this method — in particular, its stability. We, further, study the recently introduced, exciting concepts of 'exact' and 'nonstandard' finite-differences, which can lead to enhanced computational accuracy and overall algorithmic efficiency.

In **Chapter 3** we provide an introduction to the theory of light propagation inside passive bulk metamaterials and waveguides. We establish a number of crucial issues pertaining to the properties of such media such as why the sign of the refractive index becomes negative when both ε and μ are negative, to which direction does a plane electromagnetic wave refracts inside an isotropic negative-index medium, and so forth. Upon deriving the Fresnel equations for the reflection and transmission coefficients of propagating *and* evanescent waves at the interface between a normal dielectric and a negative-index medium, we study in some detail one of the most remarkable properties of negative-index metamaterials — namely the fact that a planar slab made of such a material can act as a 'perfect' lens, able to bring into focus and restore both the propagating *and* the evanescent components of an object source. Since surface waves that decay evanescently at the two interfaces of a negative-index slab play a crucial role to the manifestation of the 'perfect' lensing effect, we then proceed by studying in detail and classifying all surface plasmon polariton (SPP) modes supported by such slab structures. We, also, study the oscillatory modes propagating in these waveguides and identify that some of them can attain zero group velocity, exist alone, and be efficiently excited — results that will be proven useful in the discussions that will be presented later, in Chapter 6.

In **Chapter 4**, the fundamental physical mechanisms related to propagation of electromagnetic waves in plasmonic and negative refractive index waveguides are discussed. Initially a method for calculating modal dispersion in arbitrary planar waveguides is presented and three important types of waveguide modes are highlighted. The optical properties of metals and negative refractive index metamaterials are then detailed. Finally, the propagation characteristics of light in a number of plasmonic and negative-index waveguide geometries and their applicability to stopped light are detailed.

Chapter 5 is concerned with how to create ultra-low- or zero-loss metamaterials over a continuous range of frequencies. Key published works are initially reviewed on how to construct metamaterials exhibiting negative electric permittivity or magnetic permeability in the microwave regime. We then introduce a methodology, utilizing equivalent electric circuit configurations with multiple degrees of freedom, which is shown to result in ultra-low-loss metamaterials exhibiting negative magnetic permeability. It is explained how with such metamaterial configurations one is able to obtain metamaterial magnetism with large figures-of-merit (defined as the ratio of the real part of μ to imaginary part of μ).

Chapter 6 introduces a, by now widely popular, method for decelerating light using metamaterial waveguide heterostructures. In particular, it is shown that an electromagnetic pulse propagating along an adiabatically, axially varying waveguide with a negative-index core (or cladding) can be entirely stopped and stored inside it, with each frequency component stopping at a different point inside the tapered waveguide. This configuration thus leads to the spatial separation and 'trapping' of the frequency components of a pulse, forming what has come to be known as a 'trapped rainbow'.

The next two chapters, **Chapters 7** and **8**, detail in a pedagogical manner the crucial impact that dissipative losses and surface roughness have on the attainment of ultraslow and stopped light in passive metamaterial or plasmonic waveguides.

Chapter 9 introduces the concepts of Maxwell-Bloch theory and gain in active metamaterials — which are usually needed in order to compensate for the dissipative losses. We then proceed with **Chapters 10** and **11** which detail the basic background behind active and quantum plasmonics — both enduringly timely in the wider field of nanophotonics. The final chapter of the book, **Chapter 12**, is devoted to a particularly interesting subject of the previous two chapters, that is, the theory and applications of nanolasers, capable of emitting coherent light well below the diffraction limit.

All in all, the book will be of interest to anyone interested in computational electrodynamics and quantum optics, metamaterials, nanophotonics, slow and stopped light, and (nano)lasers. It includes topics (such as the study of surface roughnesses) and techniques (such as the nonstandard FDTD method) that are usually not covered, but which are indispensable for analyzing and understanding the performance of realistic devices. It is our hope that young students and mature researchers alike will have something interesting and useful to find in its pages, and we welcome feedback for the future editions of the book.

Kosmas L. Tsakmakidis
Konstantinos G. Baskourelos
Marek S. Wartak

Chapter 1

Vectorial Field Theory and Modelling of 3D Dielectric Waveguides

1.1 Introduction

The basic principles and underlying field theory of two- (2D) and three-dimensional (3D) dielectric waveguides are introduced in this chapter. First, we summarise Maxwell's equations, the vectorial wave equations and the associated boundary conditions for the electromagnetic fields. We, then, concisely describe several analytic methods, including the effective index method and Marcatili's method, commonly deployed for prompt investigations of 3D waveguiding structures. Finally, we describe in some detail the rigorous semi- (SV) and fully-vectorial (FV) finite-difference frequency-domain (FDFD) mode-solving techniques for the acquisition of the electromagnetic eigenmodes in a dielectric waveguide with an arbitrary 2D cross-section.

1.2 Maxwell's Equations of Electromagnetism

For a homogeneous, isotropic, non-dispersive, linear and achiral medium the electric field $\mathbf{E}$ (V/m), magnetic field $\mathbf{H}$ (A/m), electric flux density $\mathbf{D}$ (Cb/m^2) and magnetic flux $\mathbf{B}$ (A/m^2) are related, in the time-domain, through the following macroscopic constitutive relationships [Born and Wolf (1999)], [Jackson (1975)]:

$$\mathbf{D} = \varepsilon\mathbf{E}, \tag{1.1}$$

$$\mathbf{B} = \mu\mathbf{H}, \tag{1.2}$$

where the electric permittivity ε and the magnetic permeability μ are defined as:

$$\varepsilon = \varepsilon_0\varepsilon_r, \tag{1.3}$$

$$\mu = \mu_0\mu_r. \tag{1.4}$$

In these expressions, ε_0 and μ_0 are, respectively, the electric permittivity and magnetic permeability of vacuum. Moreover, with ε_r and μ_r we denote the relative (to the vacuum) permittivity and permeability of the material. As we shall see later, in Chapter 5, the relative permeability μ_r of several artificial, non-magnetic,

1

dielectrics may exceed 1, or even become negative (e.g. $\mu_r = -1$). Denoting the velocity of light in vacuum with c, one may then obtain:

$$\varepsilon_0 = \frac{1}{c^2 \mu_0} \simeq 8.85 \cdot 10^{-12} \ \text{F/m}, \tag{1.5}$$

$$\mu_0 = 4\pi \ \cdot 10^{-7} \ \text{H/m}. \tag{1.6}$$

It should also be noted here that in a conductive dielectric material the current density $\mathbf{J}$ (A/m^2) relates to the electric field $\mathbf{E}$ through the following relationship:

$$\mathbf{J} = \sigma\mathbf{E}, \tag{1.7}$$

where σ is the conductivity of the material.

Armed with the above relations, we may now proceed to analytically describe the interactions between the electric and magnetic fields in such media. In general, the electromagnetic fields satisfy the following set of coupled, first-order, differential equations, introduced by Maxwell more than a century ago [Maxwell (1891)]:

$$\nabla \times \mathbf{E} = -\frac{\partial \mathbf{B}}{\partial t}, \tag{1.8}$$

$$\nabla \times \mathbf{B} = \frac{\partial \mathbf{D}}{\partial t} + \mathbf{J}. \tag{1.9}$$

By means of the vector identity $\nabla \cdot (\nabla \times \mathbf{A}) = 0$ which holds for an arbitrary vector $\mathbf{A}$, we can rewrite Eqs. (1.8) and (1.9) in the following form:

$$\nabla \cdot \mathbf{B} = 0, \tag{1.10}$$

$$\nabla \cdot \mathbf{D} = \varrho. \tag{1.11}$$

In deriving Eq. (1.11), it was assumed that the current density $\mathbf{J}$ is related to the charge density ϱ (Cb/m^2) as follows:

$$\nabla \cdot \mathbf{J} = -\frac{\partial \varrho}{\partial t}. \tag{1.12}$$

1.3   The Wave Equations

Let us now assume that an electromagnetic field oscillates at a single angular frequency ω (rad/m). Then, the electric and magnetic field, as well as the electric and magnetic flux density, each designated with a vector $\mathbf{A}$, may be expressed as:

$$\mathbf{A}(\mathbf{r}, t) = \text{Re}\left[\bar{\mathbf{A}}(\mathbf{r}) \exp(j\omega t)\right]. \tag{1.13}$$

In the following, and for the sake of simplicity, we will denote the phasors $\bar{\mathbf{E}}$, $\bar{\mathbf{H}}$, $\bar{\mathbf{D}}$ and $\bar{\mathbf{B}}$ as $\mathbf{E}$, $\mathbf{H}$, $\mathbf{D}$ and $\mathbf{B}$ respectively. Based on Eq. (1.13), we can rewrite Eq. (1.8) to Eq. (1.11) in the frequency domain as:

$$\nabla \times \mathbf{E} = -j\omega\mathbf{B} = -j\omega\mu_0\mathbf{H}, \tag{1.14}$$

$$\nabla \times \mathbf{H} = j\omega\mathbf{D} = j\omega\varepsilon\mathbf{E}, \tag{1.15}$$

$$\nabla \cdot \mathbf{H} = \mathbf{0}, \tag{1.16}$$

$$\nabla \cdot (\varepsilon_r \mathbf{E}) = \mathbf{0}, \tag{1.17}$$

where it is assumed that $\mu_r = 1$ and $\varrho = 0$.

1.3.1 *Vectorial Wave Equation for the Electric Field E*

We, now, derive in some detail the fully-vectorial (FV) wave equation for the electric field. This equation will be of particular relevance when, later in this chapter, we shall be concerned with the various mode-solving techniques used in the analysis of 3D dielectric waveguides. Applying the vectorial rotation operator to Eq. (1.14), we obtain:

$$\nabla \times (\nabla \times \mathbf{E}) = -j\omega\mu_0 \nabla \times \mathbf{H}. \tag{1.18}$$

Using the vectorial formula:

$$\nabla \times (\nabla \mathbf{A}) = \nabla(\nabla \cdot \mathbf{A}) - \nabla^2 \mathbf{A}, \tag{1.19}$$

with the symbol ∇^2 denoting the 3D Laplacian operator, we can rewrite the left-hand side of Eq. (1.18) as:

$$\nabla(\nabla \cdot \mathbf{E}) - \nabla^2 \mathbf{E}. \tag{1.20}$$

We note at this point that Eq. (1.17) can also be written as:

$$\nabla \cdot (\varepsilon_r \mathbf{E}) = \nabla\varepsilon_r \cdot \mathbf{E} + \varepsilon_r \nabla \cdot \mathbf{E}, \tag{1.21}$$

from where one may obtain:

$$\nabla \cdot \mathbf{E} = -\frac{\nabla\varepsilon_r}{\varepsilon_r} \cdot \mathbf{E}. \tag{1.22}$$

Therefore, the left-hand side of Eq. (1.18) becomes:

$$\nabla\left(\frac{\nabla\varepsilon_r}{\varepsilon_r} \cdot \mathbf{E}\right) - \nabla^2 \mathbf{E}. \tag{1.23}$$

On the other hand, by means of Eq. (1.15), the right-hand side of Eq. (1.18) becomes:

$$k_0^2 \varepsilon_r \mathbf{E}, \tag{1.24}$$

where k_0^2 is the wavenumber in vacuum and is expressed as:

$$k_0 = \omega\sqrt{\varepsilon_0\mu_0} = \frac{\omega}{c}. \tag{1.25}$$

Thus, for a dielectric medium with relative permittivity ε_r, the vectorial wave equation for the electric field $\mathbf{E}$ is given by:

$$\nabla^2 \mathbf{E} + \nabla\left(\frac{\nabla\varepsilon_r}{\varepsilon_r} \cdot \mathbf{E}\right) + k_0^2 \varepsilon_r \mathbf{E} = \mathbf{0}. \tag{1.26}$$

Moreover, if we assume that the wavenumber k in the aforesaid medium is:

$$k = k_0 n = k_0\sqrt{\varepsilon_r} = \omega\sqrt{\varepsilon_0\varepsilon_r\mu_0} = \omega\sqrt{\varepsilon\mu_0}, \tag{1.27}$$

we may rewrite Eq. (1.26) as:

$$\nabla^2 \mathbf{E} + \nabla\left(\frac{\nabla\varepsilon_r}{\varepsilon_r} \cdot \mathbf{E}\right) + k^2 \varepsilon_r \mathbf{E} = \mathbf{0}. \tag{1.28}$$

For a homogeneous medium, i.e. one in which the relative permittivity ε_r is constant, the vectorial wave equation can be reduced to the well-known, scalar, Helmholtz equation for the **E**-field:

$$\nabla^2 \mathbf{E} + k^2 \mathbf{E} = \mathbf{0}. \tag{1.29}$$

1.3.2 *Vectorial Wave Equation for the Magnetic Field H*

In the analysis of dielectric waveguides, the fully-vectorial wave equation for the magnetic field is sometimes preferred over the corresponding equation for the electric field, owing to the absence of field discontinuities at the dielectric interfaces that could hinder high-accuracy calculations. Here, we show how the FV wave equation for the **H**-field can be obtained from Eq. (1.14) to Eq. (1.17). To this endeavour, we start by applying the vectorial rotation operator $\nabla\times$ to Eq. (1.15):

$$\nabla \times (\nabla \times \mathbf{H}) = j\omega\varepsilon_0 \nabla \times (\varepsilon_r \mathbf{E}). \tag{1.30}$$

We, now, successively have:

$$\begin{aligned} \nabla(\nabla \cdot \mathbf{H}) - \nabla^2 \mathbf{H} &= j\omega\varepsilon_0(\nabla\varepsilon_r \times \mathbf{E} + \varepsilon_r \nabla \times \mathbf{E}) \\ &= j\omega\varepsilon_0(\nabla\varepsilon_r \times \mathbf{E}) + j\omega\varepsilon_0\varepsilon_r(-j\omega\mu_0\mathbf{H}) \\ &= j\omega\varepsilon_0(\nabla\varepsilon_r \times \mathbf{E}) + k_0^2\varepsilon_r\mathbf{H}. \end{aligned} \tag{1.31}$$

Based on Eqs. (1.15) and (1.16), one may further obtain:

$$\mathbf{E} = \frac{1}{j\omega\varepsilon_0\varepsilon_r}\nabla \times \mathbf{H}, \tag{1.32}$$

which, inserted into Eq. (1.31), leads to

$$\nabla^2\mathbf{H} + \frac{\nabla\varepsilon_r}{\varepsilon_r} \times (\nabla \times \mathbf{H}) + k_0^2\varepsilon_r\mathbf{H} = \mathbf{0}. \tag{1.33}$$

Using Eq. (1.27), we can rewrite Eq. (1.33) as:

$$\nabla^2\mathbf{H} + \frac{\nabla\varepsilon_r}{\varepsilon_r} \times (\nabla \times \mathbf{H}) + k^2\mathbf{H} = \mathbf{0}. \tag{1.34}$$

For a homogeneous medium, i.e. one in which the relative permittivity ε_r is constant, the vectorial wave equation can be reduced to the well-known, scalar, Helmholtz equation for the **H**-field:

$$\nabla^2\mathbf{H} + k^2\mathbf{H} = \mathbf{0}. \tag{1.35}$$

For a dielectric waveguide whose structure is uniform in the z direction (i.e. longitudinally invariant in z) the derivative of an electromagnetic field component with respect to z may be written as:

$$\frac{\partial}{\partial z} = -j\beta, \tag{1.36}$$

where β is the longitudinal propagation constant of the field. The ratio of the propagation constant β to the wavenumber in vacuum, k_0, is usually referred to as the effective index:

$$n_{eff} = \frac{\beta}{k_0}. \tag{1.37}$$

The physical meaning of the propagation constant β is the phase accumulated (or rotated) per unit propagation distance. Therefore, the effective index, n_{eff}, can be interpreted as the ratio of a phase rotation in a medium to the phase rotation in vacuum.

We may summarise the Helmholtz equations for the electric and magnetic field as:

$$\nabla_\perp^2 \mathbf{U} + (k^2 - \beta^2)\mathbf{U} = \mathbf{0}, \tag{1.38}$$

or

$$\nabla_\perp^2 \mathbf{U} + k_0^2(\varepsilon_r - n_{eff}^2)\mathbf{U} = \mathbf{0}, \tag{1.39}$$

where $\mathbf{U} = \mathbf{E}$ or $\mathbf{H}$, and $\nabla_\perp^2 = \frac{\partial^2}{\partial x^2} + \frac{\partial^2}{\partial y^2}$.

1.4 Boundary Conditions for the Electromagnetic Fields

The boundary conditions required for the electromagnetic fields, in the case of two adjoining dielectric media 1 and 2, are summarised as follows.

(a) Tangential components of the electric fields are continuous, such that:

$$E_{1t} = E_{2t}. \tag{1.40}$$

(b) When no current flows on the media interface, tangential components of the magnetic fields are continuous, such that:

$$H_{1t} = H_{2t}. \tag{1.41}$$

By contrast, when a current flows on the interface, the magnetic field is discontinuous. The magnetic components at each side of the interface are related to the current density as follows:

$$H_{1t} - H_{2t} = J_S. \tag{1.42}$$

Since the magnetic field and the current are perpendicular to each other, the vectorial representation is:

$$\mathbf{n} \times (\mathbf{H}_1 - \mathbf{H}_1) = \mathbf{J}_S, \tag{1.43}$$

where the vector $\mathbf{n}$ is the unit normal vector at the dielectrics' interface.

(c) When there is no charge on the interface, the normal components of the electric flux densities are continuous, such that:

$$D_{1n} = D_{2n}. \tag{1.44}$$

By contrast, when there are charges on the surface, the electric flux densities are discontinuous and are related to the charge density, ρ_S, as follows:

$$D_{1n} - D_{2n} = \rho_S \, . \tag{1.45}$$

(d) Normal components of the magnetic flux densities are continuous, such that:

$$B_{1n} = B_{2n} \, . \tag{1.46}$$

Before discussing the numerical, mode-solving, techniques for the study of dielectric waveguides it is instructive at this point to describe the corresponding analytic methodologies. First, we highlight an exact method for the analysis of a three-layer, asymmetric, slab waveguide. We shall return to this method later, in Chapter 6, when we shall discuss about the mode theory of generalized waveguides with negative refractive index. Here, we shall instead examine two further common methods, namely the effective index method and Marcatili's method, which are commonly deployed for the mode analysis of rectangular dielectric waveguides with 2D cross-section.

1.5 Exact Analysis of a Three-Layer Slab Dielectric Waveguide

Let us consider a three-layer, planar, dielectric slab waveguide [Snyder and Love (1983)] with refractive indices n_1, n_2 and n_3, as illustrated in Fig. 1.1. The direction of propagation is along the $+z$ direction. The structure is uniform in the y and z directions (i.e. $\partial/\partial y = 0$ and $\partial/\partial z = -j\beta$, see Eq. (1.36)). Region 2 is the core layer that has refractive index higher than those of the cladding layers. Since media 1 and 3 are semi-infinite (and uniform) and the tangential electromagnetic field components are, indeed, connected at the interfaces between adjacent media, as mentioned in the previous paragraph, it suffices to start our analysis with the scalar Helmholtz equations (1.38) or (1.39) applied to each dielectric region.

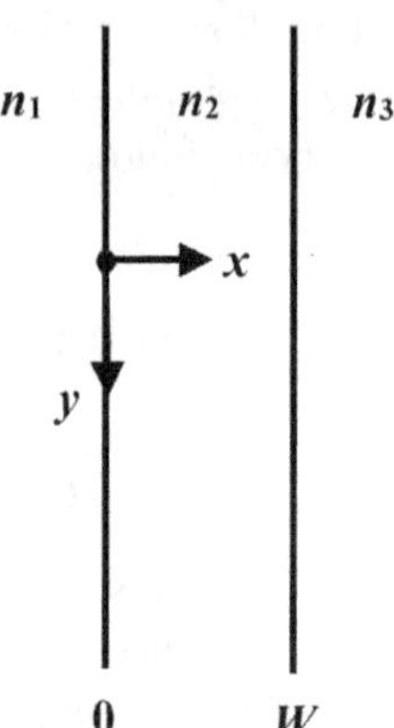

Figure 1.1: Schematic illustration of a three-layer, asymmetric, dielectric waveguide.

Next, we discuss the two types of modes that propagate in this structure. To this end, we note from Eqs. (1.26) and (1.33) that the transverse electric and magnetic field components *decouple* owing to the absence of variation in the materials' refractive indices along the y direction. It, then, swiftly turns out directly from Maxwell's equations (1.14) and (1.15) that we may distinguish between two different electromagnetic modes: the transverse electric mode (TE mode), in which the electric field is not in the longitudinal direction ($E_z = 0$) but in the transverse direction ($E_y \neq 0$), and the transverse magnetic mode (TM mode), in which the magnetic field is not in the longitudinal direction ($H_z = 0$) but in the transverse direction ($H_y \neq 0$). In this paragraph, we shall focus on the somewhat more involved TM case. From Maxwell's equations it turns out that, in this case, it is $H_x = H_z = E_y$. The two electric field components are given by:

$$E_x = \frac{\beta}{\omega \varepsilon_0 \varepsilon_r} H_y, \tag{1.47}$$

$$E_z = -\frac{j}{\omega \varepsilon_0 \varepsilon_r} \frac{\partial H_y}{\partial x}, \tag{1.48}$$

and the sole magnetic field component, H_y, fulfils the scalar wave equation:

$$\frac{d^2 H_y}{dx^2} + k_0^2 (\varepsilon_r - n_{eff}^2) H_y = 0. \tag{1.49}$$

The principal magnetic field component in the three dielectric regions can, therefore, be expressed as:

$$H_y = \begin{cases} C_1 \exp(\gamma_1 x), & x \leq 0 \\ C_2 \exp(\gamma_2 x + \alpha), & 0 \leq x \leq W \\ C_3 \exp(-\gamma_3(x - W)), & x \geq W \end{cases} \tag{1.50}$$

where $\gamma_1 = k_0(n_{eff}^2 - n_1^2)$, $\gamma_2 = k_0(n_2^2 - n_{eff}^2)$ and $\gamma_3 = k_0(n_{eff}^2 - n_3^2)$.

Imposing the boundary conditions on the tangential field components at $x = 0$ and $x = W$, results in the following characteristic equation:

$$\gamma_2 W = \arctan\left(\frac{\varepsilon_{r2}}{\varepsilon_{r1}} \frac{\gamma_1}{\gamma_2}\right) + \arctan\left(\frac{\varepsilon_{r2}}{\varepsilon_{r3}} \frac{\gamma_3}{\gamma_2}\right) + m\pi, \quad m = 0, 1, 2, \dots. \tag{1.51}$$

Using the identity:

$$\arctan\left(\frac{y}{x}\right) = \frac{\pi}{2} - \arctan\left(\frac{x}{y}\right), \tag{1.52}$$

one may also obtain:

$$\gamma_2 W = -\arctan\left(\frac{\varepsilon_{r1}}{\varepsilon_{r2}} \frac{\gamma_2}{\gamma_1}\right) - \arctan\left(\frac{\varepsilon_{r3}}{\varepsilon_{r2}} \frac{\gamma_2}{\gamma_3}\right) + (m+1)\pi, \quad m = 0, 1, 2, \dots. \tag{1.53}$$

Following a similar (dual) analysis for the TE modes, for which $E_x = E_z = H_y = 0$, one discovers that the so obtained characteristic equation is akin to Eqs. (1.51) and (1.53), with the sole difference being that the ratio of the relative permittivities appearing in the argument of the inverted tangents is, now, removed.

1.6   Effective Index Method

For the analysis of dielectric waveguides that are not uniform in the transverse, y, direction (i.e. they have a 2D cross-section), the analytic method reported before cannot be efficiently applied owing to the difficulty of matching the tangential field components at the dielectric *corners*. A quasi-analytic method, known as Marcatili's method [Marcatili (1969)], is usually deployed in these situations and will be described in some detail in the next paragraph. Here, we shall focus on the effective index method, which allows one to analyse 2D cross-sectional dielectric waveguides by simply repeating the 1D slab waveguide analysis highlighted above.

Figure 1.2 shows an example of a 3D optical waveguide (uniform in z direction) and concisely illustrates the concept of the effective index method. We start by considering the scalar wave equation:

$$\frac{\partial^2 \varphi(x,y)}{\partial x^2} + \frac{\partial^2 \varphi(x,y)}{\partial y^2} + k_0^2(\varepsilon_r(x,y) - n_{eff}^2)\varphi(x,y) = 0, \tag{1.54}$$

where n_{eff} is the eigenmode's effective index to be obtained. We, then, separate the wave function, $\varphi(x,y)$ into two functions:

$$\varphi(x,y) = f(x)g(y). \tag{1.55}$$

Equation (1.55) implies that there is no coupling between the transverse (x and y) electromagnetic field components. This assumption, strictly speaking, is not true. However, it may lead to acceptable results on the condition that the dominant electric and magnetic field components are *well confined* inside the central/guiding layer (core), having negligible magnitude near the dielectric corners. As a result, the present method leads to reasonable results, typically, for the first couple of eigenmodes, or for weakly-guiding structures, for which the field-coupling at the corners is small.

Substituting Eq. (1.55) into Eq. (1.54) and dividing the resultant equation by the wave function we obtain:

$$\frac{1}{f(x)}\frac{d^2 f(x)}{dx^2} + \frac{1}{g(y)}\frac{d^2 g(y)}{dy^2} + k_0^2(\varepsilon_r(x,y) - n_{eff}^2) = 0. \tag{1.56}$$

Setting the sum of the *second* and *third* terms of Eq. (1.56) equal to $k_0^2 N^2(x)$ we arrive at:

$$\frac{1}{g(y)}\frac{d^2 g(y)}{dy^2} + k_0^2 \varepsilon_r(x,y) = k_0^2 N^2(x). \tag{1.57}$$

Inspection of Eqs. (1.56) and (1.57) reveals that the sum of the *first* and *fourth* terms is equal to $-k_0^2 N^2(x)$:

$$\frac{1}{f(x)}\frac{d^2 f(x)}{dx^2} - k_0^2 n_{eff}^2 = -k_0^2 N^2(x). \tag{1.58}$$

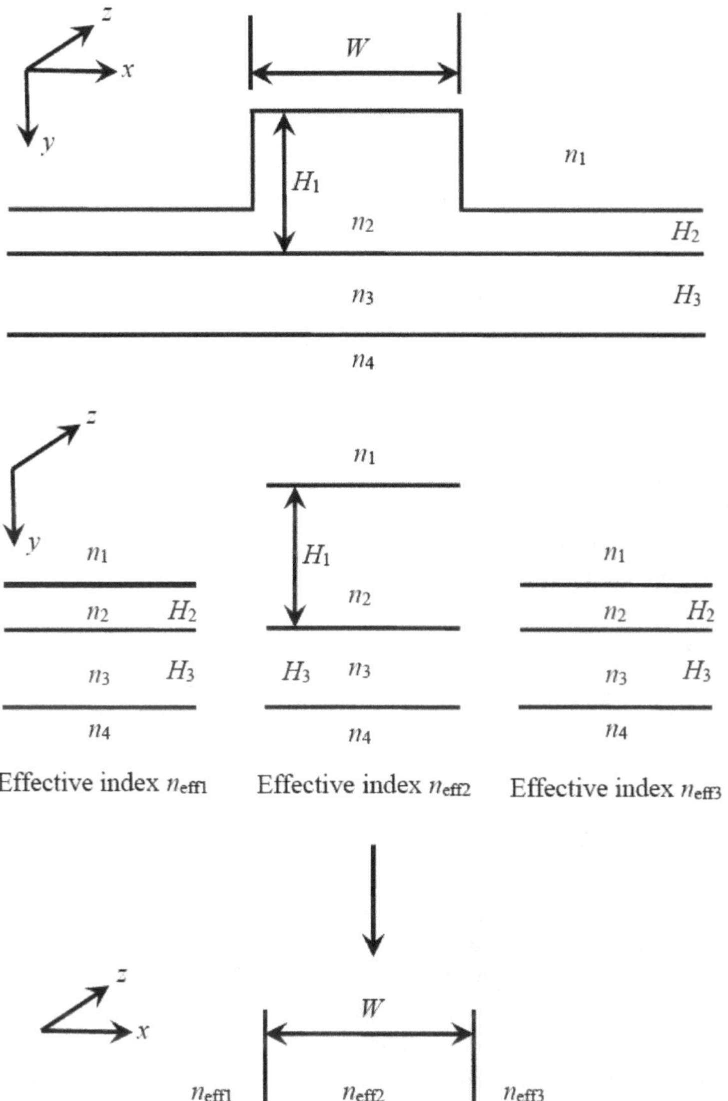

Figure 1.2: Schematic illustration of the "effective index method" concept.

Therefore, via the aforementioned procedures, we obtain the following two independent equations:

$$\frac{d^2g(y)}{dy^2} + k_0^2\big(\varepsilon_r(x,y) - N^2(x)\big)g(y) = 0, \tag{1.59}$$

and

$$\frac{d^2f(x)}{dx^2} + k_0^2\big(N^2(x) - n_{eff}^2\big)f(x) = 0. \tag{1.60}$$

Based on Eqs. (1.59) and (1.59), and on Fig. 1.2, the eigenmode's effective index calculation procedure can be summarised as follows:

(a) First, the 2D cross-sectional dielectric waveguide is replaced with a combination of 1D slab waveguides, as illustrated in Fig. 1.2.
(b) For each slab waveguide, we calculate the effective index along the y axis, based on the methodology described in Section 1.5.
(c) We derive an "equivalent" slab waveguide by placing the effective indices calculated in (b) along the x-axis.
(d) Finally, we calculate the investigated eigenmode's effective index by solving (using again the methodology described in Section 1.5) the model obtained in (c).

It is important to note here that, for the quasi TE-like eigenmode of the 3D waveguide shown in Fig. 1.2 (**E**-field parallel to the x-axis), one should first pursue the TE-mode analysis, followed by a corresponding *TM-mode* analysis, since in the final "equivalent" slab structure, the E_x-field will be perpendicular to the dielectric interfaces; hence the mode will, now, appear as a TM one. In a similar vein, for the quasi TM-like eigenmode (**E**-field parallel to the y-axis), one should first pursue the TM-mode analysis, followed by a *TE-mode analysis*, since in the "equivalent" slab waveguide, shown in the last part Fig. 1.2, the E_y-field will be parallel to the dielectric interfaces; hence the mode will appear as a TE one.

1.7   Marcatili's Method

Proposed by Marcatili, who at that time, exactly 53 years ago [Marcatili (1969)], was working at Bell Labs, the method that we shall discuss in this section, is still in wide use for prompt investigations of various types of rectangular dielectric waveguides. Figure 1.3 illustrates a cross-sectional view of a, so-called, buried optical waveguide. The core has refractive index n_1, width $2a$, and height $2b$. It is surrounded by cladding that has a refractive index n_2. Likewise the effective-index method, it is again assumed that the dominant components of the electric and magnetic fields are well-confined to the core and do not exist at all in the four shaded regions, shown in Fig. 1.3. Therefore, the continuity conditions for the tangential components need only be imposed at the interfaces of regions 1 and 2, 1 and 3, 1 and 4, and 1 and 5. The inherent limitations of the aforementioned assumption are essentially the same with the ones highlighted in Section 1.6.

Let us focus on the E_{pq}^x eigenmode, which has E_x and H_y as principal field components. Here, p and q are integers that, respectively, correspond to the number of peaks of optical power in the x and y directions. Thus, unlike ordinary mode-orders, which begin from 0 (for e.g. see Eqs. (1.51) and (1.53)), the present ones begin from 1.

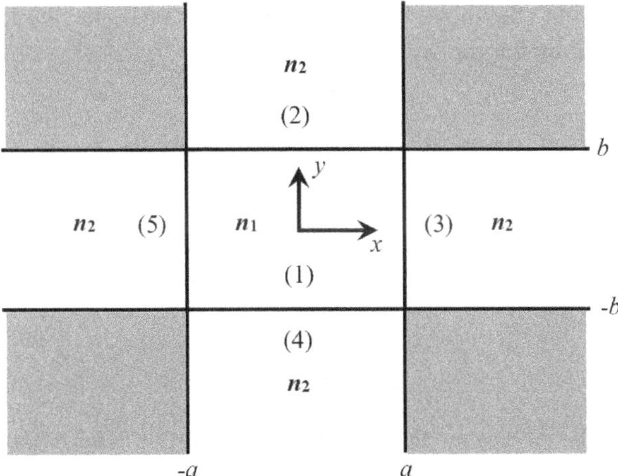

Figure 1.3: Geometry considered for the analysis of Marcatili's method.

The electric field of the E^x_{pq} eigenmode is assumed to be polarised in the x direction, which results in $E_y = 0$. Since the structure of the investigated waveguide is assumed to be invariant in the z direction, the derivative with respect to z is replaced by $-j\beta$ (see Eq. (1.36)). Then, after some algebraic manipulations starting from Maxwell's equations (1.8)–(1.10), one may obtain:

$$H_x = \frac{1}{\omega\mu_0\beta}\frac{\partial^2 E_x}{\partial x\partial y}, \tag{1.61}$$

$$H_y = \frac{1}{\omega\mu_0\beta}\left(\beta^2 E_x - \frac{\partial^2 E_x}{\partial x^2}\right), \tag{1.62}$$

$$H_z = \frac{1}{j\omega\mu_0}\frac{\partial E_x}{\partial y}, \tag{1.63}$$

$$E_z = \frac{1}{j\beta}\frac{\partial E_x}{\partial x}. \tag{1.64}$$

It, further, swiftly turns out that the principal field component, E_x, satisfies the 2D scalar wave equation:

$$\frac{\partial^2 E_x}{\partial x^2} + \frac{\partial^2 E_y}{\partial y^2} + k_0^2(\varepsilon_r - n^2_{eff})E_x = 0. \tag{1.65}$$

Since the dominant electric field component is a solution of Eq. (1.65), it should be described by the following expressions throughout all dielectric regions:

$$H_y = \begin{cases} C_1\cos(k_x x + a_1)\cos(k_y y + a_2), & x \in [-a, a],\ y \in [-b, b] \\ C_2\cos(k_x x + a_1)\exp(-\gamma_y(y - b)), & x \in [-a, a],\ y \in [b, \infty) \\ C_3\exp(-\gamma_x(x - a))\cos(k_y y + a_2), & x \in [a, \infty),\ y \in [-b, b] \\ C_4\cos(k_x x + a_1)\exp(\gamma_y(y + b)), & x \in [-a, a],\ y \in (-\infty, b] \\ C_5\exp(\gamma_x(x + a))\cos(k_y y + a_2), & x \in (-\infty, -a],\ y \in [-b, b] \end{cases} \tag{1.66}$$

Upon substituting the above functions into the wave equation (1.65), we obtain the following relations for the various wavenumbers:

$$k_x^2 + k_y^2 + \beta^2 = k_0^2 n_1^2, \tag{1.67}$$

$$k_x^2 - \gamma_y^2 + \beta^2 = k_0^2 n_2^2, \tag{1.68}$$

$$- \gamma_x^2 + k_y^2 + \beta^2 = k_0^2 n_2^2. \tag{1.69}$$

Subtracting Eq. (1.67) from Eq. (1.69), and Eq. (1.67) from Eq. (1.68), we arrive at the following expressions for the two decay constants:

$$\gamma_x^2 = k_0^2(n_1^2 - n_2^2) - k_x^2, \tag{1.70}$$

$$\gamma_y^2 = k_0^2(n_1^2 - n_2^2) - k_y^2. \tag{1.71}$$

The next step in the presented methodology is to impose the boundary conditions, specified by Eqs. (1.40) and (1.41), on the tangential E_x and H_z field components. By doing so at the $y = b$ boundary, we obtain the following characteristic equation:

$$k_y b + \alpha_2 = \arctan\left(\frac{\gamma_y}{k_y}\right) + m_1 \pi, \quad m_1 = 0, 1, 2, \ldots. \tag{1.72}$$

In a similar vein, by imposing Eqs. (1.40) and (1.41) on E_x and H_z at the $y = -b$ dielectric interface, we get:

$$k_y b - \alpha_2 = \arctan\left(\frac{\gamma_y}{k_y}\right) + m_2 \pi, \quad m_2 = 0, 1, 2, \ldots. \tag{1.73}$$

By adding the two last equations, one arrives at:

$$k_y b = \arctan\left(\frac{\gamma_y}{k_y}\right) + \frac{1}{2}(m - 1)\pi, \quad m = 1, 2, \ldots. \tag{1.74}$$

It, now, remains to implement the same conditions at the remaining two boundaries, namely $x = a$ and $x = -a$. By doing so at the first one, we obtain after some mathematical manipulations:

$$k_x a + \alpha_1 = \arctan\left(\frac{(k_0^2 n_1^2 - k_y^2)\gamma_x}{(k_0^2 n_2^2 - k_y^2)k_x}\right) + n_1 \pi, \quad n_1 = 0, 1, 2, \ldots. \tag{1.75}$$

On the other hand, from the boundary $x = -a$, we have:

$$k_x a - \alpha_1 = \arctan\left(\frac{(k_0^2 n_1^2 - k_y^2)\gamma_x}{(k_0^2 n_2^2 - k_y^2)k_x}\right) + n_2 \pi, \quad n_2 = 0, 1, 2, \ldots. \tag{1.76}$$

Thus, upon adding Eqs. (1.75) and (1.76), we finally get:

$$k_x a = \arctan\left(\frac{(k_0^2 n_1^2 - k_y^2)\gamma_x}{(k_0^2 n_2^2 - k_y^2)k_x}\right) + \frac{1}{2}(n - 1)\pi, \quad n = 1, 2, \ldots. \tag{1.77}$$

Since for most practical cases it is $k_0 n_{1,2} \gg k_y$, Eq. (1.77) further simplifies to:

$$k_x a = \arctan\left(\frac{\varepsilon_{r1}}{\varepsilon_{r2}^2}\frac{\gamma_x}{k_x}\right) + \frac{1}{2}(n - 1)\pi, \quad n = 1, 2, \ldots. \tag{1.78}$$

We may, now, summarise Marcatili's methodology for calculating the longitudinal propagation constant β of an eigenmode supported by a rectangular waveguide, as follows: First, we determine the value of k_x by numerically solving Eq. (1.78). Similarly, we obtain k_y by making use of Eqs. (1.71) and (1.74). Finally, we immediately obtain the value of β from Eq. (1.67).

It is noteworthy that while Eq. (1.74) corresponds to the characteristic equation of a TE mode in a three-layer slab waveguide parallel to the x axis, Eq. (1.78) corresponds to the characteristic equation of a TM mode in a three-layer slab waveguide parallel to the y axis. The situation is reminiscent of that described in the last paragraph of Section 1.6.

Following closely a dual analysis, one may yet obtain analogous results for the other mode, E_{pq}^y, in which the *magnetic* field is assumed to be polarised in the x direction, resulting in $H_y = 0$.

1.8   Finite-Difference Frequency-Domain Mode-Solvers

The introduction of semi-vectorial finite-difference (SV-FD) mode-solving techniques by Stern [Stern (1988a,b, 1991)] led to the development of numerically efficient waveguide analysis methods, which take polarisation into consideration and providing fairly accurate results. Therefore, they are widely used in the computer-aided design (CAD) analyses of dielectric waveguiding structures with (rectangular) 2D cross-section. Nonetheless, for increased accuracy, particularly when high-index contrast or circular cross-sectional structures (e.g. fibres) are considered, one inevitably has to resort to fully-vectorial computations, which take field-coupling into account as well.

In this section we will start by deriving the vectorial wave equations for each transverse electromagnetic field component. As a next step, the so-called semi-vectorial wave equations are obtained by one's ignoring the coupling terms. We shall proceed by formulating the finite-difference approximations for both types of wave equations. At this point we describe the structure of the resultant sparse matrix and highlight the procedure for solving the accompanying eigenvalue matrix equation [Xu *et al.* (1994)]. In the last part of the section, exemplary results will be presented and compared with published work.

1.8.1   *Vectorial Wave Equations for the E-Field Components*

As we have already shown, the vectorial equation for the electric field $\mathbf{E}$ is (see Eq. (1.26)):

$$\nabla^2 \mathbf{E} + \nabla\left(\frac{\nabla \varepsilon_r}{\varepsilon_r} \cdot \mathbf{E}\right) + k_0^2 \varepsilon_r \mathbf{E} = \mathbf{0}\,. \tag{1.79}$$

Let us consider a structure uniform in the z direction. In this case, the derivative of the relative permittivity with respect to z will be zero:

$$\frac{\partial \varepsilon_r}{\partial z} = 0. \tag{1.80}$$

Thus, the second term in Eq. (1.79) can be written as:

$$\nabla \left(\frac{\nabla \varepsilon_r}{\varepsilon_r} \cdot \mathbf{E} \right) = \nabla \left(\frac{1}{\varepsilon_r} \frac{\partial \varepsilon_r}{\partial x} E_x + \frac{1}{\varepsilon_r} \frac{\partial \varepsilon_r}{\partial y} E_y \right). \tag{1.81}$$

Upon substituting Eq. (1.81) into Eq. (1.79), we separate Eq. (1.79) into the x and y components. Thus, we obtain the following vectorial wave equations for the E_x and E_y field components:

$$\frac{\partial^2 E_x}{\partial x^2} + \frac{\partial}{\partial x} \left(\frac{1}{\varepsilon_r} \frac{\partial \varepsilon_r}{\partial x} E_x \right) + \frac{\partial^2 E_x}{\partial y^2} + \frac{\partial^2 E_x}{\partial z^2} + k_0^2 \varepsilon_r E_x + \frac{\partial}{\partial x} \left(\frac{1}{\varepsilon_r} \frac{\partial \varepsilon_r}{\partial y} E_y \right) = 0, \tag{1.82}$$

$$\frac{\partial^2 E_y}{\partial x^2} + \frac{\partial^2 E_y}{\partial y^2} + \frac{\partial}{\partial y} \left(\frac{1}{\varepsilon_r} \frac{\partial \varepsilon_r}{\partial y} E_y \right) + \frac{\partial^2 E_x}{\partial z^2} + k_0^2 \varepsilon_r E_x + \frac{\partial}{\partial y} \left(\frac{1}{\varepsilon_r} \frac{\partial \varepsilon_r}{\partial x} E_x \right) = 0. \tag{1.83}$$

Furthermore, we note that:

$$\frac{\partial}{\partial x} \left(\frac{1}{\varepsilon_r} \frac{\partial}{\partial x} (\varepsilon_r E_x) \right) = \frac{\partial^2 E_x}{\partial x^2} + \frac{\partial}{\partial x} \left(\frac{1}{\varepsilon_r} \frac{\partial \varepsilon_r}{\partial x} E_x \right), \tag{1.84}$$

and:

$$\frac{\partial}{\partial y} \left(\frac{1}{\varepsilon_r} \frac{\partial}{\partial y} (\varepsilon_r E_y) \right) = \frac{\partial^2 E_x}{\partial y^2} + \frac{\partial}{\partial y} \left(\frac{1}{\varepsilon_r} \frac{\partial \varepsilon_r}{\partial y} E_y \right). \tag{1.85}$$

Considering also Eq. (1.36), we may hence rewrite Eqs. (1.82) and (1.83) in the following compact form:

$$\frac{\partial}{\partial x} \left(\frac{1}{\varepsilon_r} \frac{\partial}{\partial x} (\varepsilon_r E_x) \right) + \frac{\partial^2 E_x}{\partial y^2} + (k_0^2 \varepsilon_r - \beta^2) E_x + \frac{\partial}{\partial x} \left(\frac{1}{\varepsilon_r} \frac{\partial \varepsilon_r}{\partial y} E_y \right) = 0, \tag{1.86}$$

$$\frac{\partial^2 E_y}{\partial x^2} + \frac{\partial}{\partial y} \left(\frac{1}{\varepsilon_r} \frac{\partial}{\partial y} (\varepsilon_r E_y) \right) + (k_0^2 \varepsilon_r - \beta^2) E_y + \frac{\partial}{\partial y} \left(\frac{1}{\varepsilon_r} \frac{\partial \varepsilon_r}{\partial x} E_x \right) = 0. \tag{1.87}$$

The last term in each of Eqs. (1.86) and (1.87) corresponds to the coupling between the x-directed electric field component, E_x, and the y-directed electric field component, E_y.

1.8.2   *Vectorial Wave Equations for the H-Field Components*

Let us next derive the corresponding components of the vectorial wave equation for the magnetic field $\mathbf{H}$. As it has already been shown, the vectorial equation for the magnetic field, is (see Eq. (1.33)):

$$\nabla^2 \mathbf{H} + \frac{\nabla \varepsilon_r}{\varepsilon_r} \times (\nabla \times \mathbf{H}) + k_0^2 \varepsilon_r \mathbf{H} = \mathbf{0}. \tag{1.88}$$

Here, the second term of Eq. (1.88) is investigated in detail. To this end, we recall that the considered structures are uniform in the z direction and that Eq. (1.80) holds. Hence, if $\mathbf{i}, \mathbf{j}, \mathbf{k}$ are, respectively, the unit vectors in the x, y and z directions, we may obtain:

$$\nabla \varepsilon_r \times (\nabla \times \mathbf{H}) = \begin{vmatrix} \mathbf{i} & \mathbf{j} & \mathbf{k} \\ \frac{\partial \varepsilon_r}{\partial x} & \frac{\partial \varepsilon_r}{\partial y} & 0 \\ (\nabla \times \mathbf{H})_x & (\nabla \times \mathbf{H})_y & (\nabla \times \mathbf{H})_z \end{vmatrix}$$

$$= \frac{\partial \varepsilon_r}{\partial y} (\nabla \times \mathbf{H})_z \, \mathbf{i} - \frac{\partial \varepsilon_r}{\partial x} (\nabla \times \mathbf{H})_z \, \mathbf{j}$$

$$+ \left(\frac{\partial \varepsilon_r}{\partial x} (\nabla \times \mathbf{H})_y - \frac{\partial \varepsilon_r}{\partial y} (\nabla \times \mathbf{H})_x \right) \mathbf{k}, \tag{1.89}$$

where we have used the following expressions:

$$(\nabla \times \mathbf{H})_x = \frac{\partial H_z}{\partial y} - \frac{\partial H_y}{\partial z}, \tag{1.90}$$

$$(\nabla \times \mathbf{H})_y = \frac{\partial H_x}{\partial z} - \frac{\partial H_z}{\partial x}, \tag{1.91}$$

$$(\nabla \times \mathbf{H})_z = \frac{\partial H_y}{\partial x} - \frac{\partial H_x}{\partial y}. \tag{1.92}$$

Substitution of Eqs. (1.90)–(1.92) into Eq. (1.89), results in:

$$\nabla \varepsilon_r \times (\nabla \times \mathbf{H}) = \frac{\partial \varepsilon_r}{\partial y} \left(\frac{\partial H_y}{\partial x} - \frac{\partial H_x}{\partial y} \right) \mathbf{i} - \frac{\partial \varepsilon_r}{\partial x} \left(\frac{\partial H_y}{\partial x} - \frac{\partial H_x}{\partial y} \right) \mathbf{j}$$

$$+ \left[\frac{\partial \varepsilon_r}{\partial x} \left(\frac{\partial H_x}{\partial z} - \frac{\partial H_z}{\partial x} \right) \frac{\partial \varepsilon_r}{\partial y} \left(\frac{\partial H_z}{\partial y} - \frac{\partial H_y}{\partial z} \right) \right] \mathbf{k}. \tag{1.93}$$

Substituting Eq. (1.93) into Eq. (1.88) and separating the result into the x and y components, one obtains the following vectorial wave equations for each magnetic field component:

$$\frac{\partial^2 H_x}{\partial x^2} + \frac{\partial^2 H_x}{\partial y^2} - \frac{1}{\varepsilon_r} \frac{\partial \varepsilon_r}{\partial y} \frac{\partial H_x}{\partial y} + \frac{\partial^2 H_x}{\partial z^2} + k_0^2 \varepsilon_r H_x + \frac{1}{\varepsilon_r} \frac{\partial \varepsilon_r}{\partial y} \frac{\partial H_y}{\partial x} = 0, \tag{1.94}$$

$$\frac{\partial^2 H_y}{\partial x^2} - \frac{1}{\varepsilon_r} \frac{\partial \varepsilon_r}{\partial x} \frac{\partial H_y}{\partial x} + \frac{\partial^2 H_y}{\partial y^2} + \frac{\partial^2 H_y}{\partial z^2} + k_0^2 \varepsilon_r H_x + \frac{1}{\varepsilon_r} \frac{\partial \varepsilon_r}{\partial x} \frac{\partial H_x}{\partial y} = 0. \tag{1.95}$$

Furthermore, because it is:

$$\varepsilon_r \frac{\partial}{\partial y} \left(\frac{1}{\varepsilon_r} \frac{\partial H_x}{\partial y} \right) = \frac{\partial^2 H_x}{\partial y^2} - \frac{1}{\varepsilon_r} \frac{\partial \varepsilon_r}{\partial y} \frac{\partial H_x}{\partial y}, \tag{1.96}$$

and:

$$\varepsilon_r \frac{\partial}{\partial x} \left(\frac{1}{\varepsilon_r} \frac{\partial H_y}{\partial x} \right) = \frac{\partial^2 H_y}{\partial x^2} - \frac{1}{\varepsilon_r} \frac{\partial \varepsilon_r}{\partial x} \frac{\partial H_y}{\partial x}, \tag{1.97}$$

we may rewrite Eqs. (1.94) and (1.95) in a more compact form, as follows:

$$\frac{\partial^2 H_x}{\partial x^2} + \varepsilon_r \frac{\partial}{\partial y}\left(\frac{1}{\varepsilon_r}\frac{\partial H_x}{\partial y}\right) + (k_0^2 \varepsilon_r - \beta^2)H_x + \frac{1}{\varepsilon_r}\frac{\partial \varepsilon_r}{\partial y}\frac{\partial H_y}{\partial x} = 0, \qquad (1.98)$$

$$\varepsilon_r \frac{\partial}{\partial x}\left(\frac{1}{\varepsilon_r}\frac{\partial H_y}{\partial x}\right) + \frac{\partial^2 H_x}{\partial y^2} + (k_0^2 \varepsilon_r - \beta^2)H_y + \frac{1}{\varepsilon_r}\frac{\partial \varepsilon_r}{\partial x}\frac{\partial H_x}{\partial y} = 0, \qquad (1.99)$$

where we have made use of Eq. (1.36).

Likewise what we mentioned in the corresponding equations for the electric field **E**, the last term in each of Eqs. (1.98) and (1.99) corresponds to the coupling between the x-directed magnetic field component, H_x, and the y-directed magnetic field component, H_y.

1.8.3   *Semi-Vectorial Wave Equations*

In the propagation equations of electromagnetic fields inside dielectric waveguides, the terms corresponding to the interaction between the x-directed electric field component, E_x, and the y-directed electric field component, E_y, i.e.:

$$\frac{\partial}{\partial x}\left(\frac{1}{\varepsilon_r}\frac{\partial \varepsilon_r}{\partial y}E_y\right) \quad \text{and} \quad \frac{\partial}{\partial y}\left(\frac{1}{\varepsilon_r}\frac{\partial \varepsilon_r}{\partial x}E_x\right)$$

in Eqs. (1.86) and (1.87), respectively, as well as the terms corresponding to the interaction between the x-directed magnetic field component, H_x, and the y-directed magnetic field component, H_y, i.e.:

$$\frac{1}{\varepsilon_r}\frac{\partial \varepsilon_r}{\partial y}\frac{\partial H_y}{\partial x} \quad \text{and} \quad \frac{1}{\varepsilon_r}\frac{\partial \varepsilon_r}{\partial x}\frac{\partial H_x}{\partial y}$$

in Eqs. (1.94) and (1.95), respectively, are quite often small. Ignoring these terms that account for the interaction, we can *decouple* the vectorial wave equations for the x- and y-directed field components and reduce them to *semi-vectorial* wave equations, which are considerably easier to solve and preserve the polarisation nature of the electromagnetic field, i.e. they take into account the discontinuities of an electric field component normal to a dielectric interface. As shown in Fig. 1.4, the semi-vectorial analyses may be divided into the quasi-TE mode analysis, in which the principal field component is E_x or H_y, and the quasi-TM mode analysis, in which the principal field component is E_y or H_x.

According to what was mentioned above, the semi-vectorial wave equations in the electric and magnetic field representation, respectively, for the quasi-TE mode are the following:

$$\frac{\partial}{\partial x}\left(\frac{1}{\varepsilon_r}\frac{\partial}{\partial x}(\varepsilon_r E_x)\right) + \frac{\partial^2 E_x}{\partial y^2} + (k_0^2 \varepsilon_r - \beta^2)E_x = 0, \qquad (1.100)$$

$$\varepsilon_r \frac{\partial}{\partial x}\left(\frac{1}{\varepsilon_r}\frac{\partial H_y}{\partial x}\right) + \frac{\partial^2 H_x}{\partial y^2} + (k_0^2 \varepsilon_r - \beta^2)H_y = 0. \qquad (1.101)$$

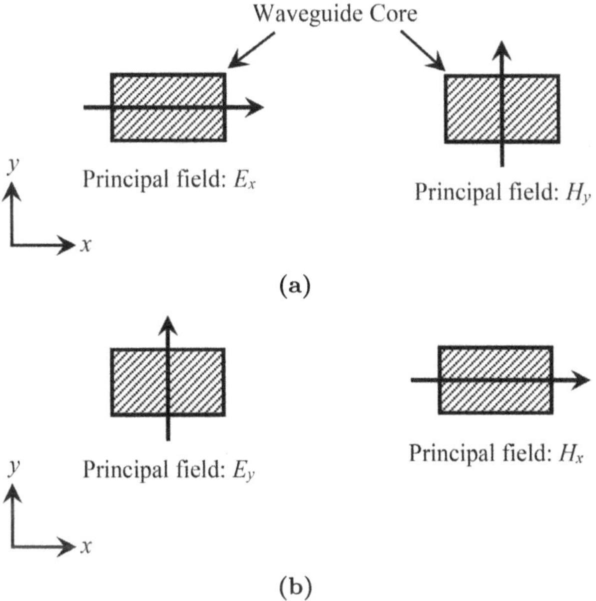

Figure 1.4: Principal electromagnetic field components for **(a)** quasi-TE, and **(b)** quasi-TM mode.

Likewise, the corresponding semi-vectorial wave equations in the electric and magnetic field representation, respectively, for the quasi-TM mode are:

$$\frac{\partial^2 E_y}{\partial x^2} + \frac{\partial}{\partial y}\left(\frac{1}{\varepsilon_r}\frac{\partial}{\partial y}(\varepsilon_r E_y)\right) + (k_0^2\varepsilon_r - \beta^2)E_y = 0, \tag{1.102}$$

$$\frac{\partial^2 H_x}{\partial x^2} + \varepsilon_r\frac{\partial}{\partial y}\left(\frac{1}{\varepsilon_r}\frac{\partial H_x}{\partial y}\right) + (k_0^2\varepsilon_r - \beta^2)H_x = 0. \tag{1.103}$$

One may notice at this point that in Eqs. (1.100)–(1.103), the derivatives of the relative permittivity, ε_r, with respect to the x and y coordinates are taken into consideration. If we, further, assume that these derivatives are zero, i.e.:

$$\frac{\partial\varepsilon_r}{\partial x} = 0 \quad \text{and} \quad \frac{\partial\varepsilon_r}{\partial y} = 0 \tag{1.104}$$

the semi-vectorial wave equations can be further reduced to the scalar wave equation:

$$\frac{\partial^2\varphi}{\partial x^2} + \frac{\partial^2\varphi}{\partial y^2} + (k_0^2\varepsilon_r - \beta^2)\varphi = 0, \tag{1.105}$$

where $\varphi = E$ or H is a wave function that designates a scalar field component.

1.8.4 *Finite-Difference Discretization*

Let us rewrite Eqs. (1.86) and (1.87) in the following, compact and more convenient, matrix form:

$$\begin{bmatrix} A_{xx} & A_{xy} \\ A_{yx} & A_{yy} \end{bmatrix} \begin{bmatrix} E_x \\ E_y \end{bmatrix} = \beta^2 \begin{bmatrix} E_x \\ E_y \end{bmatrix}, \tag{1.106}$$

where the differential operators A_{ij} ($i = x, y$ and $j = x, y$) are defined as follows [Xu *et al.* (1994)]:

$$A_{xx} E_x = \frac{\partial}{\partial x} \left(\frac{1}{n^2} \frac{\partial(n^2 E_x)}{\partial x} \right) + \frac{\partial^2 E_x}{\partial y^2} + n^2 k_0^2 E_x, \tag{1.107}$$

$$A_{yy} E_y = \frac{\partial^2 E_y}{\partial x^2} + \frac{\partial}{\partial y} \left(\frac{1}{n^2} \frac{\partial(n^2 E_y)}{\partial y} \right) + n^2 k_0^2 E_y, \tag{1.108}$$

$$A_{xy} E_y = \frac{\partial}{\partial x} \left(\frac{1}{n^2} \frac{\partial(n^2 E_y)}{\partial y} \right) - \frac{\partial^2 E_y}{\partial x \partial y}, \tag{1.109}$$

$$A_{yx} E_y = \frac{\partial}{\partial y} \left(\frac{1}{n^2} \frac{\partial(n^2 E_x)}{\partial y} \right) - \frac{\partial^2 E_x}{\partial y \partial x}. \tag{1.110}$$

Similar formulations may be applied for the magnetic field, following Eqs. (1.98)–(1.99).

We may efficiently solve Eq. (1.106) by means of a finite-difference mode-solving technique [Xu *et al.* (1994)]. In the finite-difference methodology [Smith (1985)], the continuous space is replaced by a discrete lattice structure defined in the computational domain. The fields at the lattice point $x = m\Delta x$ and $x = n\Delta y$ are represented by their discrete counterparts. The differential operators appearing in Eqs. (1.107)–(1.110) are also approximated by finite-difference approximations. The discrete form of these operators can, in fact, be found in a straightforward way by noting that at the dielectric interfaces, although the normal electric field components are discontinuous, the displacement vectors $n^2 \hat{E}_i$ ($i = x, y$) appearing in Eqs. (1.107)–(1.110) are continuous As a result, a central finite-difference scheme can be applied directly, without invoking any special treatments.

After some algebraic manipulations, the expressions for the differential operators A_{ij} ($i = x, y$ and $j = x, y$) are found to be the following [Smith (1985)], [Huang (1993)]:

$$\begin{aligned} A_{xx} \hat{E}_x = \ & \frac{T_{m+1,n} \hat{E}_x(m+1, n) + T_{m-1,n} \hat{E}_x(m-1, n)}{\Delta x^2} \\ & - \frac{(2 - R_{m+1,n} - R_{m-1,n}) \hat{E}_x(m, n)}{\Delta x^2} \\ & + \frac{\hat{E}_x(m, n+1) - 2\hat{E}_x(m, n) + \hat{E}_x(m, n-1)}{\Delta x^2} \\ & + n^2(m, n, l) k_0^2 \hat{E}_x(m, n), \end{aligned} \tag{1.111}$$

where:

$$T_{m\pm1,n} = \frac{2n^2(m\pm1,n)}{n^2(m\pm1,n)+n^2(m,n)}, \tag{1.112}$$

$$R_{m\pm1,n} = T_{m\pm1,n} - 1, \tag{1.113}$$

are, respectively, the transmission and reflection coefficients across the index interfaces between $m\Delta x$ and $(m+1)\Delta x$. For the A_{yy} operator, we have:

$$\begin{aligned}
A_{yy}\hat{E}_y =\ & \frac{T_{m,n+1}\hat{E}_y(m,n+1)+T_{m,n-1}\hat{E}_y(m,n-1)}{\Delta y^2} \\
& - \frac{(2-R_{m,n+1}-R_{m,n-1})\hat{E}_y(m,n)}{\Delta y^2} \\
& + \frac{\hat{E}_y(m+1,n)-2\hat{E}_y(m,n)+\hat{E}_x(m-1,n)}{\Delta x^2} \\
& + n^2(m,n,l)k_0^2\hat{E}_y(m,n),
\end{aligned} \tag{1.114}$$

where:

$$T_{m,n\pm1} = \frac{2n^2(m,n\pm1)}{n^2(m,n\pm1)+n^2(m,n)}, \tag{1.115}$$

$$R_{m,n\pm1} = T_{m,n\pm1} - 1, \tag{1.116}$$

are, respectively, the transmission and reflection coefficients coefficients across the index interfaces between $n\Delta y$ and $(n+1)\Delta y$. In a similar vein, the remaining two operators, A_{xy} and A_{yx}, are expressed by:

$$\begin{aligned}
A_{xy}\hat{E}_y =\ & \frac{1}{4\Delta x\Delta y}\Bigg[\left(\frac{n^2(m+1,n+1)}{n^2(m+1,n)}-1\right)\hat{E}_y(m+1,n+1) \\
& -\left(\frac{n^2(m+1,n-1)}{n^2(m+1,n)}-1\right)\hat{E}_y(m+1,n-1) \\
& -\left(\frac{n^2(m-1,n+1)}{n^2(m-1,n)}-1\right)\hat{E}_y(m-1,n+1) \\
& +\left(\frac{n^2(m-1,n-1)}{n^2(m-1,n)}-1\right)\hat{E}_y(m-1,n-1)\Bigg],
\end{aligned} \tag{1.117}$$

and:

$$\begin{aligned}
A_{yx}\hat{E}_y =\ & \frac{1}{4\Delta x\Delta y}\Bigg[\left(\frac{n^2(m+1,n+1)}{n^2(m,n+1)}-1\right)\hat{E}_x(m+1,n+1) \\
& -\left(\frac{n^2(m-1,n+1)}{n^2(m,n-1)}-1\right)\hat{E}_x(m-1,n+1) \\
& -\left(\frac{n^2(m+1,n-1)}{n^2(m,n+1)}-1\right)\hat{E}_x(m+1,n-1) \\
& +\left(\frac{n^2(m-1,n-1)}{n^2(m,n-1)}-1\right)\hat{E}_x(m-1,n-1)\Bigg],
\end{aligned} \tag{1.118}$$

Calculating Eqs. (1.111)–(1.118) for each node in the discrete computational space, results in the following *eigenvalue matrix equation*:

$$[A]\{\phi\} = \beta^2\{\phi\}, \tag{1.119}$$

where β^2 is an eigenvalue and ϕ is an eigenvector, expressed as:

$$\{\phi\} = \{\phi_1 \quad \phi_2 \quad \phi_3 \quad \cdots \quad \phi_M\}^{\mathrm{T}}, \tag{1.120}$$

with $M = m \times n$, of the square sparse matrix $[A]$.

We can calculate the longitudinal propagation constant β as well as the field distribution, by solving the eigenvalue matrix equation (1.119). In the presented, fully-vectorial finite-difference frequency-domain (FV-FDFD) mode-solving technique, the sparse matrix is non-symmetric, having only 13 diagonals. Its *bandwidth* and *dimension* are $(4m + 7)$ and $2mn$, respectively. Once matrix $[A]$ is defined, the eigenvalues and eigenvectors of its nonzero *band* can be found by means of the shifted-inverse power iteration method. This method is, further, commonly deployed for the solution of various eigenproblems in commercially available software.

1.8.5 *Numerical Examples of the FV-FDFD Mode-Solver*

The finite-difference expressions presented in this chapter are widely used in computer-aided (CAD) software available for the investigation of arbitrary dielectric heterostructures having 2D cross-sections. Here, we shall concisely discuss exemplary results obtained with the previously described FV-FDFD mode-solver. In all cases, for simplicity, a Dirichlet condition was implemented at the boundaries of our numerical space, except from the symmetry-plane, where we made use of a Neumann condition in order to halve the computational overhead.

Figure 1.5 illustrates a rib waveguide structure, also considered by the COST-216 group using 12 different methods [Working Group 1, COST-16 (1989)], and by Xu *et al.* in [Xu *et al.* (1994)]. The refractive indices for a wavelength 1.55 μm are also indicated in this figure. First, we examined the convergence of our mode-solving technique for several InP thicknesses, t, by progressively reducing the mesh size, while keeping the mesh size ratio $\Delta x/\Delta y$ fixed at the value of 2. Figure 1.6 clearly

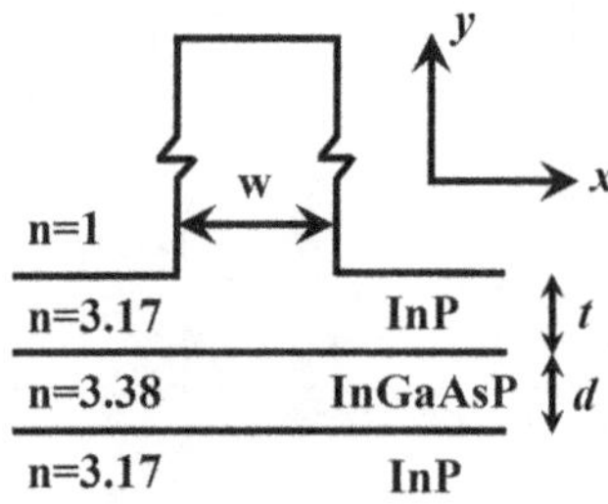

Figure 1.5: Schematic illustration of the InP-based rib waveguide under consideration. The geometrical parameters used in the computations are $w = 2.4$ μm, $d = 0.2$ μm and $t = 0.0, 0.2, 0.4$ μm.

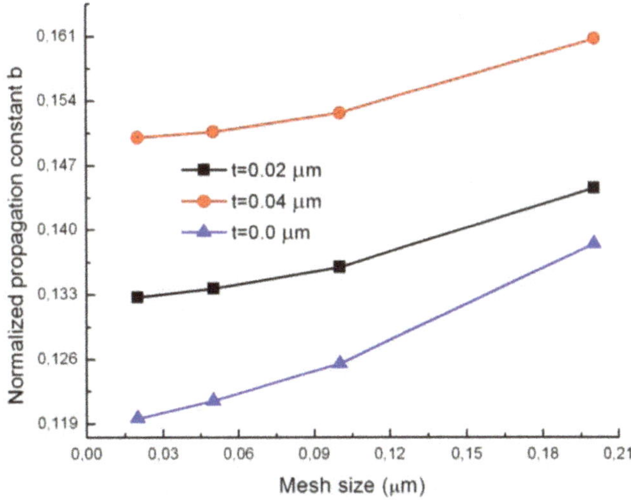

Figure 1.6: Convergence of the developed finite-difference frequency-domain mode solver for various thicknesses t of the InP layer shown in Fig. 1.5.

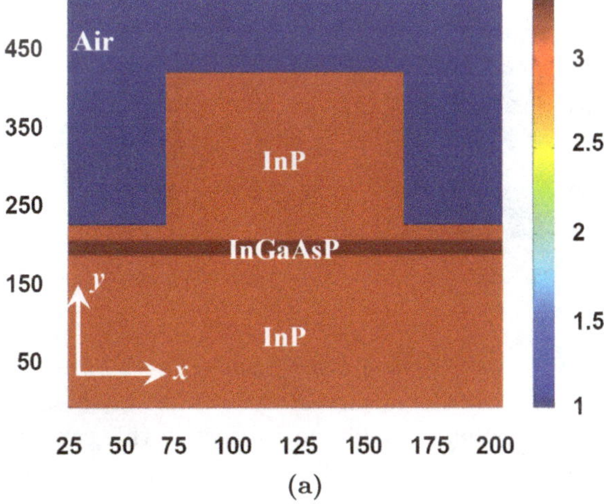

(a)

Figure 1.7: (a) Refractive index 2D distribution used in the FV-FDFD mode-solver.

illustrates that the mode-solver converges to, evidently, more accurate values of the calculated normalised propagation constant, b [Stern (1988a,b, 1991)], in all three cases. Very similar results were also reported in [Xu *et al.* (1994)].

Figure 1.7 illustrates the calculated transverse electric field components of the first TE-like mode for $t = 0.2$ μm. The refractive index distribution that was considered in our calculations is, also, shown in this figure. The mesh sizes utilised in these investigations were $\Delta x = 0.05$ μm and $\Delta y = 0.025$ μm. The results in all cases were obtained in approximately 11 seconds on an Intel Pentium 4 (2.80 GHz),

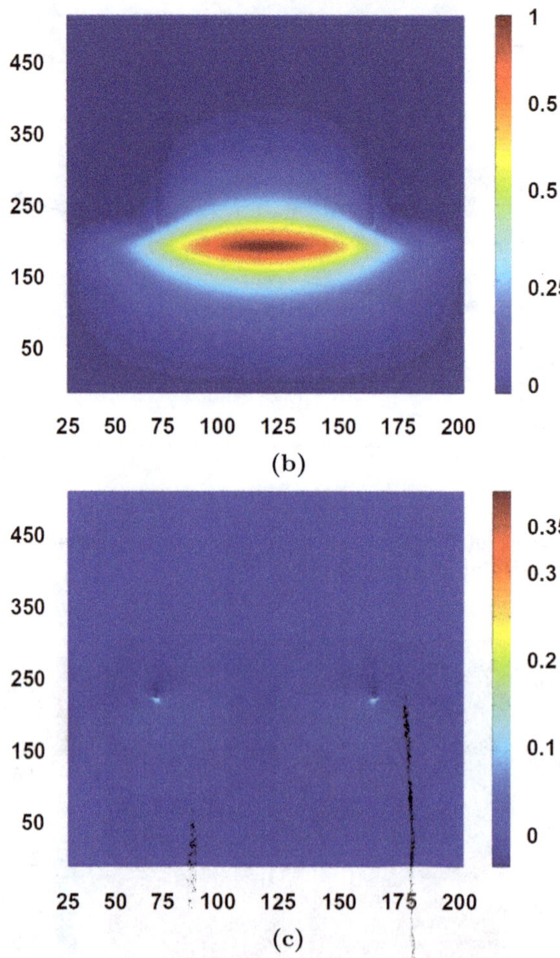

Figure 1.7: Computed **(b)** dominant E_x, and **(c)** weak E_y field component.

personal workstation. One may readily notice that in this case, the mode is rather weakly-guided, since it is not well-confined below the central rib. Moreover, we note that the weak transverse electric field component is mainly concentrated at the dielectric corners, as expected, somewhat resembling a surface wave. The obtained normalised propagation constant is, indeed, considerably small (approximately 0.1336) and compares very favourably with the value 0.1365 calculated in [Xu *et al.* (1994)] for the same mesh sizes. The minor discrepancy is attributed to the finite, approximately 2 μm rather than infinite, as used in [Xu *et al.* (1994)], height of the central rib that we selected for our present calculations.

In Fig. 1.8 we report the 2D spatial distribution of the dominant, E_x, electric field component of the fundamental TE-like eigenmode, for $t = 0.4$ and 0.0 μm. For the first case, it is seen that the mode is somewhat better guided and confined, compared to the previous case, as judged by observing the field immediately below

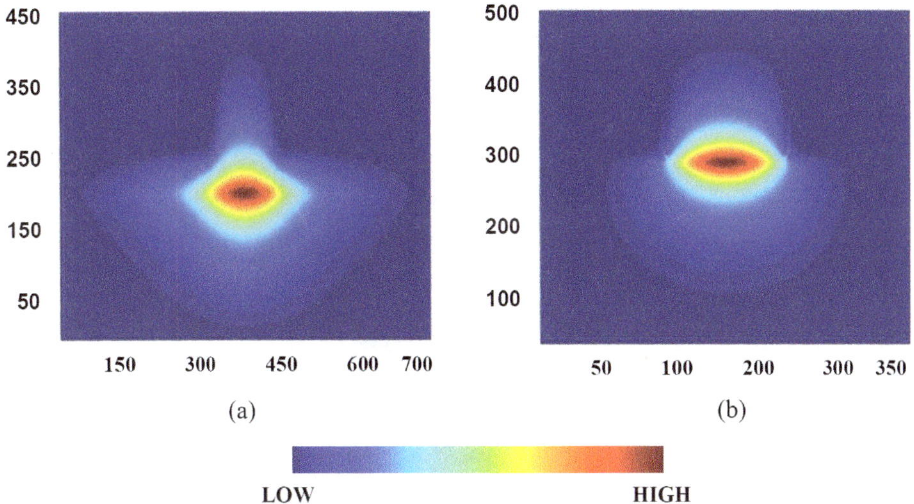

Figure 1.8: Computed 2D spatial distribution of the E_x field component for **(a)** $t = 0.4$ μm and **(b)** $t = 0.0$ μm, for the structure shown in Fig. 1.5.

the dielectric corners. The discontinuity of the dominant electric component at the two interfaces of the rib with vacuum, is clearly observed in Fig. 1.8. The computed normalised constant is $b \simeq 0.1507$ (Fig. 1.8(a)), which is again found to be in excellent agreement with the corresponding value of 0.1511, reported in [Xu *et al.* (1994)]. Very similar values were also found in [Working Group 1, COST-16 (1989)]. Finally, for $t = 0.0$ μm it is readily inferred that the mode exhibits the weakest guidance compared to both previous cases, since it considerably extends to both sides of the InGaAsP layer, while having a normalised propagation constant equal to just 0.1215 (the corresponding value in [Xu *et al.* (1994)] is 0.1228).

Chapter 2

Conventional and Nonstandard Finite-Difference Time-Domain Method

2.1 Introduction

In the analytic investigations of dielectric waveguides that we have been concerned with so far, we have assumed that the structures were axially uniform (see Eq. (1.36)). Moreover, our analysis involved steady-state solutions of the wave equations, in which the time-derivative $\partial/\partial t$ was replaced by $j\omega$ (see Eqs. (1.13)–(1.15)). For the investigation of transient phenomena, axially varying, irregularly shaped or other types of more involved (e.g. nonlinear) structures, one has to resort to full 3D, time-domain methodologies. A rigorous and well-established such algorithm, based directly on Maxwell's equations, is the Finite-Difference Time-Domain (FDTD) method [Yee (1966); Berenger (2008); Taflove and Hagness (2000)]. Upon establishing the necessary nomenclature, we shall here examine in detail several aspects of this algorithm, including its numerical stability and the induced numerical dispersion and anisotropy errors that occur within it, which occasionally can contaminate the simulation outcomes and seriously compromise the overall accuracy of the method. As a means of subduing the aforementioned algorithmic errors we also examine the recently introduced concepts of *"exact"* and *"nonstandard"* finite-differences [Mickens (1984)], their implementation within the context of the FDTD algorithm [Cole (1995)] and the significantly enhanced computational performance, in terms of memory and/or computational time, which can arise from their adoption.

2.2 The Yee Algorithm

As we show in Section 1.2, Maxwell's equations of electromagnetism for a homogeneous, isotropic, lossless, non-dispersive, linear, achiral and stationary medium take the form:

$$-\mu_0 \frac{\partial \mathbf{H}}{\partial t} = \nabla \times \mathbf{E}, \tag{2.1}$$

$$\varepsilon_0 \varepsilon_r \frac{\partial \mathbf{E}}{\partial t} = \nabla \times \mathbf{B}. \tag{2.2}$$

25

We can rewrite these equations in terms of the six electromagnetic field components, as follows:

$$-\mu_0 \frac{\partial H_x}{\partial t} = \frac{\partial E_z}{\partial y} - \frac{\partial E_y}{\partial z}, \tag{2.3}$$

$$-\mu_0 \frac{\partial H_y}{\partial t} = \frac{\partial E_x}{\partial z} - \frac{\partial E_z}{\partial x}, \tag{2.4}$$

$$-\mu_0 \frac{\partial H_z}{\partial t} = \frac{\partial E_y}{\partial x} - \frac{\partial E_x}{\partial y}, \tag{2.5}$$

$$\varepsilon_0 \varepsilon_r \frac{\partial E_x}{\partial t} = \frac{\partial H_z}{\partial y} - \frac{\partial H_y}{\partial z}, \tag{2.6}$$

$$\varepsilon_0 \varepsilon_r \frac{\partial E_y}{\partial t} = \frac{\partial H_x}{\partial z} - \frac{\partial H_z}{\partial x}, \tag{2.7}$$

$$\varepsilon_0 \varepsilon_r \frac{\partial E_z}{\partial t} = \frac{\partial H_y}{\partial x} - \frac{\partial H_x}{\partial y}. \tag{2.8}$$

In order to discretise these equations using finite differences, we assume that the electromagnetic field components are located in three dimensions according to the arrangement shown in Fig. 2.1, known also as the "Yee cube" after the name of the scientist who first proposed the present algorithm [Yee (1966)]. Further, we assume that the electric field components are calculated at integer time increments Δt, i.e. at $n\Delta t$ ($n = 0, 1, 2, \ldots$), whereas the magnetic field components at half-integer time intervals, $(n - \frac{1}{2})\Delta t$. With these assumptions in place, we may self-consistently discretise each one of Eqs. (2.3)–(2.8). For instance, the left-hand side of Eq. (2.3) becomes:

$$-\frac{\mu_0}{\Delta t}\left[H_x^{n+\frac{1}{2}}\left(i, j + \frac{1}{2}, k + \frac{1}{2}\right) - H_x^{n-\frac{1}{2}}\left(i, j + \frac{1}{2}, k + \frac{1}{2}\right)\right], \tag{2.9}$$

whereas for the right-hand side one obtains:

$$i - \frac{1}{2}\frac{1}{\Delta y}\left[E_z^n\left(i, j + 1, k + \frac{1}{2}\right) - E_z^n\left(i, j, k + \frac{1}{2}\right)\right]$$
$$- \frac{1}{\Delta z}\left[E_y^n\left(i, j + \frac{1}{2}, k + 1\right) - E_y^n\left(i, j + \frac{1}{2}, k\right)\right]. \tag{2.10}$$

We, thus, arrive at the following updating equation for the H_x field component:

$$H_x^{n+\frac{1}{2}}\left(i, j + \frac{1}{2}, k + \frac{1}{2}\right) = H_x^{n-\frac{1}{2}}\left(i, j + \frac{1}{2}, k + \frac{1}{2}\right)$$
$$- \frac{\mu_0}{\Delta y \Delta t}\left[E_z^n\left(i, j + 1, k + \frac{1}{2}\right) - E_z^n\left(i, j, k + \frac{1}{2}\right)\right]$$
$$- \frac{\mu_0}{\Delta z \Delta t}\left[E_y^n\left(i, j + \frac{1}{2}, k + 1\right) - E_y^n\left(i, j + \frac{1}{2}, k\right)\right]. \tag{2.11}$$

Similar updating equations are obtained for the remaining five electromagnetic field components. We, hence, have an algorithm for calculating all components, at every

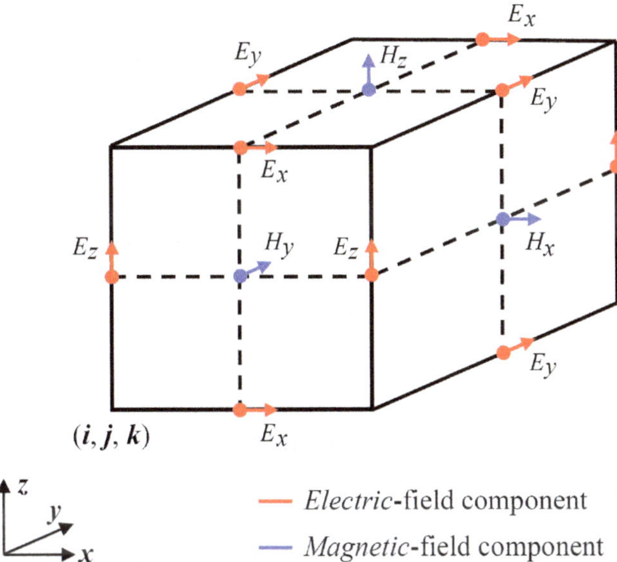

Figure 2.1: Three-dimensional spatial arrangement of the six electromagnetic field components ("Yee cube") for the finite-difference time-domain (FDTD) numerical solution of Maxwell's equations.

discrete point and at each time step. Starting from the electric field components, which at $n = 0$ are assumed to be zero throughout the computational space, we calculate the magnetic field components (e.g., for the new value of the H_x component, we use Eq. (2.11); then, based on the new values of the **H**-field, we find the new values of the **E**-field components, and so on. This process enables the calculation of the full time evolution of the electromagnetic fields in the defined computational space. We note that, in contrast to the mode-solving techniques analysed in the previous chapter (see Eq. (1.106)), the FDTD algorithm performs a *direct* solution of Maxwell's equations, i.e. it is not based on the wave equations and it does not involve complex computations of eigenvalues/vectors of sparse matrixes.

2.3 Stability of the Yee Algorithm

A point of attention in the correct implementation of the FDTD algorithm and, in fact, of every explicit algorithm is that the time step, Δt, cannot be arbitrarily large; it has to be upper-bounded by an explicit figure, which depends solely on the spatial discretisation steps and the defined dielectric structure; otherwise the algorithm will become unstable (i.e., the computed fields will unphysically start increasing with time, tending to infinity). In this section we show how one may analytically calculate the aforementioned upper limit of the time step, known also as the "Courant–Friedrichs–Lewy (CFL) limit" [Taflove and Hagness (2000)], [Smith (1985)].

The algebraic calculations are greatly simplified if we consider the (equivalent) scalar wave equation. For the sake of clarity we shall examine the 1D case. We, thus, start with the following equation:

$$\frac{\partial^2 \varphi}{\partial x^2} - \varepsilon\mu\frac{\partial^2 \varphi}{\partial t^2} = 0, \tag{2.12}$$

where φ is an electric or magnetic field component. Assuming without loosing generality that the oscillation amplitude of the component is unity and that the propagation constant along the x-direction is β_x, it is a simple matter to see that the first part (a) of the following expression for φ satisfies Eq. (2.12):

$$\varphi(x,t) = \exp(j\beta_x x)\exp(\alpha t) \tag{2.13a}$$
$$= \exp(j\beta_x l\Delta x)\exp(\alpha n\Delta t) \tag{2.13b}$$
$$= \exp(j\beta_x l\Delta x)\xi^n, \tag{2.13c}$$

where $\xi = \exp(\alpha n\Delta t)$, $l, n = 0, 1, 2, \ldots$ and Δx and Δt being the spatial and temporal increments, respectively.

We note from Eq. (2.13) that if the solution φ is to be stable, that is not growing unphysically with time, then the variable ξ must satisfy the restriction:

$$|\xi| \leq 1. \tag{2.14}$$

We mentioned above Eq. (2.13a) satisfies Eq. (2.12). Obviously, the same must hold true for Eq. (2.13c). Upon substituting it into Eq. (2.12) we, thus, obtain:

$$\frac{1}{\Delta x^2}\left[\exp\left(j\beta_x(l+1)\Delta x\right)\xi^n - 2\exp\left(j\beta_x l\Delta x\right)\xi^n + \exp\left(j\beta_x(l-1)\Delta x\right)\xi^n\right]$$
$$- \frac{\varepsilon\mu}{\Delta t^2}\left[\exp\left(j\beta_x l\Delta x\right)\xi^{n+1} - 2\exp\left(j\beta_x l\Delta x\right)\xi^n + \exp\left(j\beta_x l\Delta x\right)\xi^{n-1}\right] = 0. \tag{2.15}$$

Upon eliminating the common $\exp\left(j\beta_x l\Delta x\right)\xi^n$ term and after performing some further algebraic manipulations we arrive at:

$$\xi^2 - 2\xi + 1 - \frac{\Delta t^2}{\varepsilon\mu}\left[-\frac{4}{\Delta x^2}\sin^2\left(\frac{\beta_x\Delta x}{2}\right)\right]\xi = 0, \tag{2.16}$$

which can be written as:

$$\xi^2 - 2U\xi + 1 = 0, \tag{2.17}$$

with U being:

$$U = -\frac{2\Delta t^2}{\varepsilon\mu}\frac{1}{\Delta x^2}\sin^2\left(\frac{\beta_x\Delta x}{2}\right) + 1. \tag{2.18}$$

From the two roots of Eq. (2.17), $\xi_{1,2} = U \pm \sqrt{U^2 - 1}$, because of the restriction of Eq. (2.14) and the fact that it is always $\sin^2(\ldots) \geq 0$, we obtain the following relation:

$$U = -\frac{2\Delta t^2}{\varepsilon\mu}\frac{1}{\Delta x^2}\sin^2\left(\frac{\beta_x\Delta x}{2}\right) + 1 \leq 1. \tag{2.19}$$

We, now, consider the following two cases: First, the case where $U < 1$. Then, we immediately see that $|\xi_2| > 1$, and hence the algorithm will run unstably. The second remaining case, given Eq. (2.19) is that where:

$$-1 \le U \le 1. \tag{2.20}$$

We may promptly find that in this case, $\xi_{1,2}$ (which are, now, complex numbers) satisfy: $|\xi_1| = |\xi_2| = 1$; hence, the algorithm will be stable.

We conclude that the finite-difference algorithm will run stably only when Eq. (2.20) is satisfied. In the light of Eq. (2.19) we see that the right-hand side of this relation is always satisfied. For its left-hand side we have:

$$-1 \le -\frac{2\Delta t^2}{\varepsilon\mu}\frac{1}{\Delta x^2}\sin^2\left(\frac{\beta_x \Delta x}{2}\right) + 1. \tag{2.21}$$

It should, therefore, be enough to identify the range of values for Δt, for which the following relation holds true:

$$1 \ge \frac{2\Delta t^2}{\varepsilon\mu}\frac{1}{\Delta x^2} + 1, \tag{2.22}$$

from whence we immediately find:

$$\Delta t \le \sqrt{\varepsilon\mu}\left(\frac{1}{\Delta x^2}\right)^{-\frac{1}{2}} = \frac{\sqrt{\varepsilon\mu}}{c_0}\left(\frac{1}{\Delta x^2}\right)^{-\frac{1}{2}} = \frac{1}{\nu}\left(\frac{1}{\Delta x^2}\right)^{-\frac{1}{2}}, \tag{2.23}$$

where ν is the speed of the electromagnetic wave in the medium.

This equation defines the upper bound in the range of values that the time step of our one-dimensional finite-difference algorithm is allowed to take. Following a similar course of analysis we find that in three dimensions the corresponding expression for Δt is:

$$\Delta t \le \frac{1}{\nu}\left(\frac{1}{\Delta x^2} + \frac{1}{\Delta y^2} + \frac{1}{\Delta z^2}\right)^{-\frac{1}{2}}. \tag{2.24}$$

Although this criterion was derived from a finite-difference discretisation of the wave equation, it swiftly turns out (as one may intuitively expect) that exactly the same restriction is obeyed by the time step in the Maxwell-equations' based Yee (FDTD) algorithm.

2.4 "Exact" and "Nonstandard" Finite-Differences

To introduce the concepts of "exact" [Mickens (1989, 1993, 1997, 1996)] and "nonstandard" [Mickens (1984)] finite-differences and how these are applied within FDTD's time-marching algorithm, let us once more start by examining the finite-difference discretisation of the one-dimensional wave Eq. (2.12). Following the example of Eqs. (2.9)–(2.10), it is an easy task to see that a second-order time and space discretisation $O(\Delta t^2, \Delta x^2)$ of this (second-order) equation leads to:

$$\left(\frac{1}{\Delta t^2}\hat{d}_t^2 - v(x)^2\frac{1}{\Delta x^2}\hat{d}_x^2\right)\tilde{\varphi}(x,t) = 0, \tag{2.25}$$

where $\hat{d}_l^2 \tilde{\varphi}(l) = \tilde{\varphi}(l + \Delta l) - 2\tilde{\varphi}(l) + \tilde{\varphi}(l - \Delta l)$, $l = t, x$ with $\tilde{\varphi}$ and v denoting the numerical field component and phase velocity, respectively.

Let us, now, see how we can construct a more accurate compared to Eq. (2.25) finite-difference approximation to Eq. (2.12), without directly resorting to higher-order implementations. To this end, we start by considering an exact solution to the wave Eq. (2.12). Such a solution can, e.g., be:

$$\varphi(x, t) = ae^{i(kx - \omega t)} + be^{i(kx - \omega t)}, \tag{2.26}$$

where $k = 2\pi/\lambda$ is the wavenumber and ω the wave angular frequency.

Assuming that φ in this expression is also a solution to the numerical, finite-difference, algorithm, we may calculate the first-order central finite difference $\hat{d}_x e^{ikx}$, where $\hat{d}_x \tilde{\varphi}(x) = \tilde{\varphi}(x + \frac{\Delta x}{2}) - \tilde{\varphi}(x - \frac{\Delta x}{2})$, as:

$$\hat{d}_x e^{ikx} = 2ie^{ikx} \sin\left(\frac{k\Delta x}{2}\right). \tag{2.27}$$

We, now, note that for $\varphi(x) = e^{ikx}$, by defining:

$$d_x \varphi(x) = \frac{\hat{d}_x \varphi(x)}{s(\Delta x)}, \tag{2.28}$$

with $s(k, \Delta x) = \frac{2}{k} \sin\left(\frac{k\Delta x}{2}\right)$, we have *exactly*:

$$d_x \varphi(x) = \varphi'(x), \quad \text{for } \varphi \in \{\sin(kx), \cos(kx), e^{\pm ikx}\}.$$

Thus, by using the "nonstandard" stipulation $s(\Delta x)$, instead of the usual Δx, in the denominator of Eq. (2.28), we have obtained an *exact* finite-difference approximation to $\varphi'(x)$. To obtain an "exact" finite-difference algorithm for Eq. (2.12), we only need to make the replacements $\Delta t \to s(\omega, \Delta t)$ and $\Delta x \to s(k, \Delta x)$ in Eq. (2.25):

$$\left[\frac{1}{4\sin^2\left(\frac{\omega\Delta t}{2}\right)/\omega^2} \hat{d}_t^2 - v(x)^2 \frac{1}{4\sin^2\left(\frac{k\Delta x}{2}\right)/k^2} \hat{d}_x^2 \right] \tilde{\varphi}(x, t) = 0. \tag{2.29}$$

It should be herein noted that even with these "nonstandard" stipulations used in Eq. (2.29), the algorithm may still lead to numerical errors, even at the "targeted" frequency ω. This is because in our finite (temporally and spatially) computational domain we can never have a strictly monochromatic wave, i.e. it is not possible to define a source that will contain and excite only a single ω frequency and wavelength λ. Moreover, the phase velocity of the numerical wave will not be equal to the phase velocity of the actual wave, owing to the presence of numerical dispersion. As a result, both, at the frequency ω and in a region around it numerical errors will arise. However, as we shall see later by means of numerical examples, these errors are still significantly smaller than those that arise within the corresponding conventional algorithm.

2.5 Generalization to Three Dimensions

2.5.1 Standard Wave-Equation Finite-Difference Algorithm

We, now, consider how the above concepts can be applied to the wave equation in three dimensions. First, we examine the standard three-dimensional finite-difference approximation to the wave equation:

$$\left(\partial_{tt} - \upsilon(\mathbf{r})^2 \nabla^2\right)\varphi(\mathbf{r}, t) = 0, \tag{2.30}$$

where $\mathbf{r} = (r, y, z)$ is the position vector and $\nabla^2 = \partial_{xx} + \partial_{yy} + \partial_{zz}$ the 3D Laplacian operator. Extending Eq. (2.25) to three dimensions, it is an easy task to see that we arrive at the following relation:

$$\left(\frac{1}{\Delta t^2}\hat{d}_t^2 - \upsilon(\mathbf{r})^2 \frac{1}{\Delta x^2}\hat{D}_1^2\right)\tilde{\varphi}(x, t) = 0, \tag{2.31}$$

where $\hat{D}_1^2 = \hat{d}_x^2 + \hat{d}_y^2 + \hat{d}_z^2$, and for simplicity we have assumed that $\Delta x = \Delta y = \Delta z$, i.e. a cubic spatial cell.

In order to work out the inherent numerical error of the finite-difference Eq. (2.31), we may insert a known, analytic solution of the actual wave Eq. (2.30) to Eq. (2.31) and calculate the resulting residual term(s). Such a solution is, e.g., $\varphi_0(\mathbf{r}, t) = e^{i(\mathbf{k}\cdot\mathbf{r} - \omega t)}$, where $k = (k_x, k_y, k_z)$, and upon inserting it into Eq. (2.31) we obtain:

$$\left(\frac{1}{\Delta t^2}\hat{d}_t^2 - \upsilon(\mathbf{r})^2 \frac{1}{\Delta x^2}\hat{D}_1^2\right)\tilde{\varphi}_0(\mathbf{r}, t)$$

$$= 2\underbrace{\left[\left(\cos(\omega\Delta t) - 1\right) - \upsilon(\mathbf{r})\frac{\Delta t^2}{\Delta x^2}\sum_{l=x,y,z}\left(\cos(k_l\Delta x) - 1\right)\right]}_{\text{algorithmic error, } \tilde{\varepsilon}}\tilde{\varphi}_0(\mathbf{r}, t). \tag{2.32}$$

Ideally, we would like the error $\tilde{\varepsilon}$ to be zero for every direction of wave propagation, i.e. for every (k_x, k_y, k_z) triad, but inspection of Eq. (2.32) reveals that this is never possible. Thus, an *anisotropy* error (in addition to the induced numerical dispersion error), originating from the particular finite-difference approximation, $\hat{D}_1^2$, to the 3D Laplacian operator ∇^2 will always be present in Eq. (2.31). In the following section we investigate a method for constructing finite-difference approximations to ∇^2 with significantly reduced anisotropy error. This method is, then, used in the formation of high-accuracy finite-difference operators that can be used within FDTD's time-marching algorithm.

2.5.2 Non-Standard Finite-Difference Time-Domain Algorithm

Critical for the NS-FDTD formulae [Cole (1995)], [Kashiwa *et al.* (2002)] is the introduction of generalised 3D Laplacian operators, $D_i^2, i = 1, 2, 3$, shown in Fig. 2.2a.

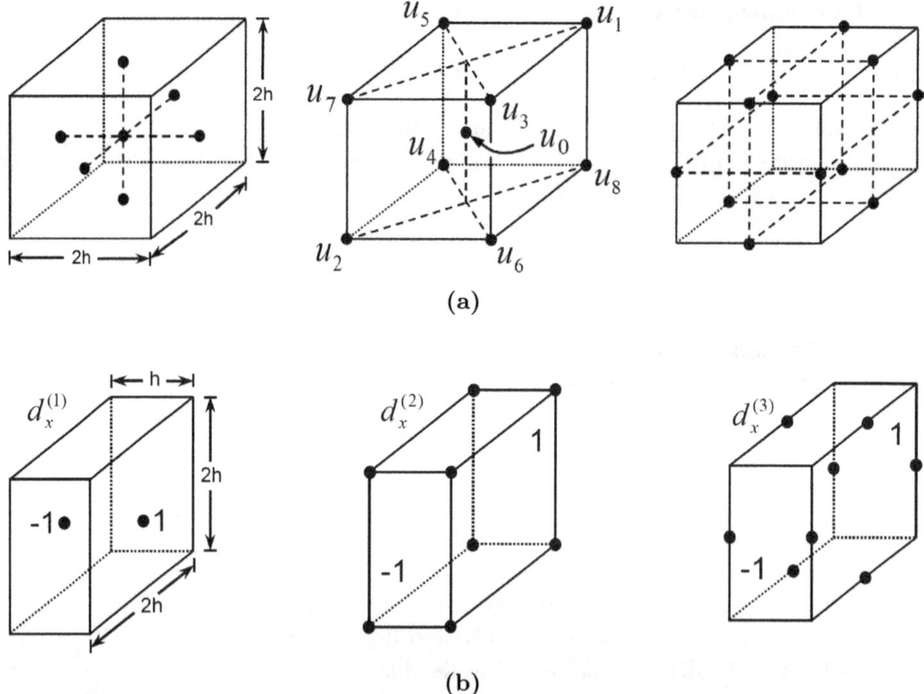

Figure 2.2: (a) Spatial arrangement of the generalized 3D Laplacian operators. (b) The elements of the nonstandard difference operator $\tilde{d}_x$. A suitable weighting component is shown for the filled points in each side of the element.

Focusing on the derivation of D_2^2, we start by introducing the parameters:

$$\hat{\xi} = h\frac{\partial}{\partial x}, \quad \hat{\eta} = h\frac{\partial}{\partial y} \quad \text{and} \quad \hat{\mu} = h\frac{\partial}{\partial z},$$

where h is a real number representing the cell size and:

$$(\hat{\xi}^2 + \hat{\eta}^2 + \hat{\mu}^2)u(\bar{r}) = h^2\nabla^2 u(\bar{r}) \tag{2.33}$$

with $u(\bar{r})$ being an electromagnetic field component in 3D space.

Denoting with $u_i^2, i = 1, 2, 3$ the values of the function $u(\bar{r})$ at the points shown in Fig. 2.2a, we have:

$$u_1 = e^{(\hat{\xi}+\hat{\eta}+\hat{\mu})}u_0 = \left[1 + (\hat{\xi} + \hat{\eta} + \hat{\mu}) + \frac{1}{2}\hat{\xi} + \hat{\eta} + \hat{\mu})^2 + ...\right]u_0. \tag{2.34}$$

Similarly:

$$u_2 = e^{(-\hat{\xi}-\hat{\eta}-\hat{\mu})}u_0,$$

$$u_3 = e^{(\hat{\xi}-\hat{\eta}+\hat{\mu})}u_0,$$

with with analogous expressions for the remaining points.

Using series expansions like the one in Eq. (2.34) and the identity in Eq. (2.33), we obtain:

$$S_2 = \sum_{1=1}^{8} u_i = 8u_0 + 4h^2 \nabla^2 u_0 + O(h^4), \qquad (2.35)$$

from which we can define an alternative second-order Laplacian D_2^2 as:

$$D_2^2 u_0 = \frac{S_2/4 - 2u_0}{h^2}. \qquad (2.36)$$

Likewise, the second-order operators D_1^2 and D_3^2 of Fig. 2.2a can be derived, as well as higher-order Laplacians. We can efficiently combine $D_i^2, i = 1, 2, 3$, to create an optimal Laplacian operator D_0^2, which equals a suitably weighted sum of D_i^2 and is sixth-order accurate in space for a reference frequency [Cole (2002)]. Decomposing D_0^2 into finite-difference operator products and employing nonstandard stipulations, yields the following spatial formulae that preserves the $O(h^6)$ accuracy (Fig. 2.2b):

$$\tilde{d}_x = \frac{1}{S_k(h)}\left(\zeta_1 d_x^{(1)} + \zeta_2 d_x^{(2)} + \zeta_3 d_x^{(3)}\right) \qquad (2.37)$$

with $S_k(h) = 2\sin(k\frac{h}{2})$, k being the wavevector that corresponds to the central frequency of the input source and $\zeta_i^2, i = 1, 2, 3$ positive weighting components fulfilling $\zeta_1 + \zeta_2 + \zeta_3 = 1$ for numerical consistency. For the other two directions analogous difference operators can be formed which, when inserted into the **E** or **H**-field updating equations, result in a significantly ameliorated FDTD scheme with regard to numerical dispersion and anisotropy.

Very crucial for the simulations herein is the formulation of the nonstandard uniaxial perfectly matched layer (NS-UPML) [Taflove and Hagness (2000)], [Cole (2002); Tsakmakidis *et al.* (2005)] inside which extend the waveguide layers. We make use of the UPML in order to suitably terminate our (finite) computational space. The numerical waves excited by our numerical source reach the UPML, enter it with practically zero reflection, and are progressively absorbed while they propagate inside it. For the sake of brevity, only the expression for the advancement of the B_y component (in Heaviside–Lorentz units) is given below:

$$B_y = \frac{\kappa_z - S_\omega(\Delta t)\sigma_z/(2\varepsilon_0)}{\kappa_z + S_\omega(\Delta t)\sigma_z/(2\varepsilon_0)}B_y + c_0\frac{S_\omega(\Delta t)}{\kappa_z + S_\omega(\Delta t)\sigma_z/(2\varepsilon_0)}(\tilde{d}_x E_z - \tilde{d}_z E_x)$$

$$(2.38)$$

with $S_\omega(\Delta t) = \frac{2}{\omega}\sin(\frac{\omega\Delta t}{2})$, c_0 the velocity of light in vacuum, and both finite-difference operators, now, involving eighteen instead of two points in a conventional formulation [Cole (2002)], [Tsakmakidis *et al.* (2005)].

For the acquisition of the 2D excitation profiles we recall from Chapter 1 that in rectangular dielectric waveguides there are no precise closed-form expressions for the cross-sectional shape of the modes, owing to the peculiar behaviour of the electromagnetic fields near the dielectric corners where they diverge to a small degree

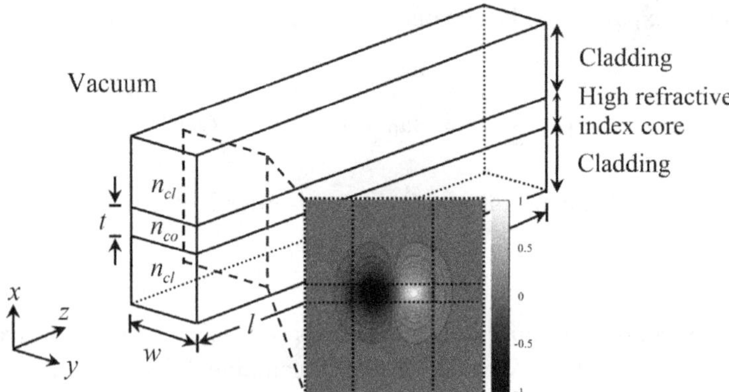

Figure 2.3: Schematic illustration of a high-index contrast (strongly guiding) planar dielectric heterostructure. Also shown is the pre-calculated excitation profile for the generation of the mode inside the rectangular waveguide.

[Sudbo (1992)]. As we show in the previous chapter, a sufficiently accurate numerical way to overcome this limitation is via computing the **E** or **H**-field eigenvectors of the vectorial wave equation that takes into account the polarisation, vector properties and discontinuity of the guided modes at the dielectric interfaces, including the corner regions [Xu *et al.* (1994)]. Upon obtaining the 2D field distributions in this manner, a proper spatial interpolation is necessary to match the Yee-grid. Such an approach is essential, particularly for FDTD in-coupling studies, and an exemplary result for the dominant E_{12}^y eigenmode is illustrated in Fig. 2.3. The associated propagation constant is obtained from the eigenvalue of the wave equation and, for sufficiently small mesh size, the error in its estimation can be made very small (e.g. less than 0.01%), as we show in discussing Figs. 1.5–1.8 in the previous chapter [Xu *et al.* (1994)].

2.6 Application of the NS-FDTD Method in the Simulation of a 3D Dielectric Waveguide

The structure on which the overall methodology was tested is a multimode rectangular waveguide having $n_{co} = 3.41$, $n_{cl} = 1.5$, width $w = 1.17$ μm and very small core thickness, as illustrated in Fig. 2.4 [Tsakmakidis *et al.* (2005)]. Our choice is motivated by the use of high index-contrast waveguides in existing photonic devices involving waveguide coupling to planar photonic crystals or various travelling-wave devices (e.g., travelling-wave photodetectors, phototransistors, etc). The length was chosen to be sufficiently large, $l = 21$ μm, to check the precision in the single-mode excitation at the far end of the waveguide. We have examined the dynamical propagation of the first three modes E^y, generated with the appropriate initial profiles. The investigated spectral range is sufficiently above the cut-off frequency of the third eigenmode (to ensure accurate computation of the propagation constant and mode profile with the mode-solver). Using the effective-index method (Section 1.6) we have calculated this frequency to be $f_{c3} = 142$ THz. For excitation we used Gaussian pulses modulating a sinusoidal carrier of frequency $f_0 = 193.5$ THz which coincided with the reference frequency in the nonstandard difference operators. The computed values of ζ_i in this study were $\zeta_1 = 0.7093$, $\zeta_2 = 0.00449$, $\zeta_3 = 0.2458$, and an exemplary part of the actual computational code is presented below:

```
Hx(i,j,k)  =   Hx(i,j,k)
               - co*ux(j,k)
               *(     a1*( Ez(i,j+1,k)  -  Ez(i,j,k)  )
                   + a2*(    Ez(i+1,j+1,k+1)  + Ez(i-1,j+1,k+1)
                          + Ez(i+1,j+1,k-1)  + Ez(i-1,j+1,k-1)
                          - Ez(i+1,j,k+1)    - Ez(i-1,j,k+1)
                          - Ez(i+1,j,k-1)    - Ez(i-1,j,k-1)  )/4
                   + a3*(    Ez(i,j+1,k+1)   + Ez(i-1,j+1,k)
                          + Ez(i+1,j+1,k)   + Ez(i,j+1,k-1)
                          - Ez(i,j,k+1)      - Ez(i-1,j,k)
                          - Ez(i+1,j,k)      - Ez(i,j,k-1)  )/4
                   - a1*( Ey(i,j,k+1)  -  Ey(i,j,k)  )
                   - a2*(    Ey(i+1,j+1,k+1)  + Ey(i+1,j-1,k+1)
                          + Ey(i-1,j+1,k+1)  + Ey(i-1,j-1,k+1)
                          - Ey(i+1,j+1,k)    - Ey(i+1,j-1,k)
                          - Ey(i-1,j+1,k)    - Ey(i-1,j-1,k)  )/4
                   - a3*(    Ey(i+1,j,k+1)   + Ey(i,j-1,k+1)
                          + Ey(i,j+1,k+1)   + Ey(i-1,j,k+1)
                          - Ey(i+1,j,k)      - Ey(i,j-1,k)
                          - Ey(i,j+1,k)      - Ey(i-1,j,k)  )/4  )
```

where "co" denotes the vacuum light speed, $a_i = \zeta_i, i = 1, 2, 3$,

and $ux = \left(\frac{1}{\omega}\sin(\frac{\omega\Delta t}{2})\right)/\left(\frac{1}{k}\sin(\frac{k\Delta x}{2})\right)$,

with ux (central film) $= 1.746516 \times 10^{-9}$, ux (cover) $= 1.743246 \times 10^{-9}$,

ux (substrate) $= 1.745135 \times 10^{-9}$, and ux (background) $= 1.732883 \times 10^{-9}$.

The corresponding piece of code for the conventional FDTD method is:

```
Hx(i,j,k)  = Hx(i,j,k)  - co*dt
                *(     ( Ez(i,j+1,k)  - Ez(i,j,k)  )/dy
                    - ( Ey(i,j,k+1)  - Ey(i,j,k)  )/dz  )
```

where dt denotes the time step, and dy, dz the dimensions of the unit cell along the y and z-axis, respectively. Note that the updating equations for the NS-FDTD method contain considerably more terms compared with the updating equation of the conventional algorithm, owing to the use of the nonstandard operators $d_m^{(i)}, i = 1, 2, 3$ and $m = x, y, z$ shown in Fig. 2.2, which "sample" additional neighbouring points in the calculations of the partial spatial derivatives compared with the conventional $d_m^{(1)}$ operator. As a result, the number of calculations per each time step do considerably increase with the implementation of the NS-FDTD methodology. However, the computational accuracy of the new algorithm increases dramatically, to the point which allows one to use much larger unit cells compared

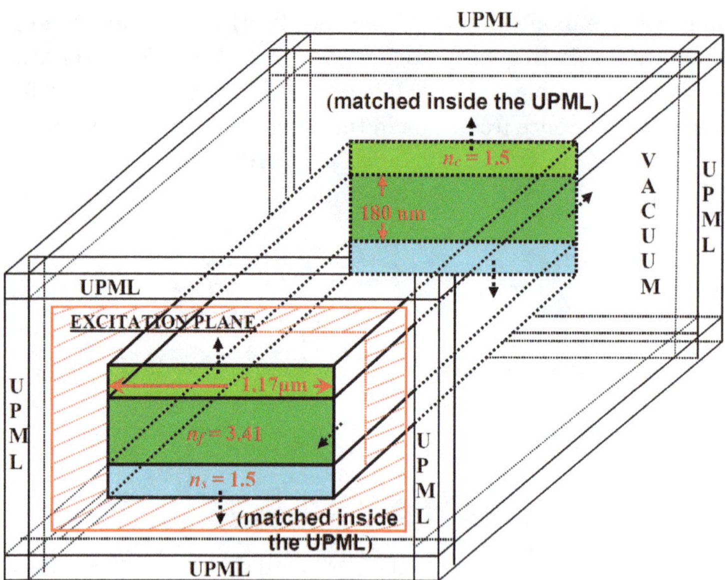

Figure 2.4: Schematic illustration of the three-dimensional dielectric waveguide simulated with the NS-FDTD method. The waveguide is laterally surrounded by air, while its top and bottom claddings extend inside the upper and bottom UPML layers, respectively. All three dielectric layers also longitudinally extend inside, both, the front and back UPML.

with the conventional algorithm, and still maintain the same (or better) levels of accuracy of the conventional algorithm. Thus, as we demonstrate in the following numerical simulations, improved accuracy or significant computational savings are achieved when the NS-FDTD strategy is deployed in the analysis of photonic structures.

Figure 2.5 presents a few snapshots from the propagation of the H_z-field component along the waveguide of Fig. 2.4. Note that at the excitation plane (at around the fiftieth unit cell in the longitudinal direction) there are, as expected, two pulses generated that travel in opposite directions. The left pulse is completely "absorbed" inside the left UPML and never enters again the main computational domain. By contrast, the right pulse is guided along the dielectric heterostructure until it reaches the right UMPL (not shown here) wherein it is also absorbed. Note, also, that the use of the proper excitation profile has allowed most of the exciting pulse's energy to be launched into the desired mode (here, into the E_{11}^y mode); otherwise, we would have observed a much more involved situation, with more than one modes being excited and propagating (with different speeds) along the waveguide.

Figure 2.6 shows the extracted longitudinal propagation constants, β, over the bandwidth of the corresponding excitation pulse, which were obtained by dividing the fast Fourier transforms (FFTs) of the pulse's time history at two fixed observation points, shown in Fig. 2.5b, along the core [Taflove and Hagness (2000)]

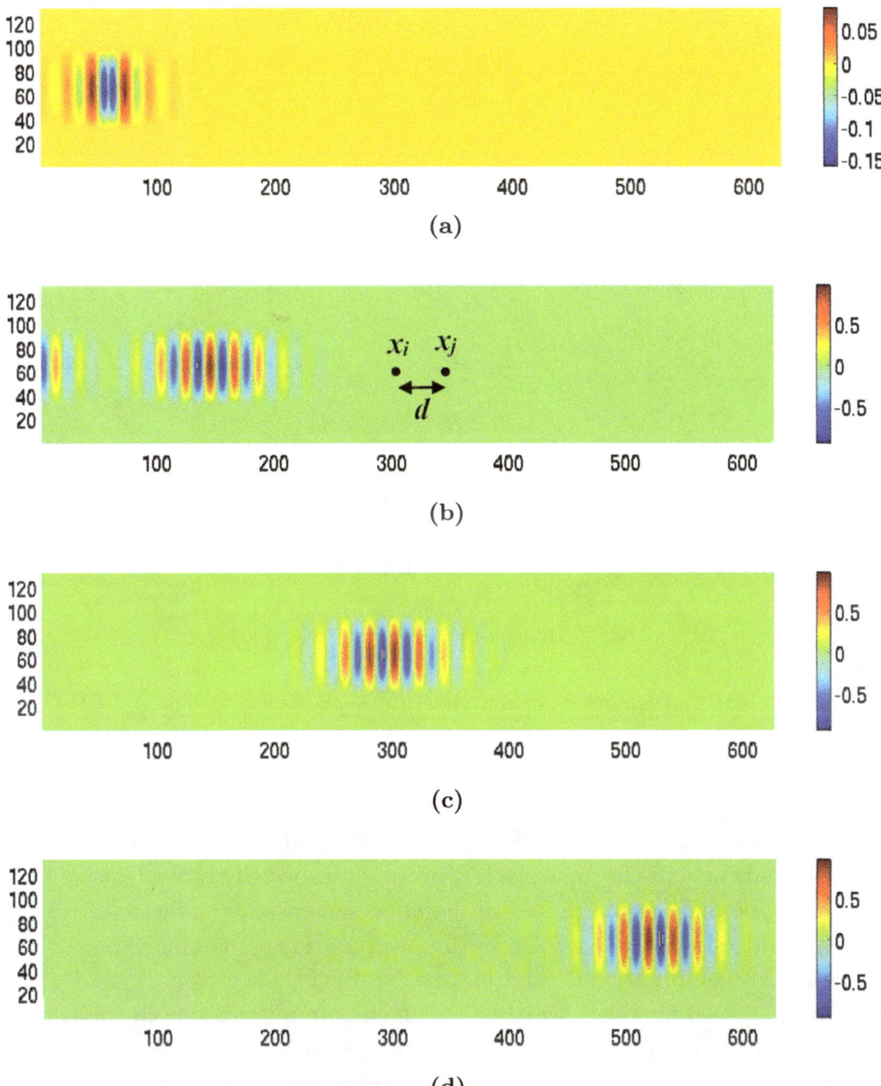

Figure 2.5: Snapshots from the propagation of the H_z-field component during the NS-FDTD simulation of the waveguide illustrated in Fig. 2.4. The plane on which the snapshots were obtained is in the middle of the waveguide core layer, lying parallel to the waveguide base. Shown here is the two-dimensional distribution of the H_z-field component on the aforementioned plane, at the time step: **(a)** $n = 1000$, **(b)** $n = 2000$, **(c)** $n = 3000$, and **(d)** $n = 4500$.

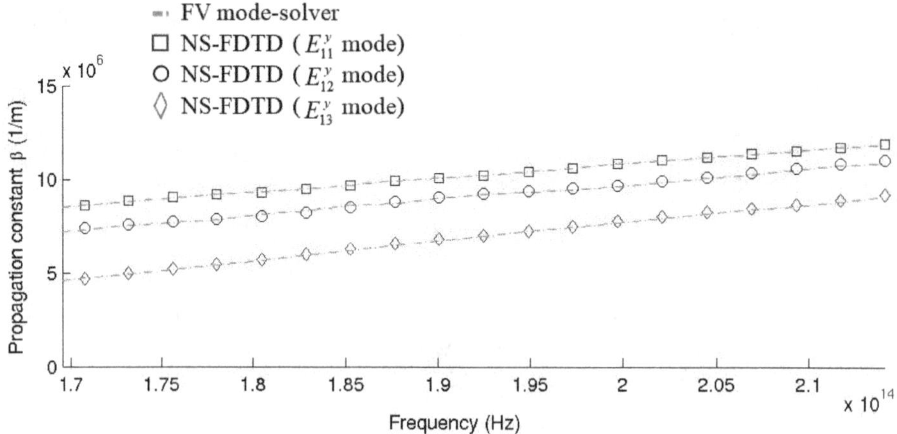

Figure 2.6: Longitudinal propagation constants of the first three E^y eigenmodes. Comparison between fully vectorial mode-solver and NS-FDTD (dashed line and symbols respectively).

Chapters 15–16, i.e. the following relation was used:

$$\beta(\omega) = \mathrm{Im}\left[\frac{1}{d}\ln\left(\frac{\mathrm{FT}\{H_z(t, x_i)\}}{\mathrm{FT}\{H_z(t, x_j)\}}\right)\right].$$

Note the excellent agreement between the values predicted by the NS-FDTD and the fully vectorial mode solvers for all three eigenmodes that confirms the precision of their excitation.

Figures 2.7a–c illustrate the relative error, with respect to the mode-solver, of the conventional and nonstandard FDTD. In each case it is found that the classical Yee scheme only attains the same levels of accuracy, especially in the high-frequency range, when the number of spatial grid points is increased by a factor of 6.5, with a corresponding significant increase in the computational time, as expected. The increase in the computational time for the NS-FDTD, however, is small due to the use of a 30% larger time-step than the maximum allowed one in the standard algorithm [Taflove and Hagness (2000); Cole (2002)]. It is also verified that optimum performance for this form of the NS-FDTD is achieved within a narrow region around f_0 [Cole (1995, 2002)] where it is always found to be more accurate than the usual Yee formulation.

These results allow one to conclude that, owing to the considerable reduction of the numerical dispersion and anisotropy errors, significantly enhanced overall computational performance is obtained when the NS-FDTD formulae is deployed. In particular, either improved accuracy or significant computational savings can be achieved when the nonstandard finite-difference concepts are incorporated within the FDTD modelling methodology of electromagnetic devices and structures.

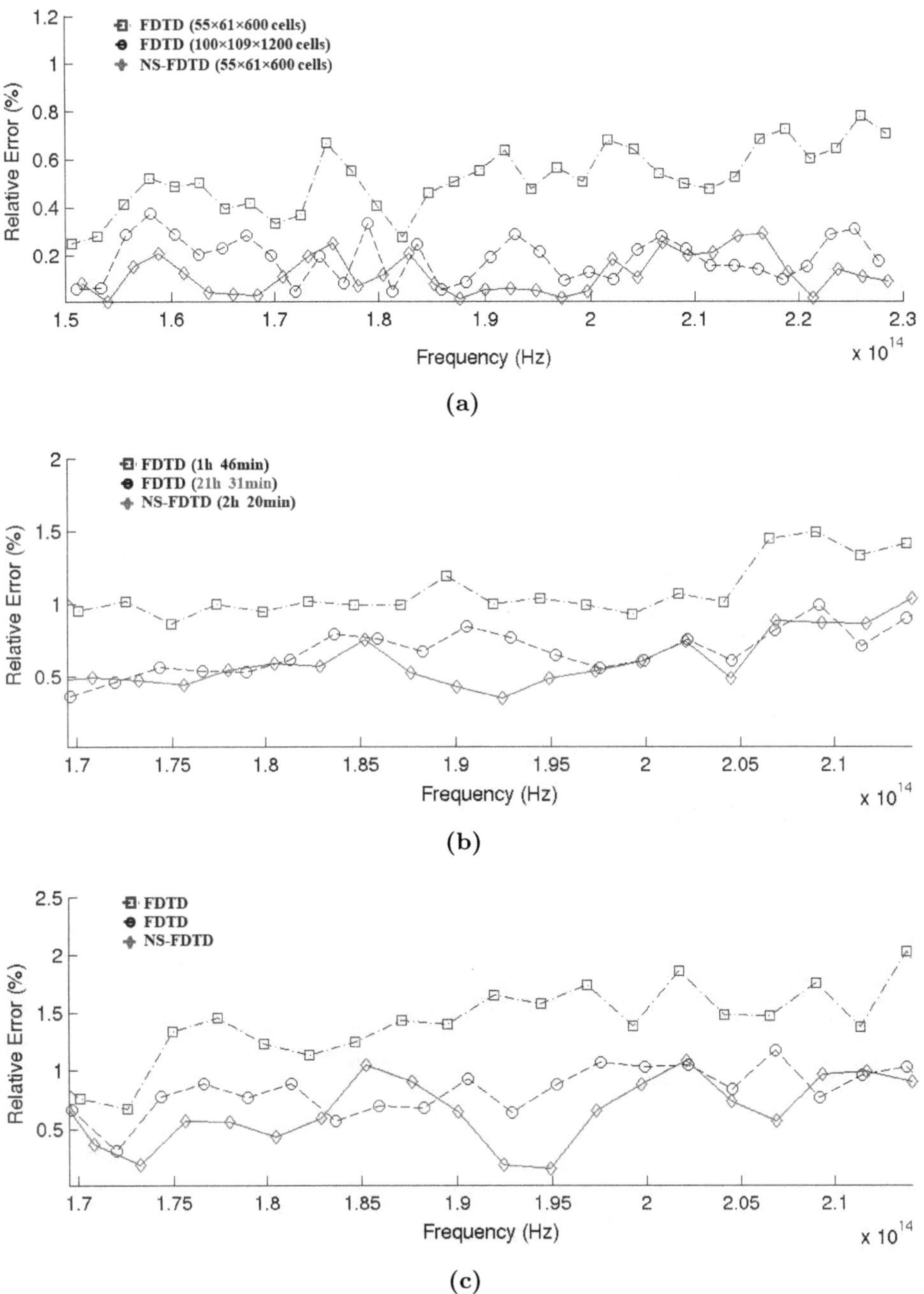

Figure 2.7: **(a)** Relative error in the NS-FDTD and FDTD-computed propagation constant of the fundamental E_{11}^{y} eigenmode. **(b)** The same calculations as in (a), but for the E_{12}^{y} eigenmode. **(c)** The corresponding calculations for the E_{13}^{y} mode supported by the same structure as in both previous cases.

Chapter 3

Light Propagation in Negative-Refractive-Index Metamaterials and Waveguides

3.1 Introduction

In the previous two chapters we investigated the propagation of light inside conventional dielectric bulk materials and waveguides. The totality of the structures that we therein examined were characterised by constitutive electromagnetic parameters (relative electric permittivity, ε_r, and magnetic permeability, μ_r) that were greater than or equal to unity, i.e. $\varepsilon_r \geq 1$ and $\mu_r \geq 1$.

The theory describing the interaction of light with and propagation within such structures is well-established, since until very recently media having $\varepsilon_r \geq 1$ and $\mu_r \geq 1$ were, indeed, considered to be the only ones that possess interesting and/or useful electromagnetic properties. For several decades, nonetheless, scientists were well-aware that more 'exotic' regions for the electromagnetic parameters of a medium can exist, such as e.g. in metals at optical frequencies, whereupon $\mathrm{Re}[\varepsilon_r] \leq 0$. However, the wider consensus was that – at least at the time – there was little to benefit from attempting to exploit such "exotic" materials, not least because usually they do not occur naturally and, also, because they were not thought of offering any particular advantage over their common dielectric counterparts. This view has, now, been revised and overturned, owing primarily to pioneering insights concerning the fundamental properties of such "exotic" materials, but also owing to technological progress that allowed for their construction and characterisation, all the way from the radio up to the optical regime.

In this chapter we will study in some detail the electromagnetic properties of materials possessing simultaneously negative electric permittivity and magnetic permeability, i.e. $\varepsilon_r \leq 0$ *and* $\mu_r \leq 0$. These materials have come to be known as metamaterials – although, at present, the same terminology is also used for a number of related materials, such as materials with only $\mathrm{Re}[\varepsilon_r]$ ($\mathrm{Re}[\mu_r]) \leq 0$ or $\mathrm{Re}[\varepsilon_r]$ ($\mathrm{Re}[\mu_r]) \simeq 0$, photonic crystals, etc – due to the fact that they allow for overcoming a number of common limitations associated with the electromagnetic properties of naturally occurring materials. Upon concisely reviewing the history behind these materials, we shall examine some of their fundamental properties, including the sign of their refractive index, Snell's law of refraction and Fresnel's formulas for

41

the reflection and transmission coefficients of a plane wave impinging at the interface between a dielectric material and a metamaterial, as well as the correct new expression for the electromagnetic energy density. We shall then study one of the most remarkable properties of such materials, namely the fact that a planar slab made of a metamaterial with $\varepsilon_r \leq 0$ and $\mu_r \leq 0$ can, in principle, act as a "perfect" lens, i.e. a lens capable of aberration-free, deep (in theory, infinite) subwavelength resolution. We shall see that this property arises ultimately from the presence of, so called, surface wave modes that are localised at and guided along the interfaces of a metamaterial with a dielectric material. We will conclude the chapter by presenting a detailed analysis into the various types of oscillatory and surface wave modes that metamaterial heterostructures can support and we will highlight some of their intriguing properties that will draw our attention in more detail on Chapter 5.

3.2   A Brief History of Metamaterials

The history of metamaterials appears to be dating back to the pioneering work of Kock [Kock (1948)] in the late 1940s. While working at Bell Labs with Sergei Schelkunoff, renowned for his "field equivalence principles" and for his work on antennas theory, Kock published a series of works wherein he proposed numerous ideas for constructing lightweight and small-volume "artificial dielectrics", used as microwave lenses in antenna systems. Amongst others, he studied the response to an incident quasi-static electromagnetic radiation of isolated or regularly-arrayed metallic particles of various shapes, such as spheres, discs, ellipsoids and prolate or oblate spheroids. He concluded that such structures *effectively* behave as a dielectric medium, whose permittivity ε and permeability μ can be purposely tuned (but not independently of each other) to an arbitrarily large or small, even negative, value by properly arranging the particles in three dimensions, i.e. the optical properties of the medium depended *solely* on the particles' geometrical set up, rather than on their own intrinsic behaviour. Kock also showed that a specially-designed structure, which recently has come to be known as "split-ring resonator" (SRR), can be used to independently *increase* the effective magnetic permeability μ, such that one can reduce or altogether eliminate the diamagnetic nature of the aforementioned composite structures. His work rose considerable interest within the engineering community of the time, with a number of works extending or elaborating on his ideas. Since then, it has been the subject of detailed coverage in standard engineering textbooks [Collin (2001)].

More than a decade later, Rotman [Rotman (1962)] also considered the quasi-static response of an array of thin conducting wires, and he showed that such a structure closely resembles, on the macroscopic level, a plasma medium. In particular, he proved that the electric permittivity ε of this artificial dielectric medium varies with frequency following a Drude-type law. Consequently, below a certain "cutoff" frequency no incident electromagnetic radiation could penetrate it. Critically, however, neither Kock nor Rotman nor, indeed, any of the early contributors investigated the properties of media exhibiting concurrently negative ε *and* μ. Partly, that was because the main motivation behind similar works at that time was to design plasma media at RF or microwave frequencies that would closely simulate the *ionosphere*, prompted by NASA's desire to secure the safe re-entrance of space-capsules into the earth's atmosphere.

Veselago [Veselago (1967, 1968)] was evidently the first to systematically consider, in the late 1960s, the possibility and some properties of "double-negative materials" (DNGMs). In particular, he showed that a negative electric permittivity ε and magnetic permeability μ would imply a reversal of almost all known electromagnetic phenomena, including the angle of refraction inside a DNGM (e.g., $\theta_t = -20°$), the Doppler effect, the sign of the refractive index n (e.g., $n = -1$) and the right-handedness of the **E**, **H** and **k** vector-triad, from where the designation of such materials as "left-handed" origins. Moreover, Veselago also proved that the phase velocity (v_{ph}) of a plane wave propagating inside a left-handed metamaterial (LH-MM) has a direction opposite to that of the group (v_g) and energy-flow velocities (v_E). However, in spite of his noteworthy findings, and apparently unaware of Kock's and other scientists' research in the same field, Veselago did not go on to materialise his theoretical conclusions. Even so, his work did not go unnoticed and he was invited several times to highlight his research at major international scientific conferences [Burstein and DeMartini (1974)].

At present, the realm of "artificial dielectrics" or "meta-materials" (from the Greek word "meta", which here means "beyond") enjoys a breadth of scientific activity and exploration, having established a sound and coherent mathematical formalism [Eleftheriades and Balmain (2005); Engheta and Ziolkowski (2006); Caloz and Itoh (2006); Milonni (2005)], the predictions of which have been verified by numerous experimental and numerical-simulation works. This revived interest followed from a series of works by Professor Sir John Pendry, wherein he proposed practical means for realizing LH-MMs experimentally [Pendry *et al.* (1998, 1999)]. Moreover, building on Veselago's work, Pendry showed that a slab constructed by a LH-MM having refractive index $n = -1$ could, ideally, act as a "perfect lens", overcoming the well-known diffraction limitations. After these key insights, the physical construction of a composite LH-MM structure has been demonstrated by Shelby *et al.* [Shelby *et al.* (2001b,a)] and the possibility of achieving subwavelength resolution of an object with the same structure has been demonstrated with a series of further theoretical [Smith *et al.* (2002, 2003)] and experimental [Parazzoli *et al.* (2003); Houck *et al.* (2003)] works.

3.3 Sign of the Refractive Index and Energy Density Expression in Passive "Double Negative" Metamaterials

Let us consider a passive medium that is characterised by simultaneously negative (effective) electric permittivity, ε, and magnetic permeability, μ, e.g. $\varepsilon = \mu = -1$. Such media do not normally occur in nature but, as we shall see in the next chapter, they can indeed be constructed in the lab by suitably engineering the electromagnetic "molecules" of a medium. At first sight, one may think that the refractive index of such a "double negative" medium, which is defined as $\sqrt{(\varepsilon\mu)/(\varepsilon_0\mu_0)}$ should still be positive since the product "$\varepsilon\mu$" appearing in the expression for the refractive index remains positive. However, when one also considers the requirement that the "double negative" medium should, in general, be passive, the sign of the square root of the product "$\varepsilon\mu$" actually turns out to be negative [Ziolkowski and Heyman (2001); Kinsler and McCall (2008)].

Indeed, in general, the relative electric permittivity ε_r and magnetic permeability μ_r of a medium are complex functions of frequency, i.e., in an obvious notation, they are of the form:

$$\varepsilon_r(\omega) = \varepsilon_r'(\omega) + i\varepsilon_r''(\omega) = \rho_\varepsilon(\omega)e^{i\theta(\omega)}, \tag{3.1a}$$

$$\mu_r(\omega) = \mu_r'(\omega) + i\mu_r''(\omega) = \rho_\mu(\omega)e^{i\phi(\omega)}, \tag{3.1b}$$

and therefore:

$$\sqrt{(\varepsilon\mu)/(\varepsilon_0\mu_0)} = \sqrt{\rho_\varepsilon\rho_\mu}\,e^{i(\theta+\phi)/2}. \tag{3.2}$$

From Eq. (3.2) one may immediately recognise that if the medium is to be passive, i.e., having a positive imaginary part (for an assumed $e^{-i\omega t}$ time dependence), it should be:

$$0 \leq \frac{1}{2}(\theta + \phi) \leq \pi. \tag{3.3}$$

On the other hand, in a double negative metamaterial the real parts of the (effective) permittivity and permeability are negative, which based on on Eq. (3.1) implies that:

$$\frac{\pi}{2} < \theta < \frac{3\pi}{2}, \tag{3.4a}$$

$$\frac{\pi}{2} < \phi < \frac{3\pi}{2}, \tag{3.4b}$$

or, by adding the two parts of Eq. (3.4):

$$\frac{\pi}{2} < \frac{1}{2}(\theta + \phi) < \frac{3\pi}{2}. \tag{3.5}$$

Comparison of Eqs. (3.3) and (3.4) leads to the following relation:

$$\frac{\pi}{2} < \frac{1}{2}(\theta + \phi) < \pi, \tag{3.6}$$

from whence it is unambiguously concluded that the real part of the medium's refractive index (obtained from Eq. (3.2)) will be negative. Thus, in a medium that has simultaneously (in the same frequency region) negative real parts in its electric permittivity and magnetic permeability, the real part of the refractive index will also be negative. This fact has immediate consequences on the way an electromagnetic wave refracts inside such a medium, but also on the three-dimensional spatial arrangement of the electric field vector, $\mathbf{E}$, magnetic field vector, $\mathbf{H}$, and the wavevector, $\mathbf{k}$, as is explained in the next section.

It is interesting to note that any medium having simultaneously negative ε and μ must necessarily be dispersive or the energy density will be negative. Indeed, if the medium was not dispersive, then the cycle-averaged energy density of an electromagnetic wave propagating inside it would have been:

$$W = \frac{1}{4}\left(\varepsilon|\mathbf{E}|^2 + \mu|\mathbf{H}|^2\right), \tag{3.7}$$

which, in view of the fact that $\varepsilon < 0$ and $\mu < 0$, would have obviously been negative. Thus, for a double negative metamaterial the correct expression for the cycle-averaged energy density is that of a dispersive medium [Veselago (1967, 1968)], i.e.:

$$W = \frac{1}{4}\left[\frac{d(\varepsilon\omega)}{d\omega}|\mathbf{E}|^2 + \frac{d(\mu\omega)}{d\omega}|\mathbf{H}|^2\right], \tag{3.8}$$

which turns out to be positive for every frequency, since causality and Kramers–Kronig relations for such a metamaterial demand [Milonni (2005)] that $\frac{d(\varepsilon\omega)}{d\omega} > 0$ and $\frac{d(\mu\omega)}{d\omega} > 0$.

3.4 Refraction and E-H-k Vector Triad Inside a "Double Negative" Metamaterial

As we remarked in the previous section, the fact that the real part of the refractive index of a DNGM is negative has some immediate consequences, most notably, on the way a plane wave refracts at the interface between a dielectric medium and a DNGM.

Indeed, by a straightforward application of boundary conditions (Section 1.4) at the interface between two media with (real parts of) refractive indices n_1 and n_2, we obtain the following relations for the angles of incidence, θ_{inc}, reflection, θ_{refl}, and refraction, θ_{refr}:

$$\theta_{refl} = \theta_{inc}, \tag{3.9a}$$

$$\theta_{refr} = \mathrm{sgn}(n_2)\sin^{-1}\left(\frac{n_1}{|n_2|}\sin(\theta_{inc})\right). \tag{3.9b}$$

From the second part of of Eq. (3.9) one sees that when the second medium has negative (real part of) refractive index, then the angle with which a plane wave refracts inside this medium is "negative", i.e. the wave refracts at the same side of the normal to the media interface as the incident wave (see Fig. 3.1). Hence, remarkably, the refraction of a plane wave, incident from a conventional dielectric medium to an isotropic DNGM, is "opposite" to that of the usual case, wherein refraction occurs inside another (positive index) dielectric medium.

The fact that the refraction of a plane wave inside an isotropic DNGM is "negative" leads to the notable possibility of using a planar slab made of such a material

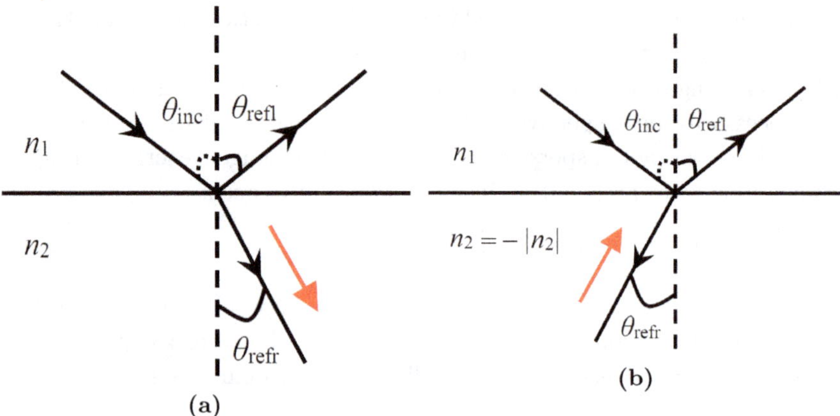

Figure 3.1: Reflection and refraction of a plane wave at the interface of a dielectric medium of refractive index n_1 with **(a)** a dielectric medium of refractive index n_2, and **(b)** a double negative metamaterial with real part of refractive index $n_2 = -|n_2|$. In both cases the solid black arrows denote the direction of power flow, while the solid red arrows reveal the direction of phase propagation inside the second medium.

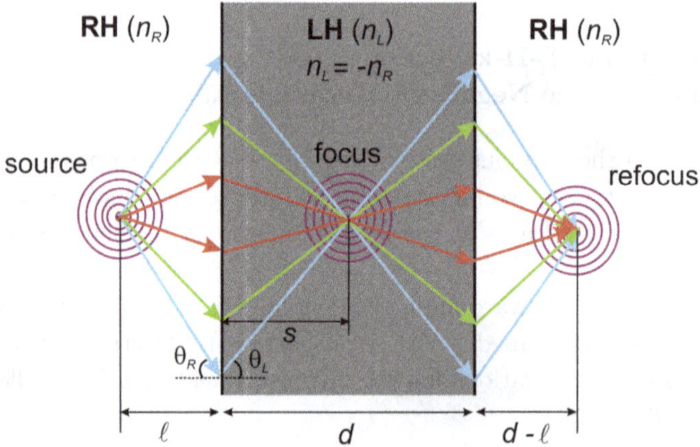

Figure 3.2: Double focusing of a source with a planar double negative (left-handed) metamaterials slab.

to bring light to a double focus, as illustrated in Fig. 3.2. Indeed, one may readily infer that a slab of thickness d and refractive index of, e.g., $n = -1$, surrounded by air, will bring all rays emanating from a source to a double focus. First, at a point inside the DNGM slab, at a distance $s = l < d$, where l is the distance of the source from the slab, and second at a point outside the slab, at a distance $d-l$. Hence, such a slab acts like a lens, and is able to bring the rays (corresponding to propagating waves) radiated by a source to a focus outside the slab, *without* reflections occurring at the media interfaces because the $n = -1 + i0$ slab is impedance-matched to free space (see Section 3.5). This result can be further verified by means of direct

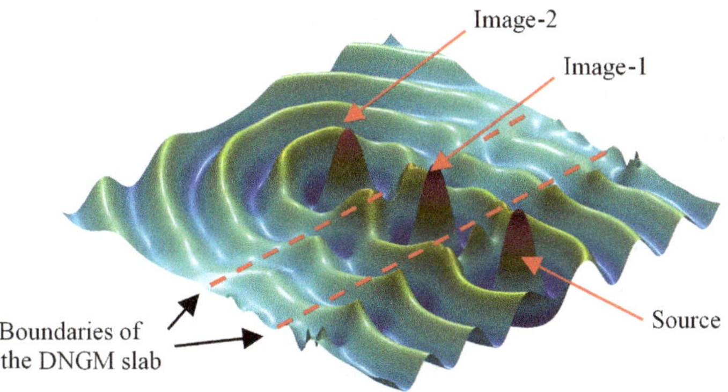

Figure 3.3: Snapshot from an FDTD simulation of an electromagnetic pulse radiated by a point source and being incident on a DNGM slab ($n = -1$) that is surrounded by air.

FDTD (numerical) solutions of Maxwell's equations. An exemplary result of such simulations is presented in Fig. 3.3 above, where we can see a snapshot from the interaction of an electromagnetic pulse with a DNGM slab, impedance matched to the surrounding medium (which is air). Note the formation of a double mirror image of the source, once inside the DNGM slab and a second time outside the slab. We can also observe that there are no reflection occurring at the two media interfaces; the only reflections observed are those which occur close to the (imperfect for DNGM) absorbing boundary conditions that are used to terminate our computational space. The small reflections that we observe in this figure, as well as the departure of the numerical wavefronts from a perfectly cylindrical shape, are entirely owing to the use of imperfect boundary conditions to simulate the extension of the DNGM slab to infinity.

It is to be noted that the aforementioned DNGM slab configuration not only is it capable of bringing outgoing light rays to a focus, but it does so without suffering from the primary (monochromatic or Seidel) aberrations (spherical, come, astigmatism, field curvature and distortion) that usually characterise conventional (e.g., spherical) lenses [Born and Wolf (1999)]. This is because the DNGM lens, being planar, may extent "infinitely" in the transverse plane, capturing all rays and focusing them at a point exactly 'in front of' each object source point, as was shown in Fig. 3.2. Thus, there will be no distortions of the object at the image/focusing plane. Moreover, one may, in principle, suitably engineer an isotropic DNGM that has a refractive index of $n = -1 + i0$ over a continuous, but finite, range of frequencies. If such a material, surrounded by air, is used (in the frequency region where $n = -1 + i0$) as a lens, then this lens will also be free of *chromatic* aberrations. Even more remarkable, however, is the fact that such a lens is not limited even by diffraction, i.e. it is capable of reconstructing at the image plane even the highly evanescent components "radiated" off by an object source, which decay rapidly away from the source and are normally completely lost. Therefore, such a configuration acts like a "perfect" lens, and will be the subject of a closer examination in Section 3.6.

Finally, one should note yet another interesting property of an isotropic DNGM, pertaining to the three-dimensional arrangement of the vectors $\mathbf{E}$, $\mathbf{H}$ and $\mathbf{k}$ inside such a material. From the first two of Maxwell's equations (Eqs. (1.8) and (1.9)), as well as from the constitutive relation given by Eqs. (1.1) and (1.2), it can readily be shown that inside every isotropic (possibly dispersive) medium, it is:

$$\mathbf{k} \times \mathbf{E} = \omega\mu(\omega)\mathbf{H}, \qquad (3.10a)$$

$$\mathbf{k} \times \mathbf{H} = \omega\varepsilon(\omega)\mathbf{E}. \qquad (3.10b)$$

We see from Eq. (3.10) that for for a conventional dielectric medium with $\varepsilon, \mu > 0$ (in a given frequency region), the vectors $\mathbf{E}$, $\mathbf{H}$ and $\mathbf{k}$ form a right-handed triad. By contrast, inside a DNGM where $\varepsilon, \mu > 0$, the vectors E, H and k form a *left*-handed triad, from where the designation of such media as "left-handed" (LH) arises. It should also be pointed out that the definition of the Poynting vector, $\mathbf{S} = \mathbf{E} \times \mathbf{H}$, remains the same for both classes of media (right- and left-handed). As a result, whereas in a right-handed medium the direction of the power flow (given by the direction of $\mathbf{S}$) is the same as the direction of the phase evolution (given by the direction of the wavevector $\mathbf{k}$), in a left-handed medium the vectors $\mathbf{k}$ and $\mathbf{S}$ are antiparallel, i.e. the phase evolves/"propagates" in the opposite direction compared with the direction of the flow of power, as illustrated in Fig. 3.1.

3.5 Fresnel's Formulas for Plane Wave Incidence at a Planar RH/LH Media Interface

It is interesting to examine how the well-known Fresnel's formulas for the reflection and transmission coefficient of a monochromatic plane wave incident at the interface between two dielectric media are modified when the second medium is assumed to be a DNGM one.

To this end, let us assume that the incident, reflected and refracted electric field is of the form:

$$\mathbf{E}_{\text{inc}} = \mathbf{E}_{0\text{inc}}e^{-i(\omega t - \mathbf{k}_{\text{inc}}\cdot\mathbf{r})}, \qquad (3.11a)$$

$$\mathbf{E}_{\text{refl}} = \mathbf{E}_{0\text{refl}}e^{-i(\omega t - \mathbf{k}_{\text{refl}}\cdot\mathbf{r})}, \qquad (3.11b)$$

$$\mathbf{E}_{\text{refr}} = \mathbf{E}_{0\text{refr}}e^{-i(\omega t - \mathbf{k}_{\text{refr}}\cdot\mathbf{r})}, \qquad (3.11c)$$

respectively, where both media were assumed to be conventional dielectrics – otherwise the direction of the wavevector $\mathbf{k}_{\text{refr}}$ inside the second, DNGM, medium would have been reversed. Then, from Eq. (3.11), the incident, reflected and refracted

B-field will, respectively, be of the form:

$$\mathbf{B}_{\mathbf{inc}} = \frac{1}{\omega}\mathbf{k}_{\mathbf{inc}} \times \mathbf{E}_{\mathbf{inc}}, \tag{3.12a}$$

$$\mathbf{B}_{\mathbf{refl}} = \frac{1}{\omega}\mathbf{k}_{\mathbf{refl}} \times \mathbf{E}_{\mathbf{refl}}, \tag{3.12b}$$

$$\mathbf{B}_{\mathbf{refr}} = \frac{1}{\omega}\mathbf{k}_{\mathbf{refr}} \times \mathbf{E}_{\mathbf{refr}}. \tag{3.12c}$$

In the case where the second medium is a DNGM one, we proceed similarly to the case where both media are normal dielectrics, by implementing the standard boundary conditions (Section 1.4) for the tangential field and wavevector components. Following such an analysis for the incident and reflected fields, with the electric field E assumed perpendicular to the plane of incidence (Fig. 3.4), yields Snell's law of reflection (Eq. (3.9)) that was highlighted above, as well as the following reflection coefficient:

$$R = \frac{E_{0refl}}{E_{0inc}} = \frac{(n_1/\mu_1)\cos\theta_{inc} - (|n_2|/|\mu_2|)\sqrt{1 - (n_1/n_2)^2\sin^2\theta_{inc}}}{(n_1/\mu_1)\cos\theta_{inc} + (|n_2|/|\mu_2|)\sqrt{1 - (n_1/n_2)^2\sin^2\theta_{inc}}}. \tag{3.13}$$

Likewise, one can obtain the following expression for the transmission coefficient of a monochromatic plane wave incident at the interface between a dielectric medium and a DNGM:

$$T = \frac{E_{0refr}}{E_{0inc}} = \frac{2(n_1/\mu_1)\cos\theta_{inc}}{2(n_1/\mu_1)\cos\theta_{inc} + (|n_2|/|\mu_2|)\sqrt{1 - (n_1/n_2)^2\sin^2\theta_{inc}}}. \tag{3.14}$$

Similar expressions to Eqs. (3.13) and (3.14) are obtained and for the case where the electric field is parallel to the plane of incidence. As expected, when the second medium is characterised by $\varepsilon_2 = -\varepsilon_1$ and $\mu_2 = -\mu_1 (\Rightarrow n_2 = -n_1)$, then we immediately obtain from Eq. (3.13) that $R = 0$, and from Eq. (3.14) $T = 1$, i.e.

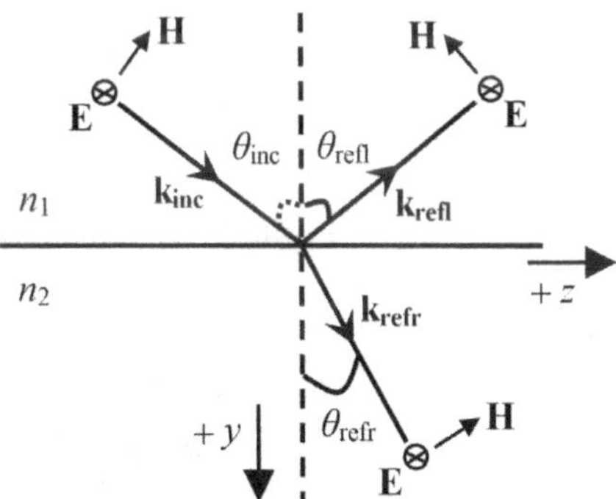

Figure 3.4: Schematic illustration of a monochromatic plane wave incident at the interface between two conventional dielectric media. Note that the vectors **E**, **H** and **k** form everywhere a right-handed triad. If the second medium is a DNGM one then, both, the sign of the angle of refraction, θ_{refr}, and the direction of the wavevector, $\mathbf{k}_{\mathbf{refr}}$, inside the second medium should be reversed.

in this case the DNGM is perfectly matched to the dielectric medium-1 and no reflection occurs.

It is interesting to note that Eqs. (3.13) and (3.14) are valid, both, when the second medium is a normal dielectric or when it is a left-handed metamaterial. This is, of course, owing to the fact that these equations equations contain the absolute values of the electromagnetic parameters of the second medium, and as such they do not lead to different results when the handness of the medium is reversed. However, it is important to realise that these equations were derived for waves that were incident to the media interface with angle of incidence $\theta_{inc} < \theta_c$, where θ_c is the critical angle (given by: $\sin\theta_c = n_2/n_1$, $n_2 < n_1$), i.e. for propagating waves.

When the incident wave impinges on the two media interface with an angle $\theta_{inc} > \theta_c$, one needs to suitably modify the expression giving the term:

$$\theta_{inc} = \sqrt{1 - (n_1/n_2)^2 \sin^2\theta_{inc}},$$

which appears in Eqs. (3.13) and (3.14). This term then becomes:

$$\cos\theta_{inc} = \pm i\sqrt{(n_1/n_2)^2 \sin^2\theta_{inc} - 1},$$

and since the refracted (transmitted) wave propagates in the $+y$ direction (see Fig. 3.4), with a spatial dependence of the form

$$\exp\left(i\frac{\omega y}{c} n_2 \cos\theta_{refr}\right),$$

with $n_2 < 0$, we should choose the minus ("$-$") sign in the aforementioned expression for $\cos\theta_{inc}$, so that the refracted wave will not diverge at infinity. By direct substitution of this expression for $\cos\theta_{inc}$ into Eq. (3.13) we obtain:

$$r\big|_{evsc} = \frac{\cos\theta_{inc} + i(\mu_1/|\mu_2|)\sqrt{\sin^2\theta_{inc} - (n_2/n_1)^2}}{\cos\theta_{inc} - i(\mu_1/|\mu_2|)\sqrt{\sin^2\theta_{inc} - (n_2/n_1)^2}} = |r|e^{i\phi}. \tag{3.15}$$

In the above equation, the term

$$\phi = 2\tan^{-1}\left[\frac{(\mu_1/|\mu_2|)}{\cos\theta_{inc}}\sqrt{\sin^2\theta_{inc} - (n_2/n_1)^2}\right]$$

is the, so-called, Goos–Hänchen phase shift that an incident ray experiences upon hitting the two media interface with an angle greater than the critical one, θ_c. We shall again return to this term later, in Chapter 5, when we will study the zigzag ray propagation inside a "slow-light" DNGM waveguide. Here, though, it is interesting to note from Eqs. (3.11) and (3.15) that, in the present case, the incident ray will be spatially-shifted in the $-z$ direction (see Fig. 3.4) hitting the media interface with an angle $\theta_{inc} > \theta_c$, contrary to the case where both media are normal dielectrics whereupon the rays are always shifted in the $+z$ direction.

3.6 Metamaterial-Enabled "Perfect" Lens

We will now turn our attention to a closer examination of one of the most remarkable properties of a DNGM slab, on which we briefly remarked on Section 3.4, namely the fact that such a slab may, in principle, work as a "perfect" lens, enabling super-resolution of an object at the image plane [Pendry (2000)].

To this end, let us start by calculating the reflection and transmission coefficient of an evanescent wave (not a plane wave, as in the previous Section 3.5) incident at a planar RH/LH interface located at $z = 0$, with the wave initially being in the RH medium (air). Assuming, without loss of generality, that the electric field is polarised along the x-axis (see Fig. 3.4), i.e. perpendicularly to the plane of incidence, the following expressions are obtained for the incident and transmitted (electric) field, respectively:

$$\mathbf{E_{inc}} = \mathbf{E_{0inc}}e^{i(k_y y + k_z^{inc} z - \omega t)}\hat{\mathbf{x}}_0, \tag{3.16}$$

with $k_z^{inc} = i\sqrt{k_y^2 - (\omega/c)^2}$,

$$\mathbf{E_{trans}} = \mathbf{E_{0trans}}e^{-i(k_y y + k_z^{trans} z - \omega t)}\hat{\mathbf{x}}_0, \tag{3.17}$$

with $k_z^{trans} = i\sqrt{k_y^2 - \varepsilon\mu\omega^2}$,

where $\varepsilon \to \varepsilon_0$, $\mu \to \mu_0$ are respectively the (negative) permittivity and permeability of the LH medium. Applying the standard boundary conditions for the tangential field components at the media interface, results in the following expressions for the reflection and transmission coefficients:

$$t = \frac{2\mu k_z^{inc}}{\mu k_z^{inc} + \mu_0 k_z^{trans}}, \tag{3.18}$$

$$r = \frac{\mu k_z^{inc} - \mu_0 k_z^{trans}}{\mu k_z^{inc} + \mu_0 k_z^{trans}}. \tag{3.19}$$

If the incident evanescent wave is initially inside the LH medium, then the new reflection and transmission coefficients are obtained, simply, by interchanging the μk_z and $\mu_0 k_z^{trans}$ terms in Eqs. (3.18) and (3.19). Furthermore, if the LH medium is assumed to be lossy, i.e. if it has electromagnetic parameters of the form:

$$\varepsilon = -\varepsilon_0(1 - i\zeta),$$

$$\mu = \mu_0(1 - i\zeta),$$

where $\zeta > 0$, then (in the limit $\zeta \to 0$) it is straightforward to show that Eqs. (3.18) and (3.19) take the form:

$$t = r = \frac{2i}{\zeta}\left(1 - \frac{\omega^2}{c^2 k_y^2}\right). \tag{3.20}$$

When *two* RH/LH interfaces are present, as is the case with a LH slab surrounded by air, an evanescent wave incident (from the air) to the slab will be reflected and transmitted multiple times at the media interfaces. A detailed analysis [Smith *et al.* (2003)] then reveals that, in this case, the overall transmission coefficient associated with the (evanescent) wave exiting the slab is given by:

$$t_{total} = \left[\left[\frac{1}{2} + \frac{1}{4} \left(\frac{\mu_0 k_z^{trans}}{\mu k_z^{inc}} + \frac{\mu k_z^{inc}}{\mu_0 k_z^{trans}} \right) \right] e^{k_z^{trans} d} \right.$$

$$\left. + \left[\frac{1}{2} - \frac{1}{4} \left(\frac{\mu_0 k_z^{trans}}{\mu k_z^{inc}} + \frac{\mu k_z^{inc}}{\mu_0 k_z^{trans}} \right) \right] e^{-k_z^{trans} d} \right]^{-1} \qquad (3.21)$$

where d is the distance between the two interfaces, i.e. the slab thickness.

With the aid of Eqs. (3.16) and 3.17) one can, now, directly infer from Eq. (3.21) that when $\varepsilon \to \varepsilon_0, \mu \to \mu_0$, the first term in Eq. (3.21) becomes equal to zero and we, therefore, obtain:

$$t_{total} = e^{k_z^{trans} d} \qquad (3.22)$$

Thus, in this case, the evanescent field that impinges upon the LH slab does not decay further inside the slab, as it usually occurs with dielectric or metallic slabs/ films, but is instead "amplified" inside the LH slab. Realizing that evanescent waves are associated with the high spatial frequencies of the field "radiated" by an object source (see Eqs. (3.16) and (3.17)), i.e. they are the part of the field that 'carries' the fine subwavelength features of an object source, Eq. (3.22) implies that the subwavelength features of an object can be recovered at the image plane. Therefore, a DNGM slab can, in principle, enable one to obtain the image of an object with "perfect" resolution, containing all the subwavelength features of an object and overcoming the usual diffraction limitations that characterise conventional lenses.

When losses are present in the DNGM slab then, as one may intuitively expect, "perfect" resolution can not be attained any more. In this case, a straightforward analysis based on Eqs. (3.20) and (3.21) reveals that the smallest feature that can be resolved at the image plane has a size $\Delta = (2\pi d)/|\ln \zeta|$, where ζ is the imaginary part of the DNGM's effective permittivity and permeability (see Eq. (3.20)). Thus, even in this case, subwavelength resolution, i.e. resolution $\Delta < \lambda$, of an image can still be obtained, so long as the losses are smaller than an upper limit $\zeta < e^{-2\pi d/\lambda}$. This remarkable feature of a DNGM slab is, ultimately, owing to the fact that the slab supports a special class of waves at its boundaries, known as "surface plasmon polaritons", to the study of which we now turn our attention.

3.7   Surface Plasmon Polaritons in Asymmetric DNGM Slab Heterostructures

As we show in the previous section, the "perfect lens" action relies critically on the amplification of an object's near field in a surface wave (SW)-like manner inside a

LH slab. Due to momentum mismatch, radiative waves cannot couple directly to the formed SWs (or surface polaritons) at the interfaces of the slab with positive-index dielectrics. Pendry, however, showed that the near field of an object, which describes its finest features and decays exponentially away form the source (evanescent), can couple to an exponentially increasing field inside the LH slab that decays similarly on the other side. The whole field pattern resembles that of a surface polariton, although the proof of its existence for the particular "perfect lens" structure ($\varepsilon = \mu = -1$) was not further elaborated in [Pendry (2000)], as well as in other more detailed analyses [Shadrivov *et al.* (2003)]. Generalisations of such analyses were investigations of asymmetric DNGM slab configurations for lensing [Ramakrishna *et al.* (2002); Wu *et al.* (2003); He *et al.* (2005)], sensing and directional coupling [Qing and Chen (2004)] applications. In both cases, the role of the coupled surface polaritons at the two interfaces of the slab waveguide was shown to be of crucial importance. For the first class of applications, the asymmetry helped to improve the limit imposed on the image resolution by the losses of the core. For the second class it improved the amplification of the evanescent waves in the device-working region, leading to enhanced performance. In this section we shall identify and classify in detail all surface plasmon polariton (SPP) eigenmodes supported by generalised asymmetric slab heterostructures. To this end, a rigorous analytical study will be pursued, which will prove that a total of 30 solutions to the involved characteristic equation giving the SPP eigenmodes can exist for all choices of the refractive index distribution, constitutive parameters ε and μ and the thickness of the core [Tsakmakidis *et al.* (2006c)]. Such an approach is essential [Prade *et al.* (1991)], particularly for the investigation of asymmetric slab configurations, because the graphical methodologies that have been proposed in the past for the modal analysis of LH waveguides [Shadrivov *et al.* (2003); Ramakrishna *et al.* (2002); Wu *et al.* (2003); He *et al.* (2005); Qing and Chen (2004)] did not reveal all SPP eigenmodes. We will see that a suitably modified form of the associated transcendental equation, derived from macroscopic electrodynamics using the well-known boundary conditions for the tangential electric and magnetic field components (Section 1.4), can obviate this limitation and allow for an analytical, unified treatment. We will confine ourselves to the discussion of the geometric dispersion (SPP effective index vs. reduced slab thickness) since, as we shall see in the following chapter, negative material parameters occur near resonances; hence most experimental realisations of LH materials considered so far were for narrow bands. In addition, all transmission (lensing) analyses of LH slab heterostructures, as well as investigations of asymmetric LH [Qing and Chen (2004)] and metallic [Prade *et al.* (1991); Burke *et al.* (1986)] films in the past, involved mainly monochromatic waves. All the important modal features, such as the number and classification of modes, number and kind of cutoffs, field enhancement, phase reversal and possible double mode-degeneracy occurrence, can be derived in a clear and conclusive way following this methodology.

The organization of this section is the following. Section 3.7.1 will make some introductory remarks regarding SPP waves at a single interface between a right-handed (RH) and a LH material. Emphasis is given on the conditions for the existence of such waves, as these are used later when the effects of retardation are taken into account. Section 3.7.2 is devoted to the discussion of the SPP eigenmodes supported by an asymmetric slab waveguide with a negative refractive index core. Following a macroscopic analysis, the DNGM waveguide is treated as a boundary value problem. Special solutions of the scalar wave equation are sought, subject to boundary conditions, to obtain the characteristic equation of the SPP eigenmodes. From the restrictions inherent in this equation, which depend on the refractive index distribution, we identify all supported SPP eigenmodes and classify them as forward or backward propagating via a closed-form expression for the total power flow P in the guide. Finally, in Section 3.7.3 we will summarise the findings of this section and we will present the main conclusions of this study.

3.7.1 *Surface Plasmon Polaritons at a Planar LH/RH Interface*

The negative permittivity and permeability of a LH medium allows for the existence of surface waves (SWs), also called surface polaritons (SPs), at the interface with a conventional dielectric. In order to investigate these solutions and derive the conditions for their existence we consider the geometry illustrated in Fig. 3.5. Here, both media are considered to be semi-infinite, homogeneous [Shadrivov *et al.* (2003)] and isotropic [Shelby *et al.* (2001b)].

Medium 1 has negative relative permittivity $\varepsilon_{r1} = -\varepsilon_{r1p} < 0$ and permeability $\mu_{r1} = -\mu_{r1p} < 0$, whereas in medium 2 we assume $\varepsilon_{r2} > 0$ and $\mu_{r2} > 0$. The coordinate axes are chosen so that the z-axis is directed along the SP propagation and the x-axis is perpendicular to the media interface.

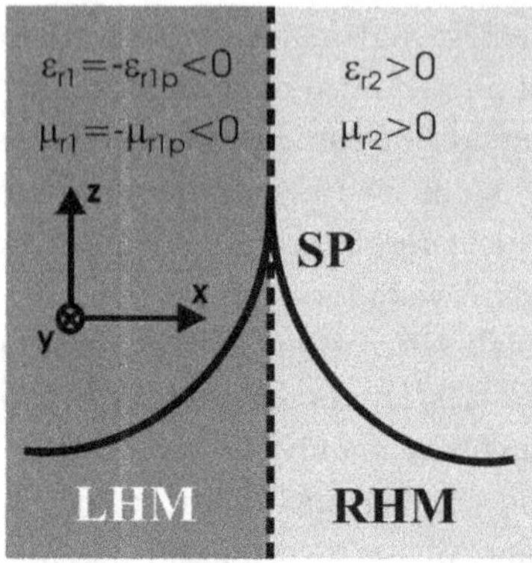

Figure 3.5: Isolated interface between a left-handed (LH) material and a right-handed (RH) material. The change in sign of the permittivity allows a *p*-polarised surface polariton (SP) to exist at this interface.

In what follows, we will examine p-polarised (transverse magnetic, TM) SP waves that exist due to the change in the sign of the permittivities. Analogous results can be obtained for s-polarised (TE) waves, following a dual analysis. The following relations describe the field components in both dielectrics (in SI units):

$$\frac{d^2 H_y}{dx^2} + (\varepsilon_r \mu_r k_0^2 - \beta^2) H_y = 0, \tag{3.23a}$$

$$E_x = \frac{\beta}{\omega \varepsilon} H_y, \tag{3.23b}$$

$$E_z = -\frac{j}{\omega \varepsilon} \frac{\partial H_y}{\partial x}. \tag{3.23c}$$

Inside the LH medium, Eq. (3.23a) becomes:

$$\frac{d^2 H_y}{dx^2} - (\beta^2 - \varepsilon_{r1p} \mu_{r1p} k_0^2) H_y = 0, \tag{3.24}$$

where β is the longitudinal propagation constant of the SP wave. Assuming

$$\beta^2 > \max\{\varepsilon_{r1p} \mu_{r1p} k_0^2, \varepsilon_{r2} \mu_{r2} k_0^2\},$$

the H_y-field component in medium 1 will be of the form $H_y = Ae^{\kappa x}$, where $\kappa = \sqrt{\beta^2 - \varepsilon_{r1p} \mu_{r1p} k_0^2}$ and A is an arbitrary constant.

From Eq. (3.23c) it follows that the E_z component will be $E_z = (jA\kappa/\omega\varepsilon_0\varepsilon_{r1p})e^{\kappa x}$. Note that the previously mentioned assumption implies a momentum mismatch between the supported SP and a radiative electromagnetic wave in the second dielectric, hence radiative waves cannot directly excite an SP at this interface.

For a bound wave to be supported by the interface, we seek H_y-solutions that decay exponentially as $x \to \pm\infty$. Therefore, we seek a solution for medium 2 of the form $H_y = Ce^{-\gamma x}$. By direct substitution of this expression into the scalar wave equation for H_y in Eq. (3.23a) we obtain:

$$\gamma = \sqrt{\beta^2 - \varepsilon_{r2} \mu_{r2} k_0^2},$$

and the E_z component given by $E_z = (jC\gamma/\omega\varepsilon_0\varepsilon_{r2})e^{-\gamma x}$.

Applying the boundary conditions associated with the tangential H_y and E_z fields at $x = 0$, , yields a characteristic or eigenvalue equation for the formed SP at the plane interface that allows us to determine the conditions for its existence. In terms of the eigenmode's effective index, $n_{eff} = \beta/k_0$, the aforementioned equation takes the form:

$$n_{eff} = \left[\frac{\varepsilon_{r1p}\varepsilon_{r2}(\mu_{r1p}\varepsilon_{r2} - \mu_{r2}\varepsilon_{r1p})}{\varepsilon_{r2}^2 - \varepsilon_{r1p}^2} \right]^{1/2}. \tag{3.25}$$

It is convenient for the subsequent discussions to rewrite Eq. (3.25) using the ratios of the permittivities, $\rho_\varepsilon = \varepsilon_{r2}/\varepsilon_{r1p}$, and permeabilities, $\rho_\mu = \mu_{r2}/\mu_{r1p}$, of the two media:

$$n_{eff} = |n_1| \left[\frac{\rho_\varepsilon(\rho_\varepsilon - \rho_\mu)}{\rho_\varepsilon^2 - 1} \right]^{1/2}. \tag{3.26}$$

Since ρ_ε is a positive quantity and (always) $n_{eff} > |n_1|$, we conclude from Eq. (3.26) that a SP at a LH/RH interface can only exist if:

$$\rho_\varepsilon \rho_\mu < 1 \quad \text{and} \quad \rho_\varepsilon > 1, \rho_\varepsilon > \rho_\mu, \tag{3.27a}$$

$$\rho_\varepsilon \rho_\mu < 1 \quad \text{and} \quad \rho_\varepsilon > 1, \rho_\varepsilon < \rho_\mu. \tag{3.27b}$$

Before closing this subsection, we wish to emphasise that these restrictions concern uncoupled ("unretarded") SPs existing at isolated LH/RH interfaces. We demonstrate in the next section that an SP eigenmode violating these constrains may exist if the interface that supports it, is brought sufficiently close to another LH/RH interface creating a new "supermode", which is not obliged to obey the two different cases in Eq. (3.27).

3.7.2 *Surface Plasmon Polaritons in Asymmetric LH Slab Waveguides*

In the following we study surface plasmon polaritons propagating along a homogeneous isotropic slab of negative permittivity and permeability bounded asymmetrically by two dielectric media, as illustrated in Fig. 3.6.

The SPP eigenmodes in the slab waveguide will be travelling along the z direction. There is no variation in the guide geometry in the z direction and by symmetry no variation in the field distributions in the y direction. The thickness of the slab is $2a$ and in all the subsequent discussions we assume, without loss of generality, that $n_2 > n_3$.

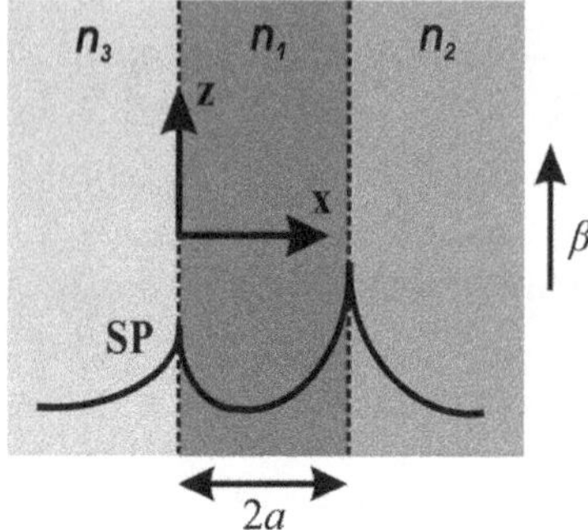

Figure 3.6: Schematic representation of the asymmetric left-handed slab heterostructure. The core is a medium with negative refractive index n_1 and thickness $2a$. Also shown are a possible field pattern of a supported SP and the direction of the longitudinal propagation constant β.

In the analysis of planar dielectric waveguides [Snyder and Love (1983)] the solution ansatz to the master equation is a monochromatic plane wave of frequency ω with a functional expression that can be symbolically written as:

$$\Psi = \Phi(x)e^{-j\beta z}e^{j\omega t}, \tag{3.28}$$

where Ψ represents an electric or magnetic field component, Φ describes its amplitude in the x-axis and β is the longitudinal component of the wavevector $\mathbf{k}$ in the slab. For TM SPPs, where the three existing field components are given by Eq. (3.23), we are seeking H_y-solutions in the three media, of the form:

$$H_y = \begin{cases} A\exp(\gamma_3 x), & x \le 0 \\ B\cosh(\kappa x) + C\sinh(\kappa x), & 0 \le x \le 2a \\ D\exp\big(-(x-2a)\gamma_2\big), & x \ge 2a \end{cases} \tag{3.29}$$

with

$$\kappa = \sqrt{\beta^2 - \varepsilon_{r1p}\mu_{r1p}k_0^2},$$

$$\gamma_2 = \sqrt{\beta^2 - \varepsilon_{r2}\mu_{r2}k_0^2},$$

$$\gamma_3 = \sqrt{\beta^2 - \varepsilon_{r3}\mu_{r3}k_0^2},$$

as in Section 3.7.1. Similarly to the single interface case we require

$$\beta^2 > \max\big\{|n_1|, n_2|, n_3\big\},$$

from which we make the ansatz to Eq. (3.29).

By matching the tangential components at $x = 0$ and $x = a$, we find:

$$B = A, \tag{3.30a}$$

$$C = -\frac{\varepsilon_{r1p}}{\varepsilon_{r3}}\frac{\gamma_3}{\kappa}, \tag{3.30b}$$

$$D = \left[\cosh(2a\kappa) - \frac{\varepsilon_{r1p}}{\varepsilon_{r3}}\frac{\gamma_3}{\kappa}\sinh(2a\kappa)\right]A \tag{3.30c}$$

and the following SPP characteristic equation is obtained:

$$\tanh(2a\kappa) = \frac{\varepsilon_{r1p}\kappa(\varepsilon_{r3}\gamma_2 + \varepsilon_{r2}\gamma_3)}{\varepsilon_{r2}\varepsilon_{r3}\kappa^2 + \varepsilon_{r1p}^2\gamma_2\gamma_3}. \tag{3.31}$$

As in classical fiber theory [Snyder and Love (1983)], it is advantageous to introduce the following reduced, dimensionless, modal parameters:

$$U = a\kappa = ak_0\sqrt{n_{eff}^2 - \varepsilon_{r1p}\mu_{r1p}}, \tag{3.32a}$$

$$W_2 = a\gamma_2 = ak_0\sqrt{n_{eff}^2 - \varepsilon_{r2}\mu_{r2}}, \tag{3.32b}$$

$$W_3 = a\gamma_3 = ak_0\sqrt{n_{eff}^2 - \varepsilon_{r3}\mu_{r3}}. \tag{3.32c}$$

With these definitions, Eq. (3.31) takes the form:

$$\tanh(2U) = \frac{\varepsilon_{r1p}U(\varepsilon_{r3}W_2 + \varepsilon_{r2}W_3)}{\varepsilon_{r2}\varepsilon_{r3}U^2 + \varepsilon_{r1p}^2 W_2 W_3}. \tag{3.33}$$

In order to determine the power propagation direction, we calculate the cycle-averaged power flow in the slab heterostructure, obtained by the integral over the guide's cross-section of the z component of the complex Poynting vector S_z:

$$P = \int_{-\infty}^{\infty} S_z \, dx = \frac{1}{2}\int_{-\infty}^{\infty} \mathrm{Re}(\mathbf{E} \times \mathbf{H}^*)_z \, dx. \tag{3.34}$$

For p-polarised SPP eigenmodes, S_z is given by:

$$S_z = \frac{1}{2}H_y^* E_x = \frac{\beta}{2\omega\varepsilon_0\varepsilon_i}|H_y|^2, \quad i = 1, 2, 3. \tag{3.35}$$

From Eqs. (3.29)–(3.31) we find the power, P_i, confined in each region to be:

$$P_3 = \left(\frac{A^2}{4\omega\varepsilon_0}\right)\frac{1}{\varepsilon_{r1p}}\frac{\beta}{\sigma_\varepsilon\gamma_3}, \tag{3.36a}$$

$$P_1 = -\left(\frac{A^2}{4\omega\varepsilon_0}\right)\frac{\beta}{\varepsilon_{r1p}}\frac{\sigma_\varepsilon\kappa^2 - \gamma_3^2}{\sigma_\varepsilon^2\kappa^2}\left(2a + \frac{\rho_\varepsilon\gamma_2}{\rho_\varepsilon^2\kappa^2 - \gamma_2^2} + \frac{\sigma_\varepsilon\gamma_3}{\sigma_\varepsilon^2\kappa^2 - \gamma_3^2}\right), \tag{3.36b}$$

$$P_2 = \left(\frac{A^2}{4\omega\varepsilon_0}\right)\frac{\beta}{\varepsilon_{r1p}\rho_\varepsilon\gamma_2}\left(\frac{\rho_\varepsilon}{\sigma_\varepsilon}\right)^2\frac{\sigma_\varepsilon^2\kappa^2 - \gamma_3^2}{\rho_\varepsilon^2\kappa^2 - \gamma_2^2} \tag{3.36c}$$

where $\sigma_\varepsilon = \varepsilon_{r3}/\varepsilon_{r1p}$ and $\rho_\varepsilon = \varepsilon_{r2}/\varepsilon_{r1p}$.

From Eqs. (3.36) we can derive a closed-form expression for the total power $P_{tot} = \sum_{i=1}^3 P_i$ in terms of the dimensionless parameters defined in Eq. (3.32) and the reduced slab thickness, ak_0:

$$P_{tot} = \left(\frac{A^2}{4\omega\varepsilon_0}\right)\frac{W_3^2 - \sigma_\varepsilon^2 U^2}{\sigma_\varepsilon^2 U^2}\frac{\left[U^2 + \varepsilon_{r1p}\mu_{r1p}(ak_0)^2\right]^{1/2}}{\varepsilon_{r1p}}$$
$$\cdot\left[2 + \frac{\rho_\varepsilon}{W_2}\frac{U^2 - W_2^2}{W_2^2 - \rho_\varepsilon^2 U^2} + \frac{\sigma_\varepsilon}{W_3}\frac{U^2 - W_3^2}{W_3^2 - \sigma_\varepsilon^2 U^2}\right]. \tag{3.37}$$

The central task at this point is the determination of the solutions to Eq. (3.33). Their existence is identified following an analytical methodology. In what follows, we discuss the features and dependence of these solutions on the thickness of the inner layer, for the various cases of the refractive index distribution. A summary of the results for each case is shown in Section 3.7.3, in Tables I–III.

Case I: $|n_1| > n_2 > n_3$

For this case, we start by defining the following two V-numbers:

$$V_2(ak_0) = \left(W_2^2 - U^2\right)^{1/2} = ak_0\left(\varepsilon_{r1p}\mu_{r1p} - \varepsilon_{r2}\mu_{r2}\right)^{1/2}, \tag{3.38a}$$

$$V_3(ak_0) = \left(W_3^2 - U^2\right)^{1/2} = ak_0\left(\varepsilon_{r1p}\mu_{r1p} - \varepsilon_{r3}\mu_{r3}\right)^{1/2}, \tag{3.38b}$$

where it is seen that the usual notation of the V-number found in fiber theory [Snyder and Love (1983)] has been properly modified to accommodate the changes in the refractive index distribution and that both parameters are functions of ak_0. We also introduce the following ratios:

$$b(n_{eff}) = \frac{U}{V_2} = \left[\frac{(n_{eff}/n_1)^2 - 1}{1 - \rho_\varepsilon \rho_\mu}\right]^{1/2}, \tag{3.39}$$

$$t = \frac{V_3}{V_2} = \left[\frac{1 - \sigma_\varepsilon \sigma_\mu}{1 - \rho_\varepsilon \rho_\mu}\right]^{1/2}, \tag{3.40}$$

obeying the restrictions $b > 0$ and $t > 1$, where $\rho_\mu = \mu_{r2}/\mu_{r1p}$ and $\sigma_\mu = \mu_{r3}/\mu_{r1p}$. Note that similar ratios to b and t are utilised in the analysis of conventional slab waveguides to denote the "normalised guide index" and the "asymmetry measure", respectively [Snyder and Love (1983)] and that b is a function of the SPP eigenmode's effective index.

For the explicit acquisition of the dispersion diagrams and the derivation of the analytical restrictions inherent in Eq. (3.33), a common strategy is to produce an inverted version of the associated characteristic equation [Burke *et al.* (1986); Kogelnik and Ramaswamy (1974)]. First, we note from Eqs. (3.38) and (3.39) that $U = bV_2$, $W_2 = V_2(b^2 + 1)^{1/2}$ and $W_3 = V_2(b^2 + t^2)^{1/2}$. With these observations, Eq. (3.33) can be rewritten in the form:

$$V_2 = \frac{1}{4b} \ln\left[\frac{(X+1)(Y+1)}{(X-1)(Y-1)}\right], \tag{3.41}$$

where $X(b) = \sigma_\varepsilon b/(b^2 + t^2)^{1/2}$ and $Y(b) = \rho_\varepsilon b/(b^2 + 1)^{1/2}$.

Since V_2, given in Eq. (3.38a), is a real number, we immediately see that Eq. (3.41) only has solutions, when the argument of the logarithm is positive, i.e. for:

$$\{X > 1 \text{ and } Y > 1\}, \tag{3.42a}$$

or

$$\{0 < X < 1 \text{ and } 0 < X < 1\}. \tag{3.42b}$$

We now examine in detail the consequences of these restrictions $X(b)$ and $Y(b)$, based on which the existing SPP eigenmodes are rigorously identified. In particular, we derive analytically the allowable range of values that b, hence the eigenmodes' effective index, can take. In pursuing this analysis, we find that it is necessary to distinguish between the following four situations that describe the possible variations in the permittivity distribution. In all four of them, it is implied that $\sigma_\varepsilon \sigma_\mu < \rho_\varepsilon \rho_\mu < 1$ from the initial assumption $|n_1| > n_2 > n_3$.

The first situation occurs for $\{\sigma_\varepsilon > 1$ and $\rho_\varepsilon > 1\}$. When $\rho_\varepsilon > \sigma_\varepsilon$, Eq. (3.42) gives the allowable values for b; that is $0 < b < b_1$ or $b > b_2$, where:

$$b_1 = \left[\frac{1}{\rho_\varepsilon^2 - 1}\right]^{1/2},\tag{3.43}$$

and

$$b_2 = \left[\frac{1 - \sigma_\varepsilon\sigma_\mu}{(\sigma_\varepsilon^2 - 1)(1 - \rho_\varepsilon\rho_\mu)}\right]^{1/2}.\tag{3.44}$$

The corresponding geometric dispersion diagram is shown in Fig. 3.7a, and the variation of the normalised power $P = P_{tot}/\left(|P_1|+|P_2|+|P_3|\right)$ with the reduced slab thickness for each solution is given in Fig. 3.7b. Three SPP eigenmodes exist in this case. The first has a lower cutoff and an upper one at $b = 0$, is forward propagating, having positive total power P, and the field intensity has a node in the core region. At the lower cutoff point this solution degenerates into the second eigenmode. This SPP has a low cutoff and no upper cutoff and is backward propagating, having negative energy velocity and, since the dielectric media are non-absorbing, negative group velocity, as well [Loudon (1970); Ruppin (2002)]. After a finite gap a third eigenmode appears, which has no cutoff and is also backward propagating having negative P for every core thickness. The corresponding fields have no node in the core region.

It is interesting to discuss the behaviour of these solutions in the extreme cases of $b \to b_1$ and $b \to b_2$. First, by letting b become equal to b_1, we recover asymptotically Eq. (3.26), which is the SPP characteristic equation at the 1-2 media interface. Such a result is expected, since from Eqs. (3.38a) and (3.41) and from Fig. 3.7a we note that for $b \to b_1$ it is $ak_0 \to \infty$, hence the two slab interfaces decouple and the SPP at an isolated interface should be recovered. For a relatively large value of ak_0, the result of plotting this SPP eigenmode is shown in the middle right inset of Fig. 3.7a, reflecting the previous conclusions. Then, assuming that $b = b_2$, we obtain:

$$n_{eff,3} = |n_1|\left[\frac{\sigma_\varepsilon(\sigma_\varepsilon - \sigma_{mu})}{\sigma_\varepsilon^2 - 1}\right]^{1/2},\tag{3.45}$$

and the SPP eigenmode at the 1-3 interface is recovered asymptotically, shown in the top inset.

When $\rho_\varepsilon < \sigma_\varepsilon$, it proves necessary to examine the intervals that the ratio ρ_ε belongs to. If $\rho_\varepsilon \in \Delta$, where $\Delta = \{(\rho_{\varepsilon,1} < \rho_\varepsilon < \rho_{\varepsilon,2}) \cap (\rho_\varepsilon > 1)\}$ and $\rho_{\varepsilon,1}$, $\rho_{\varepsilon,2}$ are the two roots of the polynomial:

$$\Pi(\rho_\varepsilon) = \left(1 - \sigma_\varepsilon\sigma_\mu\right)\rho_\varepsilon^2 + \rho_\varepsilon\rho_\mu\left(\sigma_\varepsilon^2 - 1\right) - \left[\left(1 - \sigma_\varepsilon\sigma_\mu\right) + \left(\sigma_\varepsilon^2 - 1\right)\right],\tag{3.46}$$

it is $b_1 > b_2$ and we see from Fig. 3.7c that two eigenmodes exist, both of which are backward propagating. The first SPP has only a lower cutoff at $b = 0$ and, contrary to the previous case, it concentrates asymptotically at the 1-3 interface

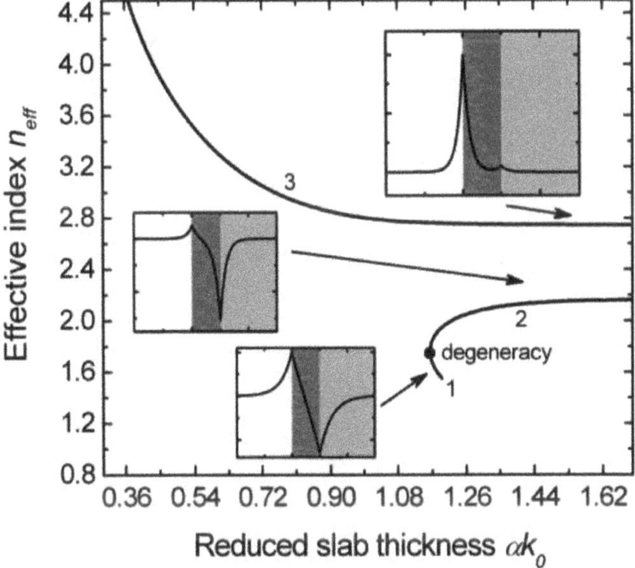

(a) n_{eff} variation for $\sigma_\varepsilon = 1.1$, $\sigma_\mu = 0.5$, $\rho_\varepsilon = 1.15$, $\rho_\mu = 0.6$.

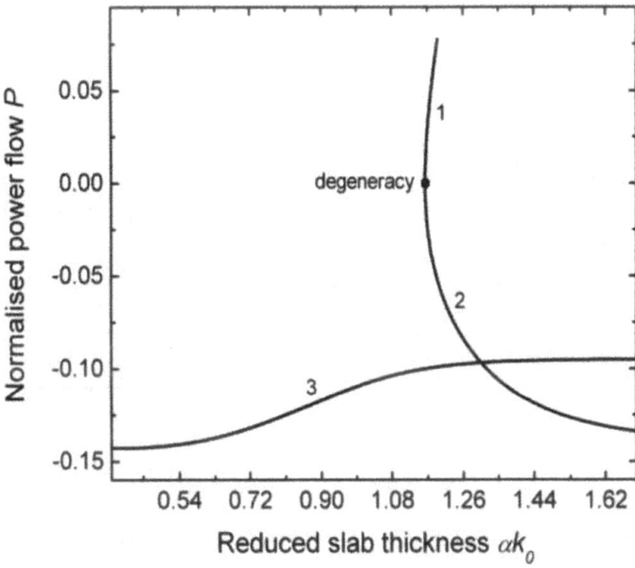

(b) P variation for $\sigma_\varepsilon = 1.1$, $\sigma_\mu = 0.5$, $\rho_\varepsilon = 1.15$, $\rho_\mu = 0.6$.

Figure 3.7: Variation of an SP eigenmode's effective index n_{eff} and normalised power P with the reduced slab thickness ak_0 in a generalised LH slab waveguide (Case I: $|n_1| > n_2 > n_3$ as indicated by the shaded background). In all cases it is assumed that $\varepsilon_r = 2$, $\mu_r = 1.2$.

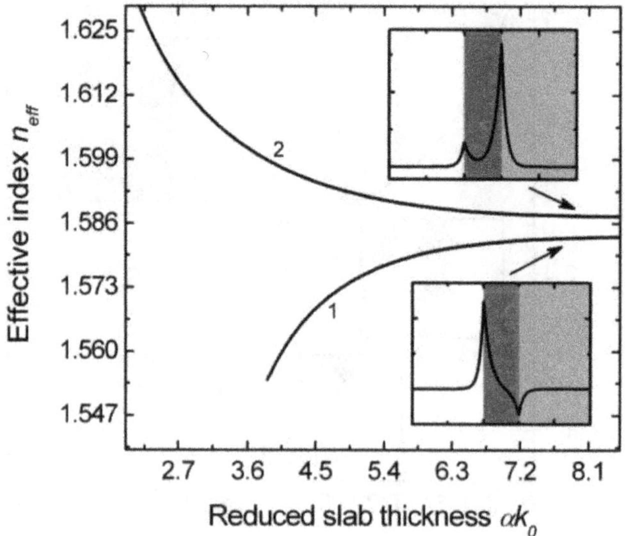

(c) n_{eff} variation for $\sigma_\varepsilon = 1.8$, $\sigma_\mu = 0.5$, $\rho_\varepsilon = 1.55$, $\rho_\mu = 0.6$.

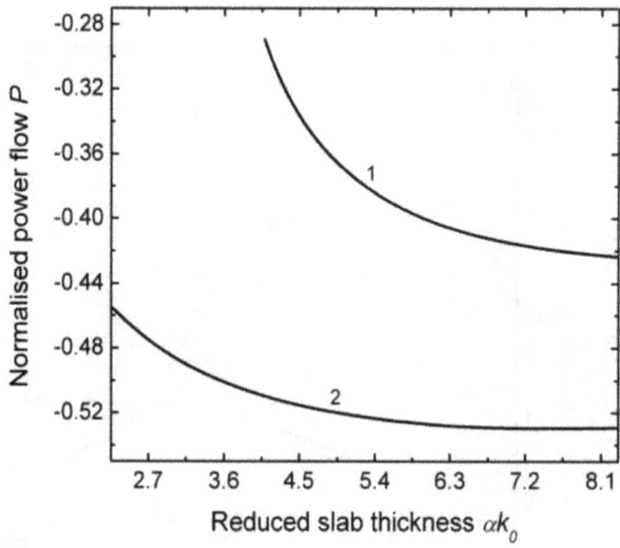

(d) P variation for $\sigma_\varepsilon = 1.8$, $\sigma_\mu = 0.5$, $\rho_\varepsilon = 1.55$, $\rho_\mu = 0.6$.

Figure 3.7: (*Continued*)

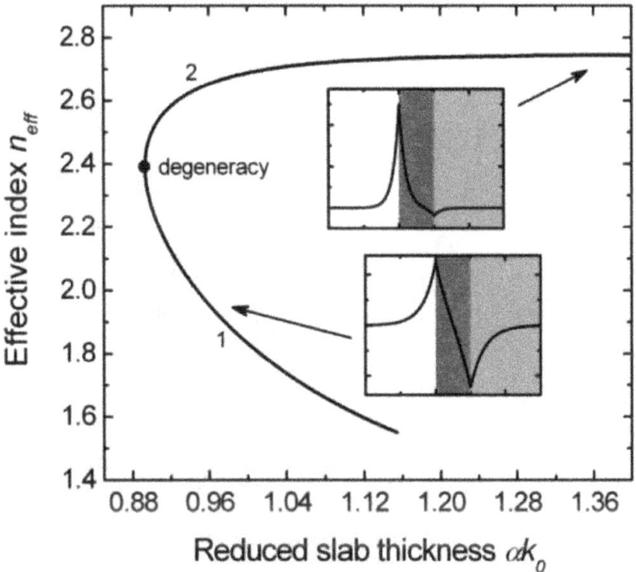

(e) n_{eff} variation for $\sigma_\varepsilon = 1.1$, $\sigma_\mu = 0.5$, $\rho_\varepsilon = 0.95$, $\rho_\mu = 0.8$.

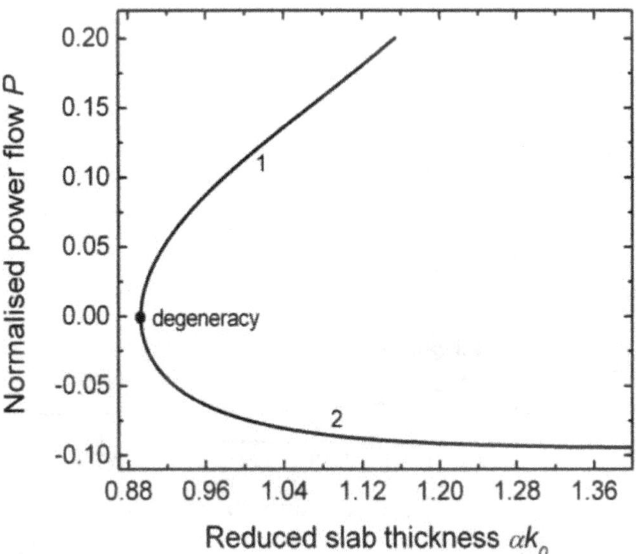

(f) P variation for $\sigma_\varepsilon = 1.1$, $\sigma_\mu = 0.5$, $\rho_\varepsilon = 0.95$, $\rho_\mu = 0.8$.

Figure 3.7: (*Continued*)

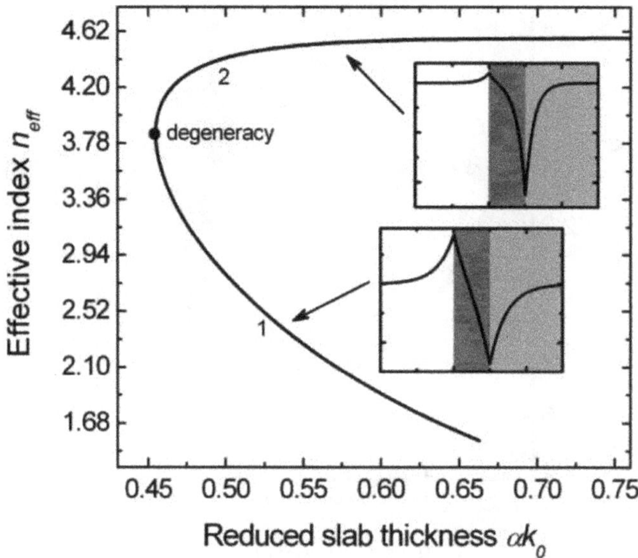

(g) n_{eff} variation for $\sigma_\varepsilon = 1.1$, $\sigma_\mu = 0.2$, $\rho_\varepsilon = 1.05$, $\rho_\mu = 0.2$.

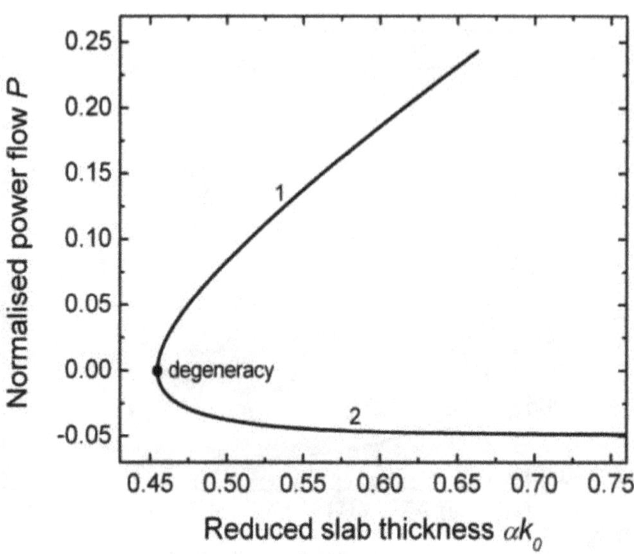

(h) P variation for $\sigma_\varepsilon = 1.1$, $\sigma_\mu = 0.2$, $\rho_\varepsilon = 1.05$, $\rho_\mu = 0.2$.

Figure 3.7: (*Continued*)

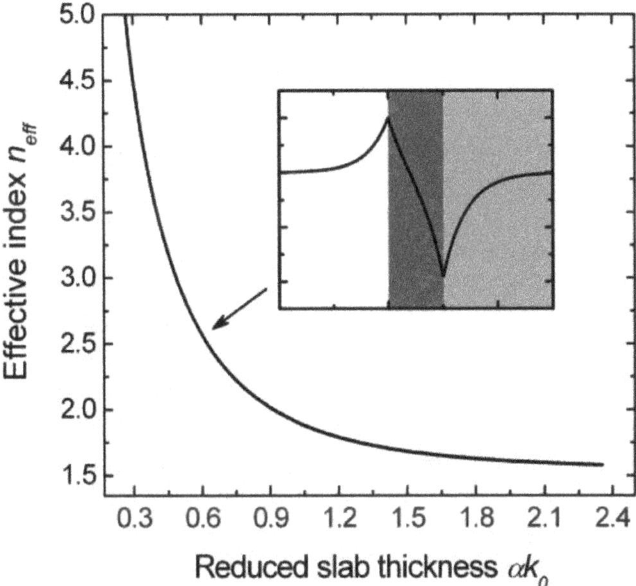

(i) n_{eff} variation for $\sigma_\varepsilon = 1.1$, $\sigma_\mu = 0.2$, $\rho_\varepsilon = 1.05$, $\rho_\mu = 0.2$.

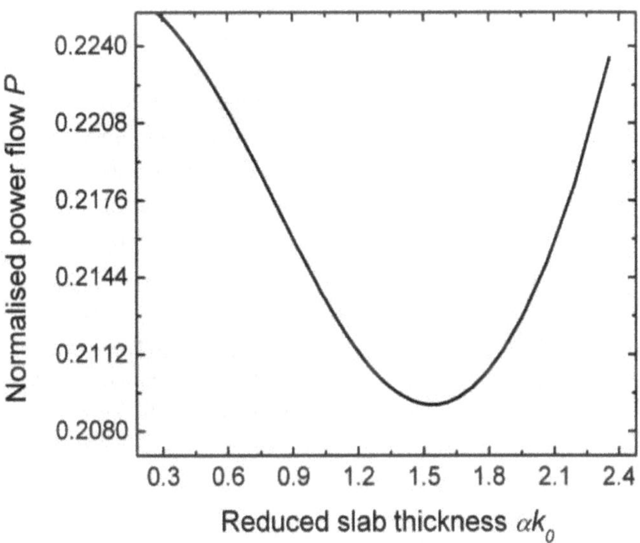

(j) P variation for $\sigma_\varepsilon = 1.1$, $\sigma_\mu = 0.2$, $\rho_\varepsilon = 1.05$, $\rho_\mu = 0.2$.

Figure 3.7: (*Continued*)

while exhibiting a phase reversal. The second (upper) SPP shows no cutoff, has positive H_y-field component throughout the slab and concentrates at the 1-2 media interface for large core thickness. For $\rho_\varepsilon \in \Delta'$, where $\Delta' = \{[(\rho_\varepsilon < \rho_{\varepsilon,1}) \cup (\rho_\varepsilon > \rho_{\varepsilon,2})] \cap (\rho_\varepsilon > 1)\}$ we have $b_1 < b_2$ and the results of the analysis hold the same as in the case $\rho_\varepsilon > \sigma_\varepsilon$, apart from the disappearance of the first SPP eigenmode that was forward propagating.

The second situation occurs for $\{\sigma_\varepsilon > 1 \text{ and } \rho_\varepsilon \leq 1\}$. From the conditions in Eq. (3.42), we find that b ranges from 0 to b_2 and the corresponding dispersion diagram is illustrated Fig. 3.7e. In this case two eigenmodes are shown to exist; for both, the H_y component has a node in the core region. The first has a low cutoff and an upper one occurring at $b = 0$ and is forward propagating. The second SPP has only a low cutoff and it concentrates asymptotically at the 1-3 interface. As in all the encountered cases that contain degeneracy, the total power P at this point equals zero, corresponding to zero group velocity. Waves with this feature are of great practical interest for optical communication and data storage applications [Milonni (2005)]. It should be noted that, for the particular values of the permittivity and permeability ratios shown in Fig. 3.7e, the 1-2 single interface does not support an SPP because the constrains in Eq. (3.27) are violated. However, the coupled SPPs at the interfaces of the slab overcome this limitation, creating the two new SPP "supermodes" for relatively large slab thickness.

In the third case, which exists for $\{\sigma_\varepsilon \leq 1 \text{ and } \rho_\varepsilon > 1\}$ it is found that the range of values for b is between 0 and b_1. The variation of the eigenmodes' effective index and power with the reduced guide thickness is shown in Figs. 3.7g,h. The conclusions for the supported SPPs are similar with those in the previous case, with the difference that the second (upper) eigenmode concentrates asymptotically at the 1-2 interface while taking negative values. In this case also, it is the 1-3 interface that violates the SPP existence conditions and, if isolated, would *not* support a bound wave.

The final situation occurs for $\{\sigma_\varepsilon \leq 1 \text{ and } \rho_\varepsilon \leq 1\}$. In this case there are no additional restrictions on b other than $b > 0$. From Figs. 3.7i,j we see that only one SPP eigenmode exists, which is forward propagating with only a higher cutoff at $b = 0$ and has opposite sign at the two interfaces of the slab.

Case II: $n_2 > n_3 > |n_1|$

For the refractive index distribution considered here, we use the following definitions for the V-numbers:

$$V_2(ak_0) = \left(U^2 - W_2^2\right)^{1/2} = ak_0\left(\varepsilon_{r2}\mu_{r2} - \varepsilon_{r1p}\mu_{r1p}\right)^{1/2}, \qquad (3.47a)$$

$$V_3(ak_0) = \left(U^2 - W_3^2\right)^{1/2} = ak_0\left(\varepsilon_{r3}\mu_{r3} - \varepsilon_{r1p}\mu_{r1p}\right)^{1/2}, \qquad (3.47b)$$

with $\rho_\varepsilon \rho_\mu > \sigma_\varepsilon \sigma_\mu > 1$. used throughout the following analysis. Then, the previously introduced b and t parameters take the form:

$$b(n_{eff}) = \frac{U}{V_2} = \left[\frac{(n_{eff}/n_1)^2 - 1}{\rho_\varepsilon \rho_\mu - 1}\right]^{1/2}, \tag{3.48}$$

$$t = \frac{V_3}{V_2} = \left[\frac{\sigma_\varepsilon \sigma_\mu - 1}{\rho_\varepsilon \rho_\mu - 1}\right]^{1/2}, \tag{3.49}$$

obeying the restrictions $b > 1$ and $t < 1$. The general form of Eq. (3.41) and the solution conditions of Eq. (3.42) remain the same, but now we have

$$W_2 = V_2\left(b^2 - 1\right)^{1/2}, \quad W_3 = V_2\left(b^2 - t^2\right)^{1/2},$$
$$X(b) = \sigma_\varepsilon b/\left(b^2 - t^2\right)^{1/2} \quad \text{and} \quad Y(b) = \rho_\varepsilon b/\left(b^2 - 1\right)^{1/2}.$$

By letting $X(b)$ and $Y(b)$ fulfill these conditions, we again find that four distinct situations arise depending on the permittivity profile.

The first situation occurs for $\{\sigma_\varepsilon > 1$ and $\rho_\varepsilon > 1\}$, and does not contain further restrictions on b. The eigenmodes' effective index and P dispersion diagrams are shown in Figs. 3.8a,b. We see that two SPPs exist; both have no node in the middle layer. The first, which is forward propagating, has a lower cutoff at $b = 1$ and also an upper cutoff point, where it degenerates into the second eigenmode. The second SPP has only a high cutoff and is backward propagating having negative total power P.

The second situation occurs for $\{\sigma_\varepsilon \geq 1$ and $\rho_\varepsilon < 1\}$ and it can be shown that b is in the range $1 < b < b_1$, where now:

$$b_1 = \left[\frac{1}{1 - \rho_\varepsilon^2}\right]^{1/2}, \tag{3.50}$$

In this case a single SPP eigenmode is shown to exist, having only a low cutoff at $b = 1$. This SPP is forward propagating and it concentrates at the 1-2 media interface for large core thickness, as illustrated in the inset of Fig. 3.8c. This result is also verified by letting $b = b_1$, where we obtain the SPP characteristic Eq. (3.26).

The third case exists for $\{\sigma_\varepsilon < 1$ and $\rho_\varepsilon \geq 1\}$ and it is found that b ranges from 1 to b_2, where now:

$$b_2 = \left[\frac{\sigma_\varepsilon \sigma_\mu - 1}{(1 - \sigma_\varepsilon^2)(\rho_\varepsilon \rho_\mu) - 1}\right]^{1/2}, \tag{3.51}$$

with the constrain $\rho_\varepsilon \rho_\mu < \left(\sigma_\varepsilon \sigma_\mu - \sigma_\varepsilon^2\right)/\left(1 - \sigma_\varepsilon^2\right)$. The variation of the effective index and total power with the reduced guide thickness are shown in Fig. 3.8e,f. The conclusions are similar with the previous case, with the difference that the supported SPP concentrates asymptotically at the 1-3 interface. For the values of ρ_ε and ρ_μ shown in Fig. 3.8e, the isolated 1-2 interface would *not* support an SPP. The final situation occurs for $\{\sigma_\varepsilon < 1$ and $\rho_\varepsilon < 1\}$.

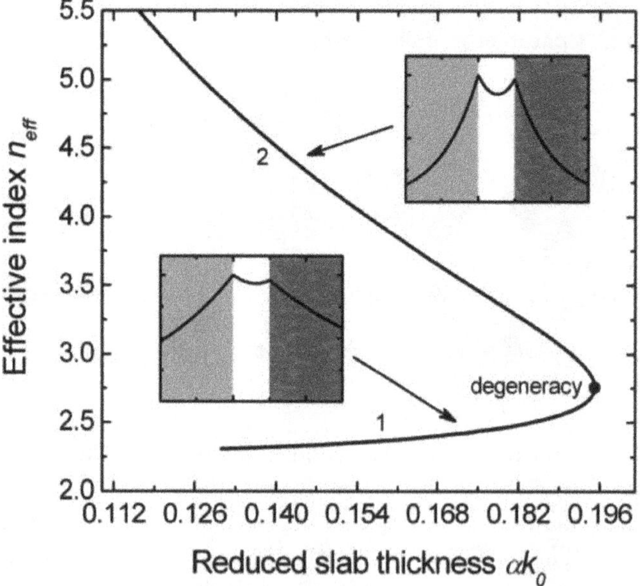

(a) n_{eff} variation for $\sigma_\varepsilon = 1.7$, $\sigma_\mu = 1.1$, $\rho_\varepsilon = 1.8$, $\rho_\mu = 1.2$.

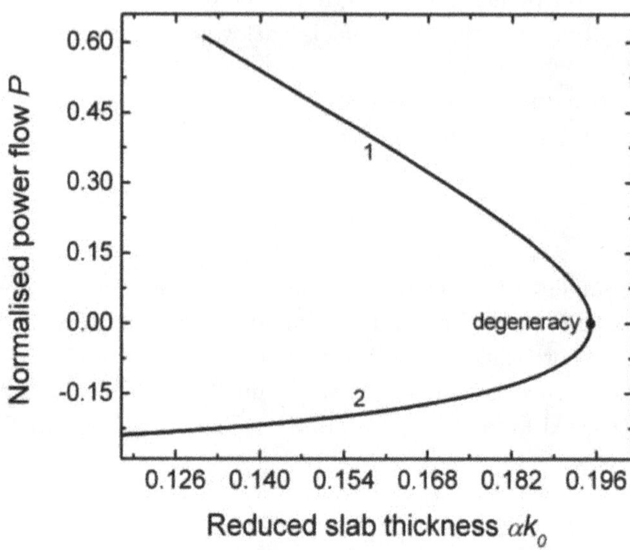

(b) P variation for $\sigma_\varepsilon = 1.7$, $\sigma_\mu = 1.1$, $\rho_\varepsilon = 1.8$, $\rho_\mu = 1.2$.

Figure 3.8: Variation of an SP eigenmode's effective index n_{eff} and normalised power P with the reduced slab thickness ak_0 in a generalised LH slab waveguide (Case II: $n_2 > n_3 > |n_1|$ as indicated by the shaded background). In all cases it is assumed that $\varepsilon_r = 2$, $\mu_r = 1.2$.

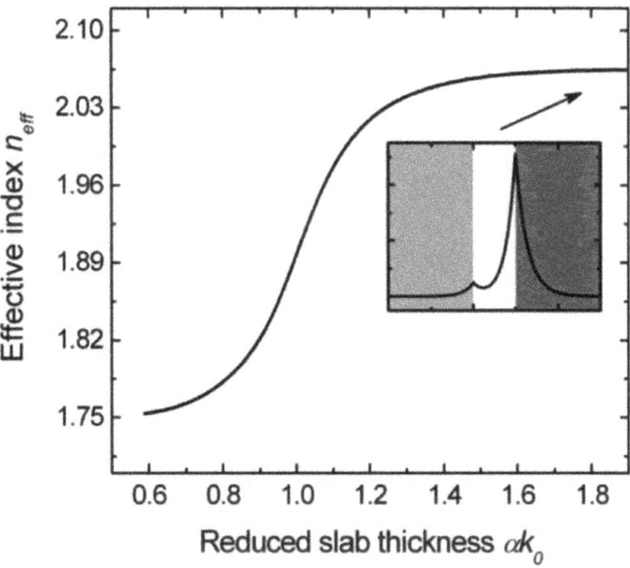

(c) n_{eff} variation for $\sigma_\varepsilon = 1.2$, $\sigma_\mu = 0.9$, $\rho_\varepsilon = 0.8$, $\rho_\mu = 1.6$.

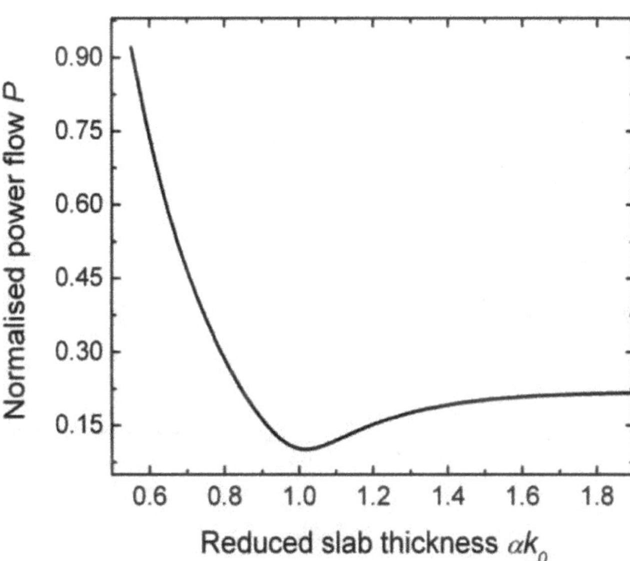

(d) P variation for $\sigma_\varepsilon = 1.2$, $\sigma_\mu = 0.9$, $\rho_\varepsilon = 0.8$, $\rho_\mu = 1.6$.

Figure 3.8: (*Continued*)

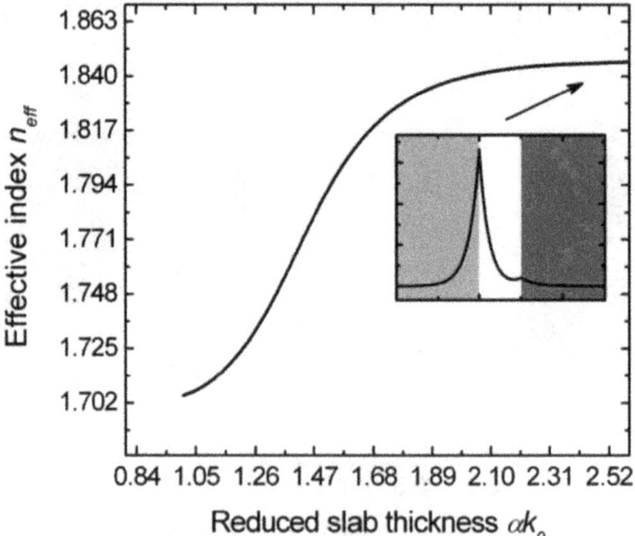

(**e**) n_{eff} variation for $\sigma_\varepsilon = 0.9$, $\sigma_\mu = 1.2$, $\rho_\varepsilon = 1.1$, $\rho_\mu = 1.1$.

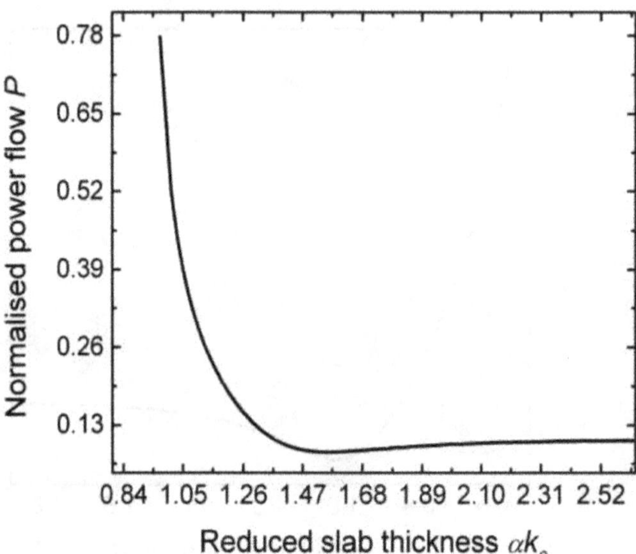

(**f**) P variation for $\sigma_\varepsilon = 0.9$, $\sigma_\mu = 1.2$, $\rho_\varepsilon = 1.1$, $\rho_\mu = 1.1$.

Figure 3.8: (*Continued*)

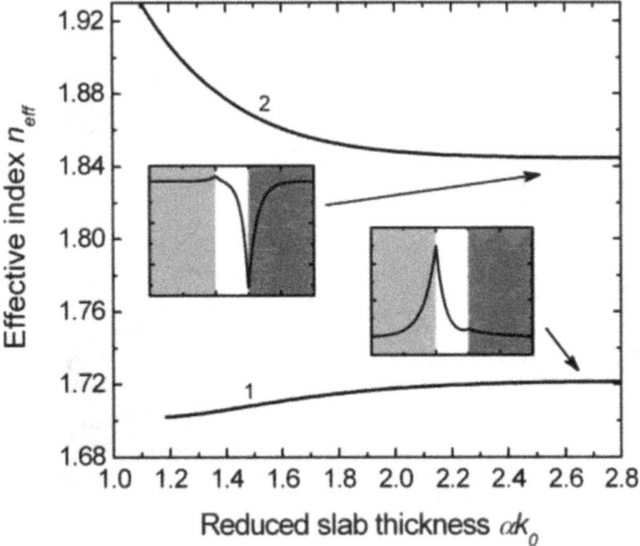

(g) n_{eff} variation for $\sigma_\varepsilon = 0.7$, $\sigma_\mu = 1.6$, $\rho_\varepsilon = 0.71$, $\rho_\mu = 1.7$.

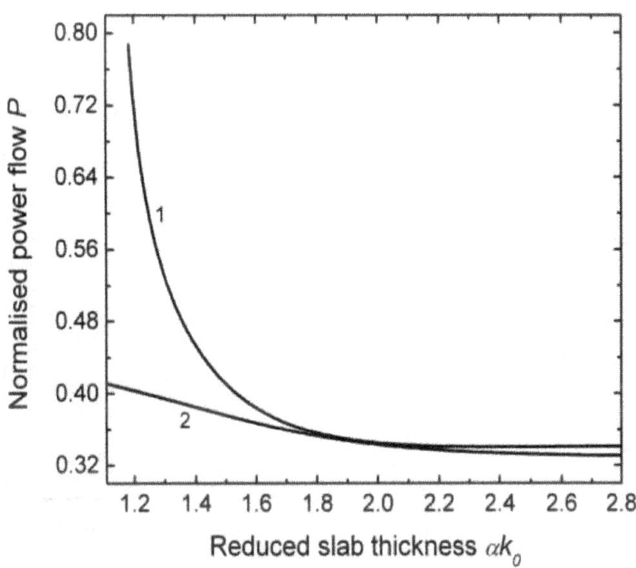

(h) P variation for $\sigma_\varepsilon = 0.7$, $\sigma_\mu = 1.6$, $\rho_\varepsilon = 0.71$, $\rho_\mu = 1.7$.

Figure 3.8: (*Continued*)

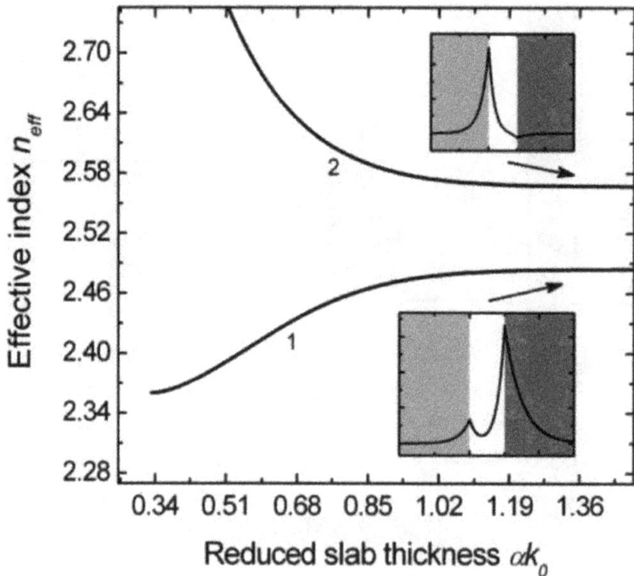

(i) n_{eff} variation for $\sigma_\varepsilon = 0.7$, $\sigma_\mu = 2.7$, $\rho_\varepsilon = 0.4$, $\rho_\mu = 5.8$.

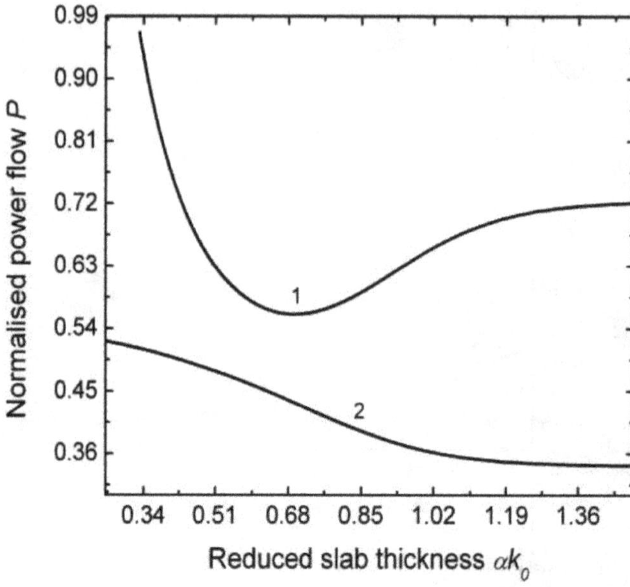

(j) P variation for $\sigma_\varepsilon = 0.7$, $\sigma_\mu = 2.7$, $\rho_\varepsilon = 0.4$, $\rho_\mu = 5.8$.

Figure 3.8: (*Continued*)

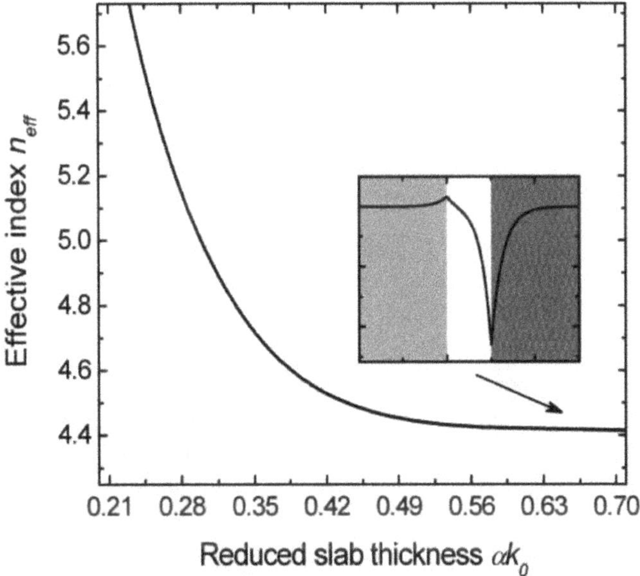

(**k**) n_{eff} variation for $\sigma_\varepsilon = 0.7$, $\sigma_\mu = 2.7$, $\rho_\varepsilon = 0.85$, $\rho_\mu = 3.5$.

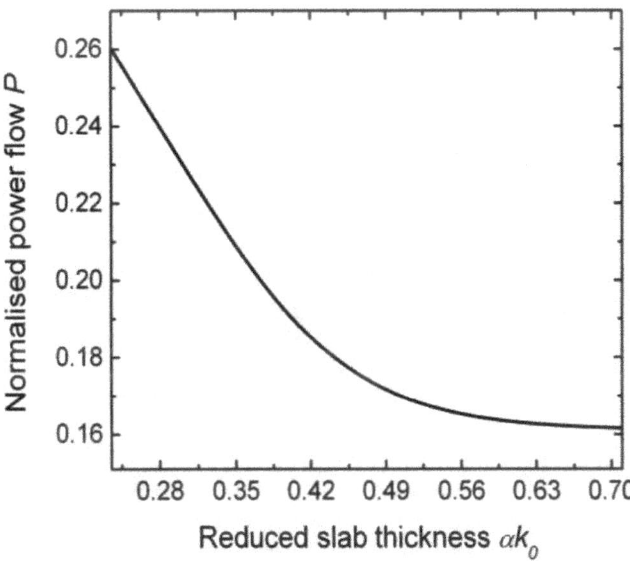

(**l**) P variation for $\sigma_\varepsilon = 0.7$, $\sigma_\mu = 2.7$, $\rho_\varepsilon = 0.85$, $\rho_\mu = 3.5$.

Figure 3.8: (*Continued*)

If $\rho_\varepsilon \rho_\mu < \left(\sigma_\varepsilon \sigma_\mu - \sigma_\varepsilon^2\right)/\left(1 - \sigma_\varepsilon^2\right)$, it proves necessary to examine the intervals that the ratio ρ_ε belongs to.

When $\rho_\varepsilon \in \Delta$, with $\Delta = \left\{ \left[(\rho_\varepsilon < \rho_{\varepsilon,1}) \cup (\rho_\varepsilon > \rho_{\varepsilon,2})\right] \cap (0 < \rho_\varepsilon < 1) \right\}$ and $\rho_{\varepsilon,1}$, $\rho_{\varepsilon,2}$ being the two roots of the polynomial:

$$\Pi(\rho_\varepsilon) = \left(\sigma_\varepsilon \sigma_\mu - 1\right)\rho_\varepsilon^2 + \rho_\varepsilon \rho_\mu \left(1 - \sigma_\varepsilon^2\right) - \left[\left(\sigma_\varepsilon \sigma_\mu - 1\right) + \left(1 - \sigma_\varepsilon^2\right)\right], \tag{3.52}$$

it is $b_1 > b_2$, and we see from Fig. 3.8g that two eigenmodes exist, both of which are forward propagating. The first SPP has only a low cutoff at $b = 1$ and concentrates asymptotically at the 1-3 interface; the corresponding field intensity has no node in the core region. The second SPP has no cutoff, exhibits phase reversal and concentrates at the 1-2 media interface for large core thickness. For $\rho_\varepsilon \in \Delta'$, where $\Delta' = \left\{(\rho_{\varepsilon,1} < \rho_\varepsilon < \rho_{\varepsilon,2}) \cap (0 < \rho_\varepsilon < 1)\right\}$ we have $b_1 < b_2$ and and again two SPP eigenmodes are shown to exist in Figs. 3.8i,j. The first has positive H_y-field amplitude across the slab. Compared with the previous two eigenmodes, the cutoff characteristics remain the same, but the order of the interfaces to which these SPPs concentrate is reversed. Finally, if $\rho_\varepsilon \rho_\mu > \left(\sigma_\varepsilon \sigma_\mu - \sigma_\varepsilon^2\right)/\left(1 - \sigma_\varepsilon^2\right)$, the constrain $b > b_1$ becomes mandatory. The dispersion diagrams for this case are shown in Figs. 3.8k,l. We see that a single SPP exists, which shows no cutoff, exhibits phase reversal, concentrates asymptotically at the 1-2 interface and has positive total power for all core thicknesses.

Case III:　$n_2 > |n_1| > n_3$

To reflect the refractive index distribution considered here, the two V-numbers are defined as:

$$V_2(ak_0) = \left(U^2 - W_2^2\right)^{1/2} = ak_0\left(\varepsilon_{r2}\mu_{r2} - \varepsilon_{r1p}\mu_{r1p}\right)^{1/2}, \tag{3.53a}$$

$$V_3(ak_0) = \left(W_3^2 - U^2\right)^{1/2} = ak_0\left(\varepsilon_{r1p}\mu_{r1p} - \varepsilon_{r3}\mu_{r3}\right)^{1/2}, \tag{3.53b}$$

with $\rho_\varepsilon \rho_\mu > 1 > \sigma_\varepsilon \sigma_\mu$ used throughout the following analysis. Accordingly, the b and t ratios take the form:

$$b(n_{eff}) = \frac{U}{V_2} = \left[\frac{(n_{eff}/n_1)^2 - 1}{\rho_\varepsilon \rho_\mu - 1}\right]^{1/2}, \tag{3.54}$$

$$t = \frac{V_3}{V_2} = \left[\frac{1 - \sigma_\varepsilon \sigma_\mu}{\rho_\varepsilon \rho_\mu - 1}\right]^{1/2}, \tag{3.55}$$

obeying the restrictions $b > 1$ and $t > 0$. Once more, the general form of Eq. (3.41) and the solution conditions of Eq. (3.42) remain the same, but now we have

$$W_2 = V_2\left(b^2 - 1\right)^{1/2}, \quad W_3 = V_2\left(b^2 + t^2\right)^{1/2},$$

$$X(b) = \sigma_\varepsilon b / \left(b^2 + t^2\right)^{1/2} \quad \text{and} \quad Y(b) = \rho_\varepsilon b / \left(b^2 - 1\right)^{1/2}.$$

By letting $X(b)$ and $Y(b)$ fulfill aforesaid conditions, the following four distinct situations arise depending on the permittivity profile.

The first situation occurs for $\{\sigma_\varepsilon > 1 \text{ and } \rho_\varepsilon \geq 1\}$.

If $\rho_\varepsilon \rho_\mu < (\sigma_\varepsilon^2 - \sigma_\varepsilon \sigma_\mu)/(\sigma_\varepsilon^2 - 1)$, the allowable range of values for b is $b > b_2$, where:

$$b_2 = \left[\frac{1 - \sigma_\varepsilon \sigma_\mu}{(\sigma_\varepsilon^2 - 1)(\rho_\varepsilon \rho_\mu) - 1} \right]^{1/2}. \tag{3.56}$$

We see from Figs. 3.9a,b. that a single eigenmode exists, which is backward propagating and has no cutoff. For large core thickness, the SPP characteristic equation at the 1-3 interface is recovered. The other interface would *not* support a SPP, if isolated. In the case where $\rho_\varepsilon \rho_\mu > (\sigma_\varepsilon^2 - \sigma_\varepsilon \sigma_\mu)/(\sigma_\varepsilon^2 - 1)$, it is seen from Figs. 3.9c,d that two SPP eigenmodes can exist; both have no node in the inner layer. The first one, which is forward propagating, has a low cutoff point at $b = 1$ and a high cutoff, where it degenerates into the second SPP. This eigenmode is backward propagating having only an upper cutoff.

The second situation is described by $\{\sigma_\varepsilon > 1 \text{ and } \rho_\varepsilon < 1\}$.

For $\rho_\varepsilon \rho_\mu < (\sigma_\varepsilon^2 - \sigma_\varepsilon \sigma_\mu)/(\sigma_\varepsilon^2 - 1)$, it proves necessary to examine the intervals that the ratio ρ_ε belong to.

When $\rho_\varepsilon \in \Delta$, with $\Delta = \left\{ \left[(\rho_\varepsilon < \rho_{\varepsilon,1}) \cup (\rho_\varepsilon > \rho_{\varepsilon,2}) \right] \cap (0 < \rho_\varepsilon < 1) \right\}$ and $\rho_{\varepsilon,1}$, $\rho_{\varepsilon,2}$ being the two roots of the polynomial:

$$\Pi(\rho_\varepsilon) = (1 - \sigma_\varepsilon \sigma_\mu)\rho_\varepsilon^2 + \rho_\varepsilon \rho_\mu(\sigma_\varepsilon^2 - 1) - \left[(1 - \sigma_\varepsilon \sigma_\mu) + (1 - \sigma_\varepsilon^2) \right], \tag{3.57}$$

it is $b_1 > b_2$, where:

$$b_1 = \left[\frac{1}{1 - \rho_\varepsilon^2} \right]^{1/2}, \tag{3.58}$$

and it can be shown that b is in the range $b_2 < b < b_1$. The corresponding dispersion diagrams are illustrated in Fig. 3.9e,f. We see that two node-less eigenmodes exist. The first one has only a low cutoff, is backward propagating and concentrates asymptotically at the 1-3 interface. The second SPP has only a low cutoff, is forward propagating and concentrates at the 1-2 interface for relatively large core thicknesses. For $\rho_\varepsilon \in \Delta'$, where $\Delta' = \left\{ (\rho_{\varepsilon,1} < \rho_\varepsilon < \rho_{\varepsilon,2}) \cap (0 < \rho_\varepsilon < 1) \right\}$ we have $b_1 < b_2$ and again two SPP eigenmodes are shown to exist in Figs. 3.9g,h. Compared with the previous two eigenmodes, the cutoff characteristics and the classification remain the same, but the order of the interfaces to which these SPPs concentrate is reversed. Also, both SPPs now exhibit phase reversal. For $\rho_\varepsilon \rho_\mu > (\sigma_\varepsilon^2 - \sigma_\varepsilon \sigma_\mu)/(\sigma_\varepsilon^2 - 1)$, we have $1 < b < b_1$ and the dispersion diagrams are shown in Figs. 3.9i,j. In this case, three distinct solutions of Eq. (3.33) are found; the H_y-field intensity for all of them has no node in the core region. The first one, which corresponds to positive total power P, has a low cutoff point at $b = 1$ and an upper cutoff, where it degenerates into the second SPP. This eigenmode has also a lower and a higher cutoff, but is

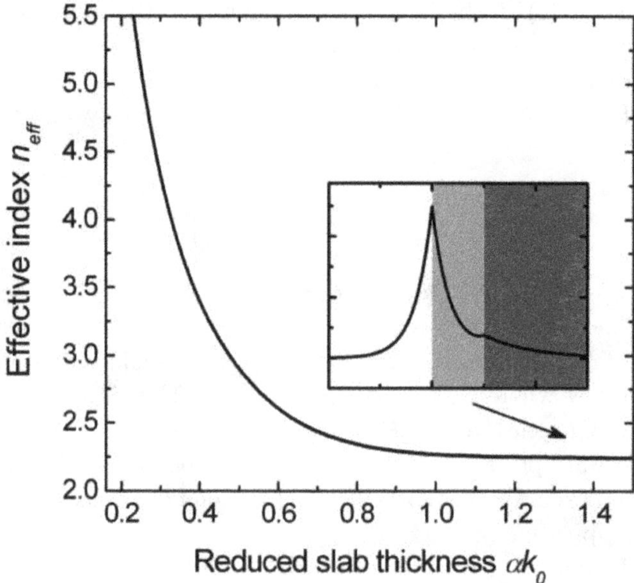

(a) n_{eff} variation for $\sigma_\varepsilon = 1.1$, $\sigma_\mu = 0.7$, $\rho_\varepsilon = 1.38$, $\rho_\mu = 1.5$.

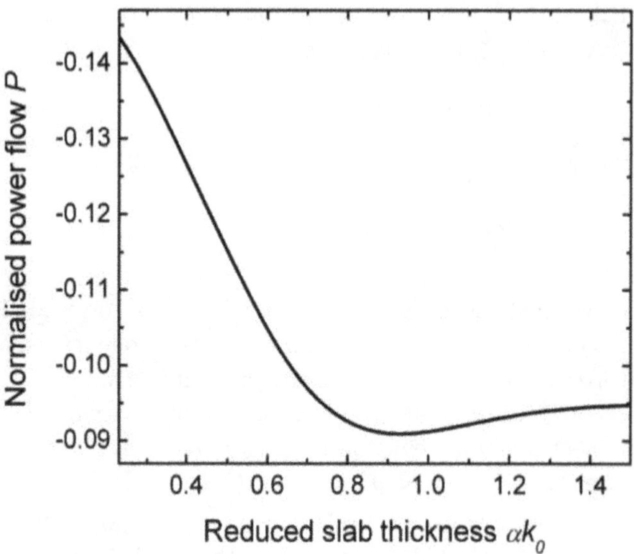

(b) P variation for $\sigma_\varepsilon = 1.1$, $\sigma_\mu = 0.7$, $\rho_\varepsilon = 1.38$, $\rho_\mu = 1.5$.

Figure 3.9: Variation of an SP eigenmode's effective index n_{eff} and normalised power P with the reduced slab thickness ak_0 in a generalised LH slab waveguide (Case III: $n_2 > |n_1| > n_3$ as indicated by the shaded background). In all cases it is assumed that $\varepsilon_r = 2$, $\mu_r = 1.2$.

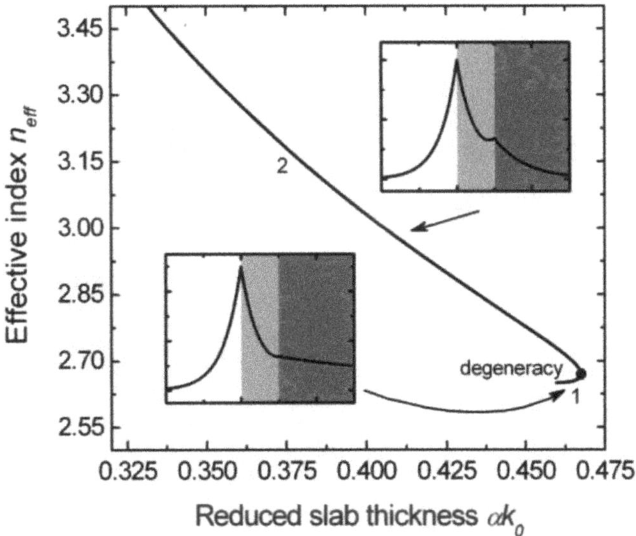

(c) n_{eff} variation for $\sigma_\varepsilon = 1.1$, $\sigma_\mu = 0.7$, $\rho_\varepsilon = 1.95$, $\rho_\mu = 1.5$.

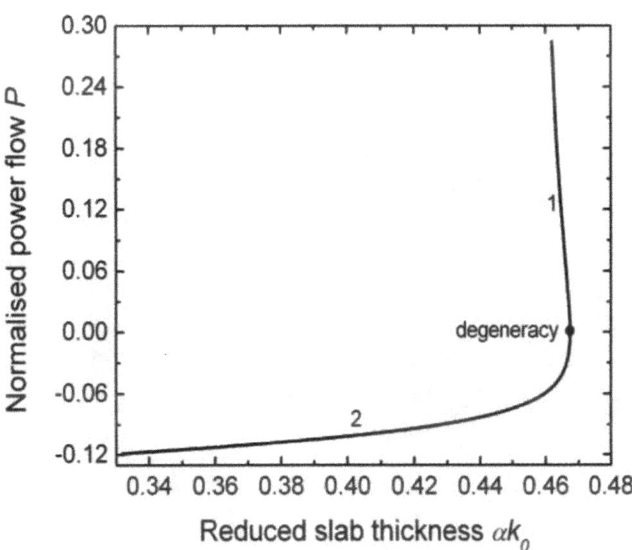

(d) P variation for $\sigma_\varepsilon = 1.1$, $\sigma_\mu = 0.7$, $\rho_\varepsilon = 1.95$, $\rho_\mu = 1.5$.

Figure 3.9: (*Continued*)

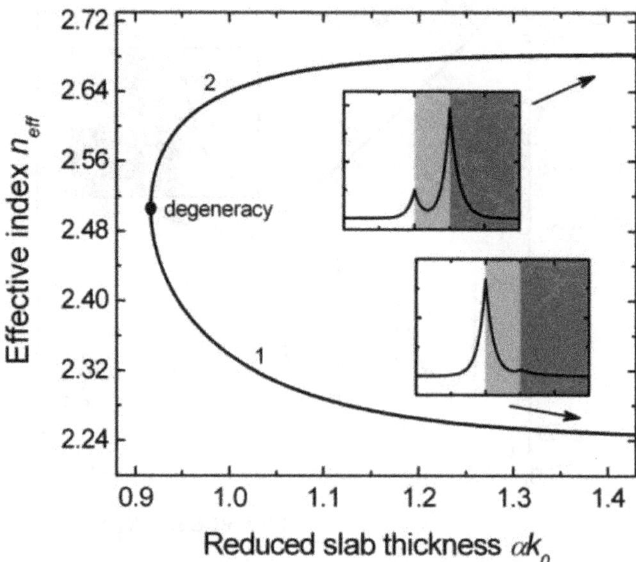

(e) n_{eff} variation for $\sigma_\varepsilon = 1.1$, $\sigma_\mu = 0.7$, $\rho_\varepsilon = 0.75$, $\rho_\mu = 2.5$.

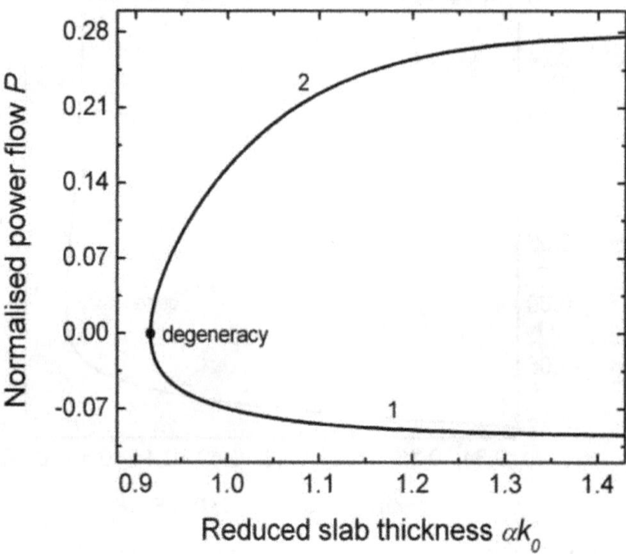

(f) P variation for $\sigma_\varepsilon = 1.1$, $\sigma_\mu = 0.7$, $\rho_\varepsilon = 0.75$, $\rho_\mu = 2.5$.

Figure 3.9: (*Continued*)

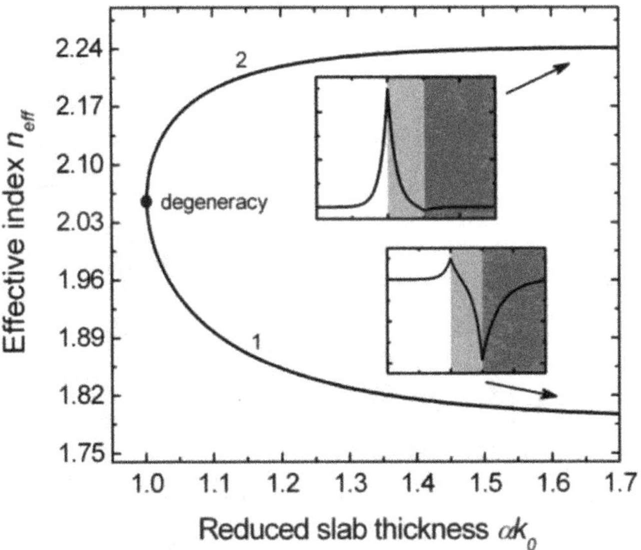

(g) n_{eff} variation for $\sigma_\varepsilon = 1.1$, $\sigma_\mu = 0.7$, $\rho_\varepsilon = 0.5$, $\rho_\mu = 2.5$.

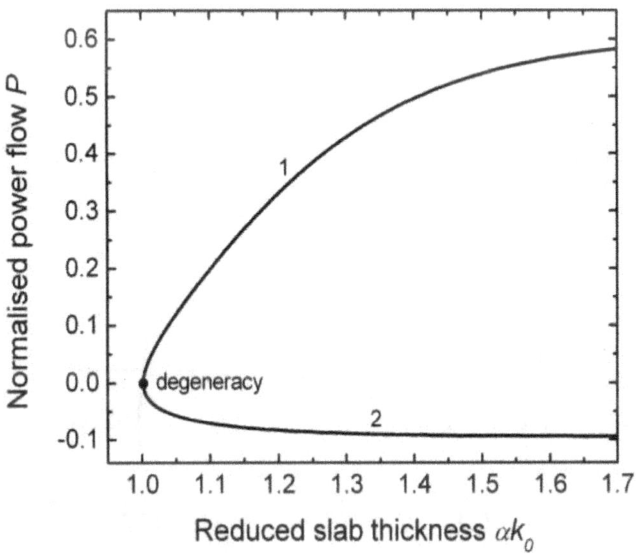

(h) P variation for $\sigma_\varepsilon = 1.1$, $\sigma_\mu = 0.7$, $\rho_\varepsilon = 0.5$, $\rho_\mu = 2.5$.

Figure 3.9: (*Continued*)

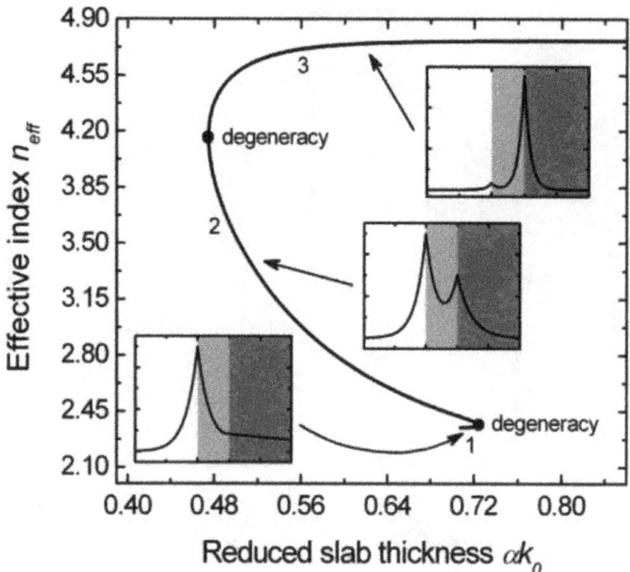

(i) n_{eff} variation for $\sigma_\varepsilon = 1.1$, $\sigma_\mu = 0.7$, $\rho_\varepsilon = 0.92$, $\rho_\mu = 2.5$.

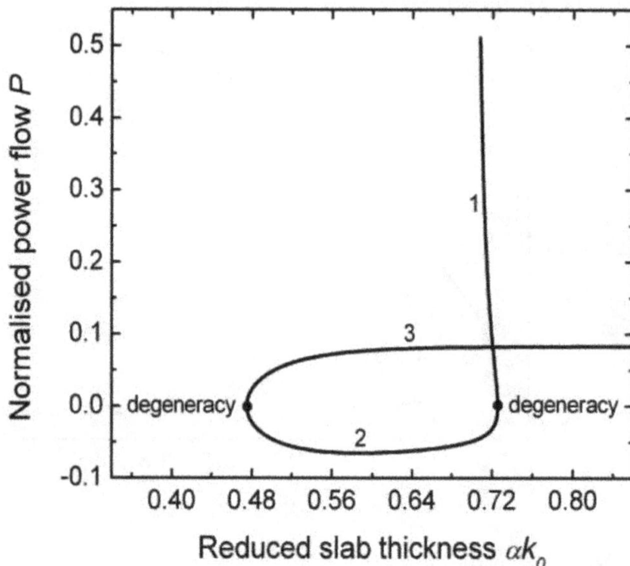

(j) P variation for $\sigma_\varepsilon = 1.1$, $\sigma_\mu = 0.7$, $\rho_\varepsilon = 0.92$, $\rho_\mu = 2.5$.

Figure 3.9: (*Continued*)

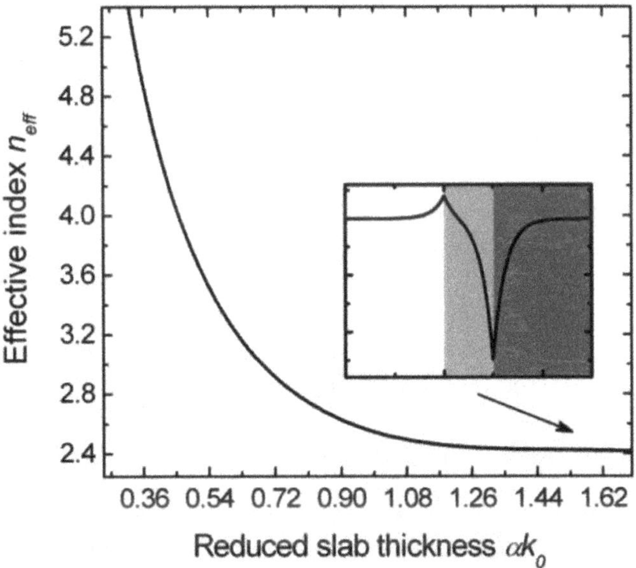

(k) n_{eff} variation for $\sigma_\varepsilon = 0.9$, $\sigma_\mu = 0.9$, $\rho_\varepsilon = 0.95$, $\rho_\mu = 1.2$.

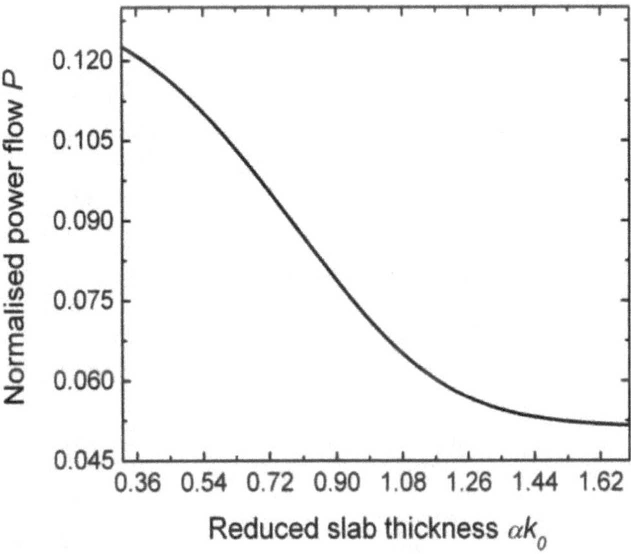

(l) P variation for $\sigma_\varepsilon = 0.9$, $\sigma_\mu = 0.9$, $\rho_\varepsilon = 0.95$, $\rho_\mu = 1.2$.

Figure 3.9: (*Continued*)

backward propagating. The third branch shows only a low cutoff, corresponds to positive P and, asymptotically, degenerates into the isolated 1-2 interface solution. This situation is the only one where two double mode-degeneracy points occur.

In the third situation, which occurs for $\{\sigma_\varepsilon < 1 \text{ and } \rho_\varepsilon > 1\}$ there are no solutions to Eq. (3.41) for any $b > 1$, hence the slab waveguide does not support SPPs.

The final situation occurs for $\{\sigma_\varepsilon \leq 1 \text{ and } \rho_\varepsilon < 1\}$ and the corresponding dispersion diagrams are shown in Figs. 3.9k,l. From the solution constrains of Eq. (3.42), we find that $b > b_1$. A single SPP exists, which does not have cutoff, is forward propagating, exhibits phase reversal and concentrates asymptotically at the 1-2 media interface. It should be noted that, for the chosen values of σ_ε and σ_μ, the isolated 1-3 interface does not support an SPP.

3.7.3 *Summary of the Identified SPP Solutions*

We have presented a systematic investigation of all solutions of the involved characteristic equation giving the p-polarised surface plasmon polariton eigenmodes in generalised slab waveguides, comprised of a negative refractive index core that is bounded by two different positive-index dielectrics. Following an analytical methodology, all SPPs are classified as forward or backward propagating, depending on the sign of the associated power flow. Contrary to the case of thin metallic films surrounded by two different dielectrics, where the SPPs depend solely on the permittivity distribution, we have shown analytically that, for the corresponding left-handed structures considered here, the refractive index distribution must also be included in the analysis.

In particular, when the absolute value of the core refractive index is greater than those of the surrounding dielectrics we identified ten TM SPP eigenmodes (see Table I); six of these solutions are backward propagating. Nine solutions existed if the core index has the smallest absolute value (Table II); one of these SPPs has antiparallel phase and group velocities. In the final case, where the core index value is between those of the claddings, eleven SPP eigenmodes were identified (Table III); four of these solutions have backwards power flow in relation to the direction of the phase velocity. The total number of 30 SPP eigenmodes, which is significantly higher than the 8 SPPs that exist in similar metallic film geometries [Prade *et al.* (1991)], is a direct result of the presence of negative magnetic permeability in the LH structures that provides additional degrees of freedom in the definition of the V-numbers.

The analytical treatment revealed features not observed before, such as the occurrence of "supermodes" in the case of a violation of the isolated interface condition, strong field enhancement and opening of gaps in the geometric dispersion diagrams owing to the asymmetry, and coexistence of three eigenmodes with double mode-degeneracy points occurring twice. We also demonstrated that the group velocity of the investigated slow waves could be decreased down to zero or become negative, by adjusting suitably the thickness of the core.

Table I: Summary of the discussion for the case $|n_1| > n_2 > n_3$.

$\sigma_\varepsilon \geq 1, \rho_\varepsilon \geq 1$	$\sigma_\varepsilon \geq 1, \rho_\varepsilon < 1$	$\sigma_\varepsilon < 1, \rho_\varepsilon \geq 1$	$\sigma_\varepsilon < 1, \rho_\varepsilon < 1$		
			$\rho_\varepsilon \in \Delta$	$\rho_\varepsilon \in \Delta'$	$\rho_\varepsilon \rho_\mu > (\sigma_\varepsilon \sigma_\mu - \sigma_\varepsilon^2)/(1 - \sigma_\varepsilon^2)$
1st: forward, 2 cutoffs. 2nd: backward, upper cutoff.	1st: forward, low cutoff.	1st: forward, low cutoff.	1st: forward, low cutoff. 2nd: forward, no cutoff.	1st: forward, low cutoff. 2nd: forward, no cutoff.	1st: forward, no cutoff.

Table II: Summary of the discussion for the case $n_2 > n_3 > |n_1|$.

$\sigma_\varepsilon > 1, \rho_\varepsilon > 1$		$\sigma_\varepsilon > 1, \rho_\varepsilon \leq 1$	$\sigma_\varepsilon \leq 1, \rho_\varepsilon > 1$	$\sigma_\varepsilon \leq 1, \rho_\varepsilon \leq 1$
$\rho_\varepsilon > \sigma_\varepsilon$	$\rho_\varepsilon \in \Delta$			
1st: forward, 2 cutoffs. 2nd: backward, low cutoff. 3rd: backward, no cutoff.	1st: backward, low cutoff. 2nd: backward, no cutoff.	1st: forward, 2 cutoffs. 2nd: backward, low cutoff.	1st: forward, 2 cutoffs. 2nd: backward, low cutoff.	1st: forward, upper cutoff.

Table III: Summary of the discussion for the case $n_2 > |n_1| > n_3$.

$\sigma_\varepsilon > 1, \rho_\varepsilon \geq 1$			$\sigma_\varepsilon > 1, \rho_\varepsilon < 1$		$\sigma_\varepsilon \leq 1, \rho_\varepsilon < 1$
$\rho_\varepsilon \rho_\mu < (\sigma_\varepsilon^2 - \sigma_\varepsilon \sigma_\mu)/(\sigma_\varepsilon^2 - 1)$	$\rho_\varepsilon \rho_\mu > (\sigma_\varepsilon^2 - \sigma_\varepsilon \sigma_\mu)/(\sigma_\varepsilon^2 - 1)$	$\rho_\varepsilon \in \Delta$	$\rho_\varepsilon \in \Delta'$	$\rho_\varepsilon \rho_\mu > (\sigma_\varepsilon^2 - \sigma_\varepsilon \sigma_\mu)/(\sigma_\varepsilon^2 - 1)$	
1st: forward, no cutoff.	1st: forward, 2 cutoffs. 2nd: backward, upper cutoff.	1st: backward, low cutoff. 2nd: forward, low cutoff.	1st: forward, low cutoff. 2nd: backward, low cutoff.	1st: forward, 2 cutoffs. 2nd: backward, 2 cutoffs. 3rd: forward, low cutoff.	1st: forward, no cutoff.

3.8　Oscillatory Guided Modes Supported by Asymmetric DNGM Slab Heterostructures

In the previous section we have seen that important potential applications of LH materials in optics and microwaves, such as the possibility to create super-resolving lenses or to improve the performance of biosensing devices, prompted detailed investigations into the properties of specific LH heterostructures [Shadrivov *et al.* (2003); Ramakrishna *et al.* (2002); Wu *et al.* (2003); He *et al.* (2005); Qing and Chen (2004)]. From this research it has been shown that LH waveguides support two classes of waves: Oscillatory or waveguide modes (OM/fast waves), which are also found in regular dielectric slab structures, and a rich variety of surface plasmon polariton modes (SPP/slow waves). We showed that up to 30 bound SPPs can exist in a generalised LH slab waveguide for all choices of opto-geometrical parameters. Although approximately a third of the SPP solutions allow for attaining *zero* group velocity – a notable characteristic of guided modes in such waveguides, to which we shall turn our attention later, in Chapter 5 – their inherent sensitivity to small variations of the media interfaces [Tsakmakidis *et al.* (2006c)] may limit their practical use. For this reason, oscillatory modes, which generally have their maxima inside the waveguide core, are more suitable for most conceived applications, including slow-light. Nonetheless, in most relevant studies, these modes have been described only qualitatively, following ad hoc graphic solution approaches [Shadrivov *et al.* (2003); Wu *et al.* (2003)]. In this manner, relevant modal properties were either not revealed or not conclusively proven and, importantly, explicit expressions for the cutoff conditions of each mode have not been established yet.

In this section we shall report on an extension of waveguide mode theory that considers an *asymmetric* three-layer slab heterostructure in which either the guiding region or the cladding region may have a negative refractive index. We derive dimensionless, closed-form expressions for the cutoff(s) of the investigated oscillatory modes that allow for the identification of their existence regions. The modes with negative energy flux that give rise to negative group velocity are identified via an explicit expression for the cycle-averaged total power flow P_{tot} in the guide. Based on this treatment we prove rigorously that, with judicious choice of parameters, there is a frequency region where the second mode can simultaneously exist *alone* and attain *zero* group velocity. Moreover, we show that the inverted (LH-RH-LH) arrangement supports similar modes as the RH-LH-RH heterostructure but with opposite power flux. To our knowledge, these are the only slab waveguide structures utilizing homogeneous, isotropic [Koschny *et al.* (2005)] media that have the potential for single-mode operation in the "slow-light regime".

In the analysis that follows we consider the geometry illustrated in Fig. 3.10, where all media are assumed to be lossless [Chen *et al.* (2004)], homogeneous and isotropic. We shall be concerned primarily with p-polarised (TM) waves, in which the magnetic field is directed along the y-axis. First, we investigate the case where the core is LH with thickness $2a$ bounded asymmetrically by two RH media that satisfy $n_2 > n_3$. The fields in the guide consist of counterpropagating forward and backward modes trapped within the core by total-internal reflection.

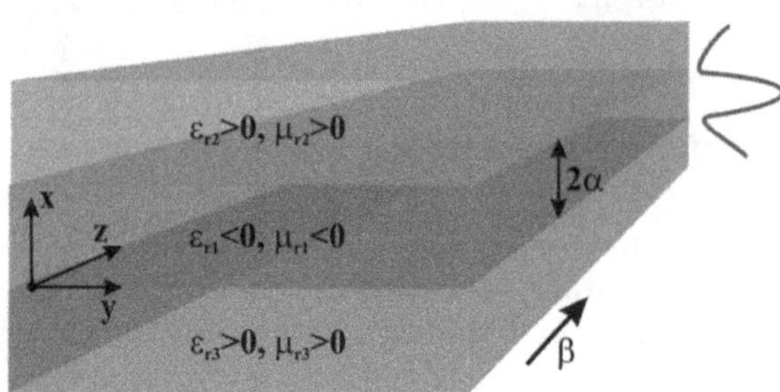

Figure 3.10: Illustration of the asymmetric left-handed planar heterostructure, supporting an oscillatory guided mode.

For wave guidance to occur the wave components of both types of modes must satisfy the total-internal reflection condition. This condition can be stated as $|n_1| > n_{eff} > n_3$, where n_{eff} is the effective index of the mode. We point out here that there is a marked contrast between the present case and that of SPP waves that we studied in the previous section: For the latter case, we required $n_{eff} > \max\{|n_1|, n_2|\}$ (see remarks after Eq. (3.29)) as a result of the somewhat different solution ansatz (see Eq. (3.29)) that we therein made to the wave equation. Matching the E_z and H_y fields at the boundaries yields the following dimensionless characteristic equation for oscillatory modes in generalised LH slab waveguides:

$$ak_0 = \frac{1}{2|n_1|\sqrt{1 - \rho_\varepsilon \rho_\mu}\, b}\left[\arctan\left(\frac{\rho_\varepsilon b}{\sqrt{1 - b^2}}\right) + \arctan\left(\frac{\sigma_\varepsilon b}{\sqrt{t^2 - b^2}}\right) + m\pi\right], \quad (3.59)$$

where, similarly to the previous section, the following ratios are introduced:

$$b(n_{eff}) = \frac{U}{V_2} = \left[\frac{1 - (n_{eff}/n_1)^2}{1 - \rho_\varepsilon \rho_\mu}\right]^{1/2}, \tag{3.60}$$

$$t = \frac{V_3}{V_2} = \left[\frac{1 - \sigma_\varepsilon \sigma_\mu}{1 - \rho_\varepsilon \rho_\mu}\right]^{1/2}, \tag{3.61}$$

with $\rho_\varepsilon = \varepsilon_{r2}/|\varepsilon_{r1}|$, $\rho_\mu = \mu_{r2}/|\mu_{r1}|$, $\sigma_\varepsilon = \varepsilon_{r3}/|\varepsilon_{r1}|$, $\sigma_\mu = \mu_{r3}/|\mu_{r1}|$, and k_0 being the free-space wavevector. In this section, the two V-parameters are defined as

$$V_2(ak_0) = ak_0\left(\varepsilon_{r1}\mu_{r1} - \varepsilon_{r2}\mu_{r2}\right)^{1/2} \quad \text{and}$$

$$V_3(ak_0) = ak_0\left(\varepsilon_{r1}\mu_{r1} - \varepsilon_{r3}\mu_{r3}\right)^{1/2}, \quad \text{whereas}$$

$$U = a\kappa = ak_0\left(\varepsilon_{r1}\mu_{r1} - n_{eff}^2\right)^{1/2},$$

with κ standing for the transverse component of the wavevector in the core. It is noteworthy that similar dispersion expressions, cast in an "inverted" form that is obtained following zigzag-ray model analysis, describe oscillatory modes in standard asymmetric optical waveguides [Kogelnik and Ramaswamy (1974)]. The parameters used in Eqs. (3.60) and (3.61) here, though, obey the restrictions $0 \le b \le 1$ and $t > 1$. The equalities for b hold at the mode-cutoff points. For each mode ($m = 0, 1, \ldots$) and refractive index distribution, the dispersion diagrams, n_{eff} versus reduced slab thickness ak_0, may be directly obtained by noting that b in Eq. (3.59) is a function of n_{eff}, which in turn increases monotonically from n_2 to $|n_1|$. In discussing the cutoff conditions, it is important to recognise the fact that several double mode-degeneracy points can, in principle, appear in the dispersion diagrams since some of the modes in the LH heterostructure will be backward propagating, having antiparallel phase (v_p) and group (v_g) velocities. The cycle-averaged total power flow P_{tot} at these (cutoff) points will vanish. It is therefore useful to derive here a general closed-form expression for P_{tot}. Integrating the z component of the complex pointing vector over the guide's cross-section [Tsakmakidis *et al.* (2006a); Snyder and Love (1983)], considering Eq. (3.59), we obtain after some algebraic manipulations:

$$P_3 = C\frac{1}{|\varepsilon_{r1}|}\frac{a\beta}{\sigma_\varepsilon W_3},$$

(3.62a)

$$P_1 = C\frac{a\beta}{|\varepsilon_{r1}|}\frac{W_3^2 + \sigma_\varepsilon^2 U^2}{\sigma_\varepsilon^2 U^2}\left[\frac{\rho_\varepsilon W_2}{W_2^2 + \rho_\varepsilon^2 U^2} + \frac{\sigma_\varepsilon W_3}{W_3^2 + \sigma_\varepsilon^2 U^2} - 2\right],$$

(3.62b)

$$P_2 = C\frac{a\beta}{|\varepsilon_{r1}|\rho_\varepsilon W_2}\frac{\rho_\varepsilon^2}{\sigma_\varepsilon^2}\frac{W_3^2 + \sigma_\varepsilon^2 U^2}{W_2^2 + \rho_\varepsilon^2 U^2},$$

(3.62c)

where P_i $(i = 1,2,3)$ the power confined in the waveguide i-layer (Fig. 3.10), β, W_j $(j = 1,2)$ the mode longitudinal propagation and decay constants, respectively, and C an arbitrary positive constant. It is inferred form Eqs. (3.62a–c) that the net power flow can become negative in the core layer, but remains positive in the cladding region. Calculating the total power $P_{tot} = \sum_{i=1}^{3} P_i$, in terms of the dimensionless parameters defined before and the reduced slab thickness, we arrive at:

$$P_{tot} = C\frac{W_3^2 + \sigma_\varepsilon^2 U^2}{\sigma_\varepsilon^2 U^2}\frac{\left(n_1^2(ak_0)^2 - U^2\right)^{1/2}}{|\varepsilon_{r1}|}$$
$$\cdot\left[\frac{\rho_\varepsilon}{W_2}\frac{V_2^2}{W_2^2 + \rho_\varepsilon^2 U^2} + \frac{\sigma_\varepsilon}{W_3}\frac{V_3^2}{W_3^2 + \sigma_\varepsilon^2 U^2} - 2\right].$$

(3.63)

To facilitate the discussion of the mode cutoff relationships, we first note from Eq. (3.59) that the upper branches $(b \to 0)$ for all modes other than the fundamental $(m \neq 0)$, show no upper cutoff thickness, since $b \to 0$ in this case yields $ak_0 \to \infty$. The situation is reminiscent of OMs in normal dielectric slab structures and is illustrated in Fig. 3.11b. However, for the fundamental mode $(m = 0)$ inspection of Eq. (3.59) reveals a somewhat unusual behaviour and a *low* cutoff point given by:

$$(ak_0)\Big|_{Low}^{m=0} = \frac{1}{2|n_1|}\left[\frac{\rho_\varepsilon}{\sqrt{1 - \rho_\varepsilon^2\rho_\mu^2}} + \frac{\sigma_\varepsilon}{\sqrt{1 - \sigma_\varepsilon^2\sigma_\mu^2}}\right].$$

(3.64)

At this transition point, depicted Fig. 3.11a, the SPP-1 mode [Tsakmakidis *et al.* (2006c)] transforms continuously to the $m = 0$ oscillatory mode, whose field pattern resembles closely that of a surface wave. In [Shadrivov *et al.* (2005)], the existence of such cutoff point for the $m = 0$ oscillatory mode in asymmetric three-layer LH heterostructures has been associated with the appearance of a complete 3D photonic bandgap. Critically, the fundamental mode also exhibits an upper cutoff that can be calculated as:

$$(ak_0)\Big|_{Upper}^{m=0} = \frac{1}{2|n_1|\sqrt{1 - \rho_\varepsilon\rho_\mu}}\left[\arctan\left(\frac{\sigma_\varepsilon}{\sqrt{t^2 - 1}}\right) + \pi\left(m + \frac{1}{2}\right)\right].$$

(3.65)

This relationship also describes the upper cutoff point of all lower branches in the mode-dispersion diagrams. An exemplary result of such curves for a "slow-light oriented" choice of optogeometric parameters is shown in Fig. 3.11b, with the insets

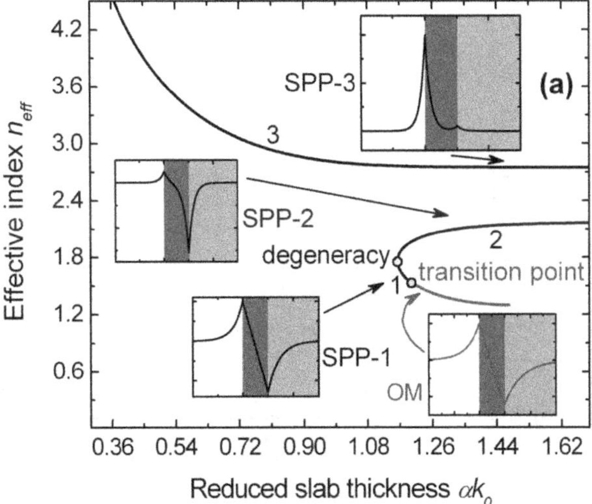

(a) n_{eff} variation for $\sigma_\varepsilon = 1.1$, $\sigma_\mu = 0.5$, $\rho_\varepsilon = 1.15$, $\rho_\mu = 0.6$, $\varepsilon_r = 2$, $\mu_r = 1.2$.

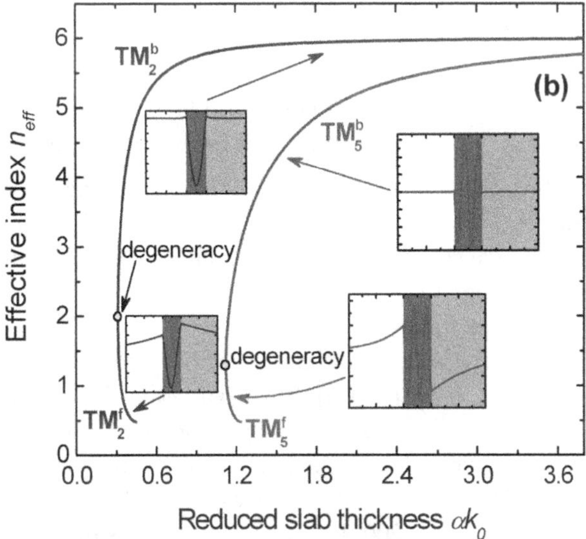

(b) n_{eff} variation for $\sigma_\varepsilon = 0.05$, $\sigma_\mu = 0.05$, $\rho_\varepsilon = 0.08$, $\rho_\mu = 0.08$, $\varepsilon_r = 6$, $\mu_r = 6$.

Figure 3.11: Variation of SPP and fundamental oscillatory mode (OM) effective index n_{eff} with the reduced slab thickness ak_0.

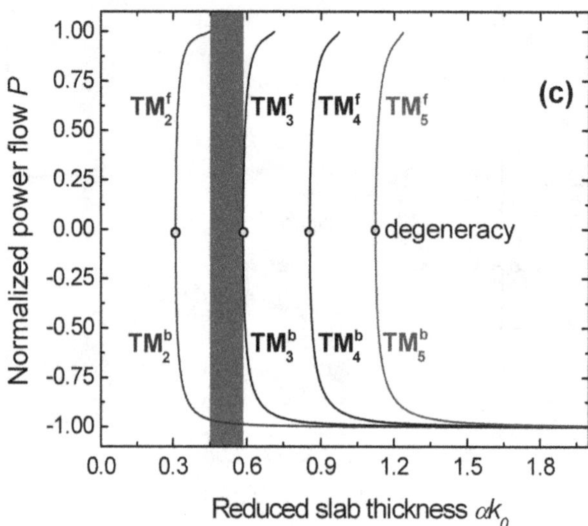

(c) Variation of oscillatory modes normalised power flow P with reduced slab thickness for the parameters defined in (b).

Figure 3.11: (*Continued*)

illustrating the corresponding field-profiles. For the sake of clarity, the dispersion curve of the fundamental TM_1 mode is not shown here, since the upper cutoff point of this mode is $\left(ak_0\right)\big|_{Upper}^{m=0} \simeq 0.18763$, i.e. well below the "degeneracy" point of the TM_2 branch, which occurs at $\left(ak_0\right)\big|_{Deg}^{m=0} \simeq 0.31047$. The proposed nomenclature for recognising the oscillatory modes, consists of a pair of letters to identify the polarisation, followed by a subscript to track the number of nodes in the core region, and a superscript to designate that the mode is forward (f) or backward (b), i.e. $TM_{m+1}^{f/b}$ for the mth-order mode.

Figure 3.11c reports the variation of the normalised power flow $P = P_{tot}/\left(|P_1| + |P_2| + |P_3|\right)$ [Tsakmakidis *et al.* (2006a); Shadrivov *et al.* (2003)] with reduced slab thickness for the previous solutions. One may notice that the two branches of the TM_2 mode merge at a critical slab thickness. At this (cutoff) point the total power, P_{tot}, of the resulting degenerate mode vanishes and the group velocity reduces to zero [Loudon (1970); Ruppin (2002)]. An exact expression for this cutoff point cannot be given except in the somewhat simplified form of $P_{tot} = 0$. However, for every mode ($m = 1, 2, \ldots$) this point can be swiftly calculated following the steps for the derivation of the dispersion diagrams outlined above and requiring $ak_0(n_{eff}) \to$ min.

Since the TM_1 and TM_2^f modes *always* show an upper cutoff, given by Eq. (3.65), we conclude that there is a unique and experimentally intriguing region, highlighted by the gray area in Fig. 3.11c, where the backward TM_2^b mode can simultaneously exist alone and allow for attaining very small or zero group velocity via adiabatically tapering to the degeneracy point. Under weaker guidance conditions, i.e. $\rho_\varepsilon \rho_\mu, \sigma_\varepsilon \sigma_\mu$

it is further possible to altogether suppress the TM_2^f branch, thereby increasing the operable width of the highlighted region from approximately 0.134 to 0.274, for the case shown in Fig. 3.11c.

It is interesting to note that similar modes, but with reversed power flow, are supported by the inverted heterostructure, i.e. one with a RH core (e.g. air) and LH claddings. Indeed, by again solving the wave equation in each waveguide region and requiring continuity of the tangential fields at the boundaries, we find that the characteristic Eq. (3.59) remains unchanged, and so does the expression for the sole magnetic field component, H_y. In a similar vein, it is found that for the new expressions of the cycle-averaged power flow in each layer, one only needs to replace $|\varepsilon_{r1}|$ in Eqs. (3.62a–c) with $-|\varepsilon_{r1}|$. Based on the explicit form of Eq. (3.59) and since the V-parameters defined above are independent of the refractive index sign distribution, it is inferred that all modal properties of the RH-LH-RH arrangement, previously analysed, are replicated by its "dual" counterpart. It appears at this time that this is the only pair of planar waveguides constructed from isotropic media that can do so. This property also provides increased flexibility in slow-light waveguiding design utilising LH materials.

In summary, on the basis of an exact, analytic appraisal, it has been shown that an asymmetric planar waveguide utilising LH media in either the core or the cladding can support single-mode operation in the slow-light regime. The investigated scheme, relying solely on sufficient decrease of slab thickness, combines the remarkably simple approach for slowing down light suggested in [Karalis *et al.* (2005)], with the use of efficiently excitable waveguide modes used in [Vlasov *et al.* (2005); Gersen *et al.* (2005)], since the profile of the TM_2 solution here closely matches that of a single-mode fibre. Moreover, the heterostructures investigated here can be designed to be monomode in the desired frequency range [Tsakmakidis *et al.* (2006b)]. The control of the group velocity is achieved solely by varying the core thickness rather than by varying the temperature or field intensity [Vlasov *et al.* (2005); Lukin and Imamoglu (2001)]. The same is true for the light in- and out-coupling, which may be satisfactorily adjusted by adiabatically tapering the size of the waveguide core. It should be stressed that the mechanism for decelerating light here does not directly rely on refractive index resonances but merely on the exchange of power between the core and cladding regions, as indicated by Eqs. (3.62)–(3.63). Hence, broadband slow light can be obtained provided that the negative material parameters are designed to exist over relatively large bandwidths [Chen *et al.* (2005)] at optical frequencies [Linden *et al.* (2004)]. This intriguing possibility is further analysed in detail in the Chapter 5.

Chapter 4

Plasmonic and Metamaterial Waveguides

4.1 Introduction

In the previous chapter, we saw how light propagates in negative-refractive-index waveguides. The analysis was therein fully a closed-form one. In the present chapter, we will go through a powerful numerical method, i.e. the transfer matrix method, for analysing both plasmonic and metamaterial waveguides, which affords for calculations of more general (e.g., more than three-layer) geometries.

Starting from Maxwell's equations the interaction of light with matter is formulated in terms of the material's polarisation and magnetisation response in Section 4.2. Next the propagation of electromagnetic waves in multilayer waveguides is considered. It is shown that the geometry of these structures gives rise to the formation of three distinct sets of modes; bound, radiative and leaky modes. The physical applicability of these states is discussed and a robust method for calculating their properties in arbitrary planar waveguides is introduced in Section 4.3.

Following this introduction to planar waveguide theory the influence of material dispersion on wave propagation is considered. A key material used in both the plasmonic and NRI waveguides is metals. The interaction of light with metals results in a number of unique phenomena such as subwavelength localisation. Many of these phenomena are related to the formation of highly localised surface waves known as surface plasmons and could enable a wide variety of exciting applications. An overview of the optical properties of metals and surface plasmons is given in Section 4.4 for both single and multilayered waveguides demonstrating the unusual propagation characteristics in these structures.

Finally the effective properties of artificial media known as metamaterials are discussed in Section 4.5. Metamaterials are a relatively new class of media that derive their optical response from the geometric design of subwavelength resonant structures from which they are composed. As a result metamaterials can be designed to exhibit optical properties not observed in natural bulk materials. One of the most remarkable being the demonstration of metamaterials with a negative refractive index. A number of studies has found that waveguides incorporating layers with

91

NRI support electromagnetic modes with unusual properties, including extremely low group velocities.

4.2   Maxwell's Equations

A wide variety of optical phenomena, including the formation of surface plasmons and optical response of metamaterials, are described by Maxwell's equations. These can be expressed as a set of four partial differential equations that, together, fully describe the classical interactions of electric and magnetic fields with their sources of charges and currents. In their differential, macroscopic form Maxwell's equations can be expressed as [Eyges (1980)]:

$$\nabla \times \mathbf{E} = -\frac{\partial \mathbf{B}}{\partial t}, \tag{4.1}$$

$$\nabla \times \mathbf{B} = \frac{\partial \mathbf{D}}{\partial t} + \mathbf{J}, \tag{4.2}$$

$$\nabla \cdot \mathbf{D} = \varrho_f, \tag{4.3}$$

$$\nabla \cdot \mathbf{B} = 0. \tag{4.4}$$

When considering electromagnetic waves in matter it is convenient to work with the equations in this form as the collective response of bound charge and current densities $(\varrho_b(\mathbf{r}, t), \mathbf{J}_b(\mathbf{r}, t))$ to external electric and magnetic induction fields $(\mathbf{E}(\mathbf{r}, t), \mathbf{B}(\mathbf{r}, t))$ is contained in auxiliary fields known as the electric displacement field $\mathbf{D}(\mathbf{r}, t)$ and the magnetic field $\mathbf{H}(\mathbf{r}, t)$. This allows for the efficient calculation of light-matter interactions over large scales and separates out the influence of any free charge $\varrho_b(\mathbf{r}, t)$ and current $\mathbf{J}_b(\mathbf{r}, t)$ densities. However, in order to apply Maxwell's equations in this form the functional dependence of $\mathbf{D}(\mathbf{E})$ on the electric field and $\mathbf{H}(\mathbf{B})$ on the magnetic induction field needs to be determined. The equations that describe this connection are known as the constitutive relations which have the following general definition:

$$\mathbf{D}(\mathbf{E}) = \varepsilon_0 \mathbf{E} + \mathbf{P}(\mathbf{E}), \tag{4.5}$$

$$\mathbf{B}(\mathbf{H}) = \mu_0^{-1} \mathbf{B} - \mathbf{M}(\mathbf{B}), \tag{4.6}$$

where $\mathbf{P}(\mathbf{E})$ is the polarisation field arising from the interaction of an external electric field with charges in the material and $\mathbf{M}(\mathbf{B})$ is magnetisation field induced by the interaction of the magnetic induction field with currents. Finally, ε_0 and μ_0 are the vacuum permittivity and permeability. Particular polarisation and magnetisation responses thus have to be specified for each material under consideration.

The dependency of polarisation and magnetisation on the electric and magnetic fields can most easily be represented in the frequency domain. Assuming a harmonic time $\exp(-i\omega t)$ for the fields, Maxwell's two curl equations become

$$\nabla \times \mathbf{E}(\omega) = i\omega \mathbf{B}(\omega), \tag{4.7}$$

$$\nabla \times \mathbf{H}(\omega) = \mathbf{J}_f - i\omega \mathbf{D}(\omega). \tag{4.8}$$

Considering the case of homogeneous isotopic media and assuming the polarisation and magnetisation are linearly dependent on the electric and magnetic fields, $\mathbf{P}(\omega)$ and $\mathbf{M}(\omega)$, can now be defined as

$$\mathbf{P}(\omega) = \varepsilon_0 \chi_e(\omega) \mathbf{E}(\omega) \tag{4.9}$$

$$\mathbf{M}(\omega) = \frac{\chi_m(\omega)}{1 + \chi_m(\omega)} \mathbf{B}(\omega), \tag{4.10}$$

where $\chi_e(\omega)$ and $\chi_m(\omega)$ are the frequency dependent electric and magnetic susceptibilities of the material. The slightly convoluted form of Eq. (4.10) is due to the traditional convention that $\mathbf{M}(\omega) = \chi_m(\omega) \mathbf{H}(\omega)$. Generally, the functional form of these susceptibilities for different materials are determined by fitting basic response models to experimental data. Particular response models often used for describing metals and metamaterials shall be discussed in the next section.

By inserting the expressions for the polarisation and magnetisation fields, Eqs. (4.9) and (4.10), into Maxwell's two curl Eqs. (4.5) and (4.6) the constitutive relations can be redefined as:

$$\mathbf{E}(\omega) = \varepsilon_0 \varepsilon_r(\omega) \mathbf{E}(\omega), \tag{4.11}$$

$$\mathbf{H}(\omega) = \mu_0^{-1} \mu_r^{-1}(\omega) \mathbf{B}(\omega), \tag{4.12}$$

where $\varepsilon_r(\omega) = 1 + \chi_e(\omega)$ and $\mu_r(\omega) = 1 + \chi_m(\omega)$ are the relative permittivity and permeability of the medium, representing the averaged response of microscopic bound charges and currents to the driving electric and magnetic fields. Henceforth, the relative permittivity and permeability shall be represented by ε and μ.

Using these new constitutive relations the wave equation for propagating electromagnetic waves in an inhomogeneous, isotropic medium in the absence of free charges and currents can be written as:

$$\nabla \times \mu^{-1}(\omega, \mathbf{r}) \big(\nabla \times \mathbf{E}(\omega, \mathbf{r}) \big) = \frac{\omega^2}{c^2} \varepsilon(\omega, \mathbf{r}) \mathbf{E}(\omega, \mathbf{r}), \tag{4.13}$$

$$\nabla \times \varepsilon^{-1}(\omega, \mathbf{r}) \big(\nabla \times \mathbf{H}(\omega, \mathbf{r}) \big) = \frac{\omega^2}{c^2} \mu(\omega, \mathbf{r}) \mathbf{H}(\omega, \mathbf{r}), \tag{4.14}$$

where $c = 1/\sqrt{\varepsilon_0 \mu_0}$ is the speed of light in vacuum.

In homogeneous media, Eqs. (4.13) and (4.14) reduce to their well known Helmholtz form describing an electromagnetic wave propagating through a medium at a reduced velocity $v(\omega) = c/n(\omega)$, where $n(\omega)$ is the frequency dependent refractive index. The dependence of velocity on refractive index results in a range of interesting optical phenomena such as the reflection and refraction of light at the interface between two materials. The angles of transmission and reflection can be calculated simply from Snell's law:

$$\frac{\sin \theta_1}{\sin \theta_2} = \frac{n_2}{n_1} \tag{4.15}$$

where θ_1 is the angle of incidence and θ_1 is the angle of transmission into material 2, the power of the transmitted and reflected waves can also be found from the Fresnel equations [Eyges (1980)]. Interestingly, for a wave propagating from a region of high refractive index to one of low index these equations predict that, for a certain range of incident angles, the wave will be completely reflected at the interface. This phenomenon is known as total internal reflection which occurs above a critical angle of incidence, dependent on the material properties as well as the polarisation of the incident wave. If a second interface is introduced so that the high index material is confined to a slab, surrounded on top and bottom by lower index cladding layers, then total internal reflection at both interfaces will restrict the wave propagation to the plane of the material layers. This structure acts as a simple waveguide, which are devices designed to control the direction of propagation of electromagnetic waves. As only certain angles will be totally internally reflected not all waves will be guided by these structures. The field configurations that are guided are called the bound modes of the waveguide. To determine the characteristics of waveguide modes the electromagnetic wave equation for waves in inhomogeneous media has to be solved. This can be performed using a number of techniques such as the finite-element-method (FEM) [Jin (2002)] or finite-difference time-domain (FDTD) method [Taflove and Hagness (2000)], which can accurately describe the interaction of fields with three-dimensional structures. For the case of simple planar waveguides, where the variation in materials is limited to just one dimension, the transcendental wave equation can be solved semi-analytically using various root finding algorithms.

Throughout this work semi-analytic calculations are used to efficiently characterise and optimise two waveguide designs for stopped light operation. FDTD simulations are then performed to test the predicted properties and to investigate the temporal dynamics of the modes. Details concerning the implementation of materials in FDTD are discussed later in this chapter while a full description of the FDTD method can be found in Chapter 2. In the next section, theory concerning the electromagnetic properties of planar waveguides and several important mode definitions shall be introduced.

4.3   Planar Waveguide Structures

Planar waveguides are characterised by a refractive index profile that varies in only one dimension. A schematic representation of a general N layer planar waveguide is shown in Fig. 4.1, where n_i and d_i are the refractive index and thickness of the nth layer, respectively.

The position of each interface y_i between layers is shown on the left, the first interface y_i between the first and second layer is at $y = 0$, subsequent interfaces are at negative y-values. The first layer, $i = 0$, is generally referred to as the superstrate and the final layer, $i = N + 1$ is the substrate layer. When EM waves interact with planar waveguides the geometry imposes constraints on the fields and the resulting allowed states are called the modes of the system. As the field is only confined in one direction it can propagate freely in the xz plane, however for convenience x is

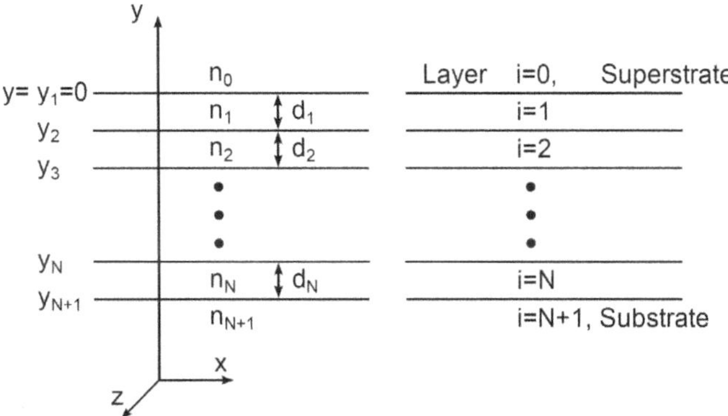

Figure 4.1: Schematic diagram of a planar multilayer waveguide. The layers are oriented in the xz plane with x chosen as the propagation direction.

here defined as the propagation direction. The modes can be further classified in terms of their field polarisation, transverse electric (TE) where the only non-zero field components are E_z, H_x and H_y, and transverse magnetic (TM) where the field components are H_z, E_x and E_y. Determining the modal characteristics involves two main steps, first the wave equation for the structure must be defined, which depends on the particular structure and the nature of the mode under investigation, secondly a robust method to find the roots of this characteristic equation is needed in order to determine the modes of the waveguide.

The problem can be most easily defined by writing down the generalised electromagnetic wave equation for a wave propagating through a spatially varying material. For simplicity the wave equation is given only for the single field component in the z-direction as the remaining field terms can be derived from this. Assuming free propagation in the x direction the z component can be written $\phi_z(y)\exp^{i(\omega t - \beta x)}$. Inserting this field ansatz into the inhomogeneous wave equations (4.13) and (4.14) yields the transcendental 1D wave equation:

$$m(y,\omega)\frac{\partial}{\partial y}m(y,\omega)^{-1}\frac{\partial}{\partial y}\phi_z(y,\beta,\omega) + \left(\frac{\omega^2}{c^2}n(y,\omega)^2\right)\phi_z(y,\beta,\omega) = 0, \qquad (4.16)$$

$$m, \phi_z = \begin{cases} \varepsilon, H_z & \text{for TM,} \\ \mu, E_z & \text{for TE,} \end{cases}$$

where $\beta = k_x$ is the component of the wavevector in the propagation direction, also referred to as the propagation constant, ω is the angular frequency, and $n(y)$, $\varepsilon(y)$, $\mu(y)$ are the spatially varying material refractive index, permittivity and permeability. In general, these are tensorial parameters to account for anisotropic materials, however, the materials considered in this text are assumed to be isotropic.

Analytic solutions to the wave equation can only be found for a few basic structures and generally the solution takes the form of a transcendental equation that can only be solved numerically. There has been a great deal of work devoted to solving the wave equation for arbitrary structures in order to quickly and efficiently find the modes and their propagation constants. For 1D waveguides the most successful method for forming the wave equation for arbitrary planar structures is the transfer matrix method (TMM) which shall now be presented.

4.3.1 *The Transfer Matrix Method*

The transfer matrix method (TMM) [Chilwell and Hodgkinson (1984)] has become established as a primary tool in the analysis of multilayer waveguides. Here, the electric and magnetic field amplitudes in adjacent waveguide layers are related via a matrix multiplication. This allows the dispersion relation to be efficiently formed for arbitrary planar waveguides.

In order to formulate the transfer matrix an initial ansatz for the field in the ith layer of the planar waveguide is required. Here the description presented by Kwon [Kwon and Shin (2004)] is used, where, assuming a harmonic time dependency of $\exp^{-i\omega t}$, the z-component of the electric or magnetic field in layer i is written as:

$$\phi_{z,i} = \phi_i \cos\left(\kappa_i(y - y_i)\right) + \phi_i' \frac{m_i}{\kappa_i} \sin\left(\kappa_i(y - y_i)\right), \tag{4.17}$$

where $\phi_{z,i}$ represents the $E_{z,i}$, $H_{z,i}$, and $m_i = \mu_i, \varepsilon_i$ for TE and TM polarisations respectively, $\kappa_i = \sqrt{k_0^2 n_i^2 - \beta^2}$ is the y-component of the wavevector in layer i. ϕ_i and ϕ_i' are the magnitudes of the $\phi_{z,i}$ field and its derivative at the start of each layer. The field coefficients in adjacent layers are connected via the field continuity conditions that follow from Maxwell's equations. At the interface between two materials of different permittivity $\varepsilon_{1,2}$ and permeability $\mu_{1,2}$ the field continuity conditions can be written as [Eyges (1980)]:

$$\varepsilon_1 E_1^\perp - \varepsilon_2 E_2^\perp = 0, \tag{4.18a}$$

$$\varepsilon_1 \mathbf{E}_1^\parallel - \varepsilon_2 \mathbf{E}_2^\parallel = \mathbf{0}, \tag{4.18b}$$

$$\mu_1 H_1^\perp - \mu_2 H_2^\perp = 0, \tag{4.18c}$$

$$\mu_1 \mathbf{E}_1^\parallel - \mu_2 \mathbf{E}_2^\parallel = \mathbf{0}. \tag{4.18d}$$

In the transfer matrix formalism these continuity conditions, along with the field phase change across each layer, are expressed in a matrix multiplication that connects the wave amplitudes at the start of layer $(i+1)$ to those at start of the previous layer i:

$$\begin{bmatrix} \phi_{i+1} \\ \phi_{i+1}' \end{bmatrix} = M_i \begin{bmatrix} \phi_i \\ \phi_i' \end{bmatrix}, \quad i = 1, 2, \ldots, N \tag{4.19}$$

where M_i is the transfer matrix with the form

$$M_i = \begin{bmatrix} \cos(\kappa_i d_i) & -\frac{m_i}{\kappa_i} \sin(\kappa_i d_i) \\ \frac{m_i}{\kappa_i} \sin(\kappa_i d_i) & \cos(\kappa_i d_i) \end{bmatrix}. \tag{4.20}$$

By repeating this multiplication the fields in the superstrate can be related to those in the substrate. The final step in utilising the TMM for waveguide analysis is to define the nature of the initial fields in the superstrate and the final fields in the substrate, where different profiles are associated with different physical modes. The three most important groups of modes and their corresponding field profiles are described in the following sections.

4.3.2 *Radiation Modes*

The first set of modes under consideration is the continuum of radiation modes. Here freely propagating electromagnetic waves are assumed to be incident upon the waveguide and the interaction with the structure results in energy being absorbed, transmitted and reflected (ATR) in varying amounts. An efficient technique for calculating the ATR spectra in arbitrary multilayer waveguides using the TMM will now be presented.

Radiation modes are characterised by forwards and backwards going waves (incident and reflected waves) in the superstrate layer and only a forward going component (the trans-mitted wave) in the substrate layer. In order to utilise the TMM formulation presented in the previous section, the field ansatz, Eq. (4.17), has to be recast into forwards and backwards propagating components:

$$
\phi_i \cos(\kappa_i y) + \phi_i' \frac{m_i}{\kappa_i} \sin(\kappa_i y) = \underbrace{\frac{1}{2}\left(\phi_i - i\phi_i'\frac{m_i}{\kappa_i}\right) e^{i\kappa_i y}}_{A_i}
$$

$$
+ \underbrace{\frac{1}{2}\left(\phi_i + i\phi_i'\frac{m_i}{\kappa_i}\right) e^{-i\kappa_i y}}_{B_i} \tag{4.21}
$$

where A_i represents the amplitude of the field propagating in the negative y-direction and B_i is the amplitude of the wave travelling in the positive y-direction.

Before the TMM can be applied, the initial field conditions in the cladding layers have to be defined. In the superstrate layer the incident wave amplitude can be chosen arbitrarily, however, the reflected amplitude can not be predefined as it depends on the interaction with the multilayer structure under consideration. In the substrate, only a single outgoing transmitted wave exists, and so the initial conditions in this region can be fully defined. The calculation thus proceeds by setting an arbitrary transmission amplitude and propagating the field coefficients back through the structure using the TMM in order to calculate the incident and reflected field components in the superstrate. Finally, the three field amplitudes are normalised with respect to the incident amplitude to find the reflection, transmission and absorption coefficients. So, the initial wave components of the outgoing transmitted wave can be defined as:

$$
A_i = 0, \tag{4.22}
$$

$$
B_i = 1, \tag{4.23}
$$

which means that the amplitude of the field in the z-direction and its derivative at the start of layer 1 have to be:

$$\phi_1 = 1, \tag{4.24}$$

$$\phi_1' = -i\frac{\kappa_1}{m_1}. \tag{4.25}$$

After multiplying through by the transfer matrix, the incident and reflected wave amplitudes in the substrate are found to be:

$$\text{incident:} \quad B_N = \phi_N + i\phi_N'\frac{m_N}{\kappa_N}, \tag{4.26}$$

$$\text{reflected:} \quad A_N = \phi_N - i\phi_N'\frac{m_N}{\kappa_N}, \tag{4.27}$$

$$\text{transmited:} \quad B_1 = \phi_1 + i\phi_1'\frac{m_1}{\kappa_1} = 1. \tag{4.28}$$

For completeness the transmitted amplitude has also been included. The next step is to write the expressions for the incident, reflected and transmitted powers in the y-direction. This can be calculated using the time averaged Poynting vector:

$$P_{av} = \langle S \rangle = \frac{1}{2}\mathrm{Re}\big(\mathbf{E} \times \mathbf{H}^*\big) \tag{4.29}$$

Applying this to the field ansatz of $\phi_z = \big(A_i e^{i\kappa_i y} + B_i e^{-i\kappa_i y}\big)e^{i(\omega t - \beta x)}$, where $\phi_z = E_z$, H_z for TE and TM polarisation respectively, expressions for the power carried by the incident, reflected and transmitted waves can be defined as:

$$P_y^{(I)} = \frac{\kappa_N^*}{2\omega m_N^*}B_N^* B_N, \tag{4.30a}$$

$$P_y^{(R)} = -\frac{\kappa_N^*}{2\omega m_N^*}A_N^* A_N, \tag{4.30b}$$

$$P_y^{(T)} = \frac{\kappa_1^*}{2\omega m_1^*}B_1^* B_1. \tag{4.30c}$$

Finally, the power reflection, transmission and absorption coefficients can thus be expressed as:

$$R = \frac{P_y^{(R)}}{P_y^{(I)}} = \frac{A_N^* A_N}{B_N^* B_N}, \tag{4.31}$$

$$T = \frac{P_y^{(T)}}{P_y^{(I)}} = \frac{\kappa_1^* m_N^*}{\kappa_N^* m_1^*}\frac{B_1^* B_1}{B_N^* B_N}, \tag{4.32}$$

$$A = 1 - (T + R). \tag{4.33}$$

This methodology allows the efficient calculation of reflection and transmission spectra for a wide variety of planar structures incorporating both lossy and dispersive media. The frequency dependence can then be found by varying the value of ω while the influence of the incidence angle is accounted for by the propagation constant β through the relation:

$$\beta = \frac{\omega}{c} n_i \sin \theta_i, \tag{4.34}$$

where θ_i is the angle of incidence and n_i is the refractive index of the superstrate.

4.3.3 *Bound Modes and the Dispersion Equation*

The next commonly considered set of modes are the discrete bound modes. These modes are characterised by exponentially decaying fields in the cladding layers (Fig. 4.2a) with the fields being, as their name suggests, bound to the waveguide. On a dispersion diagram, a plot of frequency against the wavevector in propagation direction, bound modes lie below the light line of the cladding materials and therefore can not be coupled to from free space due to the momentum mismatch. Exciting these modes requires special setups such as prism coupling, where energy is coupled into the waveguide through evanescent fields, or alternatively via end fire coupling. Different coupling schemes will be discussed in more detail in the next chapter as the method of excitation becomes very important in slow and stopped light waveguides.

In order to calculate the bound waveguide modes with the TMM the field ansatz must take an exponential form in the cladding layers:

$$\phi_i \cos(\kappa_i y) + \phi_i' \frac{m_i}{\kappa_i} \sin(\kappa_i y) \Rightarrow \alpha_i e^{\pm i \gamma_i y}, \tag{4.35}$$

where α_i is an amplitude related to the coefficients ϕ_i and ϕ_i', and γ_i is the decay constant. By writing the equations in hyperbolic form it is found that for the fields to have the desired profile in the superstrate layer the amplitude coefficients in this

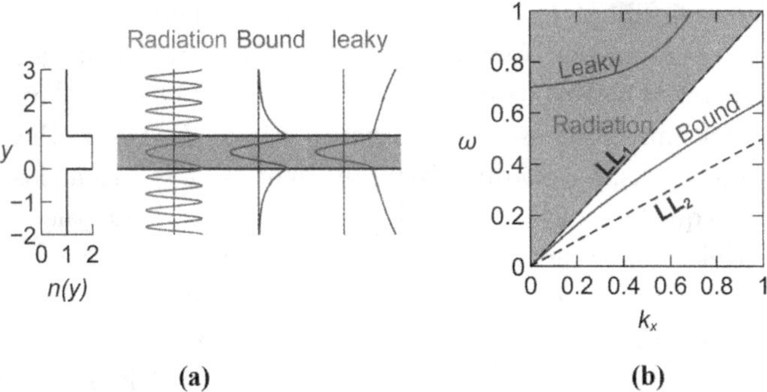

Figure 4.2: (a) Schematic field profiles of the radiation, bound and leaky modes in a simple slab waveguide consisting of a high index core (ε_2) and low index cladding layers (ε_1). (b) Representation of the three mode types on a dispersion diagram, LL_j is the light line of material j given by $\omega_j^{(LL)} = ck_x/n_j$.

layer, ϕ_i and ϕ'_i, must take the form:

$$\phi_1 = \alpha_1 = 1, \tag{4.36a}$$

$$\phi'_1 = -\alpha_1 \frac{\gamma_1}{m_1} = -\frac{\gamma_1}{m_1}, \tag{4.36b}$$

and that the decay constant be equal to $\gamma_1 = i\kappa_1 = \pm\sqrt{\beta^2 - k_0^2 n_1^2}$. Now that the fields in the superstrate have been defined the fields in the substrate can be calculated via the TMM. Again to match the bound mode profile a similar condition must be fulfilled by the substrate fields to ensure they also have an exponentially decaying profile. This condition is met when the amplitude coefficients in the substrate layer fullfill the following criteria,

$$\phi_N = \phi'_N \frac{m_N}{\gamma_N} = \alpha_N \phi_i. \tag{4.37}$$

The above is the final condition which gives the characteristic waveguide dispersion equation:

$$F(\omega, \beta^2) = \phi'_N - \frac{\gamma_N}{m_N} \phi_N = 0. \tag{4.38}$$

The dispersion equation is an eigenvalue equation where the roots correspond to the bound modes of the waveguide. Roots can either be sought on the complex-β plane for a real valued frequency, ω, these are called the complex-β (or k) solutions. Alternatively roots can be found on the complex-ω plane for a real valued propagation constant β, referred to as the complex-ω solutions. When the waveguide layers are free of loss the roots are real valued and both sets of solutions are equal. However, when loss is present the two types of solutions are no longer equivalent and can yield different predictions for modal dispersion and loss rates, particularly near regions of low group velocity. Due to this, both types of solutions will be investigated in this thesis whereas, typically, only the complex-β solutions are reported in the literature.

Except for some trivial cases, the dispersion equation is transcendental and needs to be solved numerically. A number of root finding methods have been suggested in the literature, including the Newton–Raphson (NR) method, iterative algorithms such as those reported by Kekatpure *et al.* [Kekatpure *et al.* (2009)] as well as graphical search algorithms, such as the bisection method. However, these approaches have various drawbacks, for example Newton's method requires a close initial guess value in order to quickly converge to the correct solution, while the iterative procedure used by Kekatpure *et al.* requires closed-form dispersion equations that have to be found for each waveguide structure. This becomes increasingly difficult the more layers are added. Another technique, which is growing in popularity, is the argument principle method (APM) [Kwon and Shin (2004); Delves and Lyness (1967); Smith *et al.* (1992)]. This technique is based on Cauchy's Theorem of complex integration and as such is also known as the Cauchy Integration Method (CIM). The APM has the advantage of being able to quickly locate all roots of a function inside a region of the complex plane without the need for initial guess values. However this method can only be used on functions that are single valued and free of poles, otherwise roots may be missed. In its current formulation the dispersion equation $F(\omega, \beta^2)$ is four-valued due to the indeterminate sign of the decay constants $\gamma_{1,N}$ in the cladding layers. As a result the dispersion equation can be associated with a four-sheeted Riemann surface with two branch points at $\beta^2 = \omega^2 c^{-2} - n_1^2$ and at $\omega^2 c^{-2} - n_N^2$. Different sheets of the Riemann surface can be selected by choosing the sign of γ_1 and γ_N resulting in a single valued function allowing the APM to be implemented over this plane. Certain issues remain as the single-valued function will still contain poles at $\beta^2 = \omega^2 c^{-2} - n_1^2$ and at $\omega^2 c^{-2} - n_N^2$. For the APM to be applicable the contour of integration must avoid these poles which complicates the implementation. Additionally a root corresponding to a certain mode can appear on different sheets when the frequency is varied, for example when a mode becomes cut-off, and so this change in modal behaviour might be missed [Smith and Houde-Walter (1993)]. Another issue is, that, in order to find all the modes supported by the waveguide, each sheet has to be analysed individually increasing computational time. An implementation that avoids these issues is presented in, e.g., [Pickering (2013)], extending the approach used in Kwon and Shin [Kwon and Shin (2004)]. For this work the implemented mode solver utilises a combination of the APM and NR in order to find the roots of the dispersion equation. Interestingly only one sheet of the Riemann surface contains bound modes, roots residing on the other three sheets are the less well known, leaky modes. In the next section the unusual characteristics of these modes are discussed.

4.3.4 *Leaky Modes*

Leaky modes are solutions to the dispersion equation with seemingly unphysical field profiles that increase exponentially to infinity away from the structure [Hu and Menyuk (2009)]. Mathematically these solutions arise from the square root terms in the cladding layers defining the decay coefficients $\kappa_i = \sqrt{\beta^2 - k_0^2 n_i^2}$, as both positive and negative values are possible. As a result the dispersion equation $F(\omega, \beta^2)$ becomes multivalued and can be described by a four-sheeted Riemann surface with two branch points at $\beta^2 = k_0^2 n_1^2$ and $\beta^2 = k_0^2 n_N^2$. The field profiles of modes found on each of the four sheets are shown in Figure 4.3. The physical meaning and applicability of the leaky modes has been the subject of much discussion in the literature [Hu and Menyuk (2009)]. In principle, any physically allowed field profile in the waveguide can be expanded in terms of a finite number of bound modes and a continuum of radiation modes which combined, form a complete basis set. The leaky modes are not a part of the basis set and have field profiles that can not be normalised. However, they are of interest as it was found that for certain structures the radiation continuum could be approximated by a finite number of leaky modes [Benech and Khalil (1995)]. This could reduce the complexity of modal expansion techniques used in waveguide analysis as they generally require a large discrete sample of the radiation continuum. As an example the leaky modes of

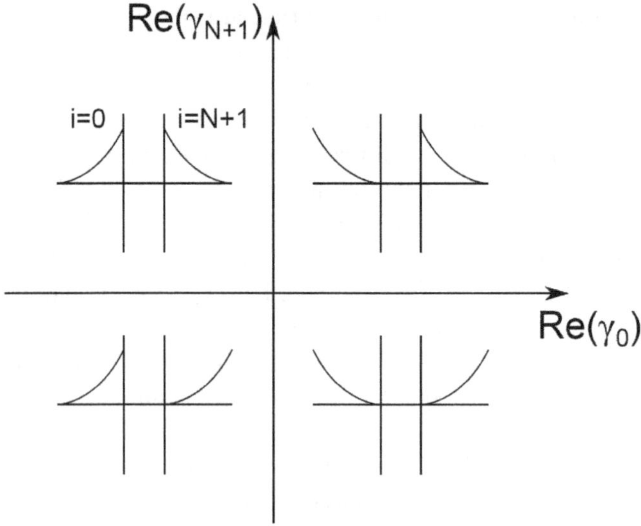

Figure 4.3: Field profiles for various combinations of the decay constants γ_{N+1} and γ_0. The four profiles shown correspond to solutions found on the different planes of the four-sheeted Reimann surface. In the top left are the bound modes while the other three quadrants correspond to the various "leaky" modes.

a thick silicon slab, bounded on both side by air half-spaces are compared to the radiation continuum shown in Fig. 4.4. It can be seen that the number of leaky modes matches the number of resonances observed in the reflection spectra and that they follow the reflection minima. Similarly, the imaginary component of the leaky mode, representing temporal attenuation, accurately predicts the width of the observed reflection resonance when modeled as a simple Lorentzian. For modes away from the light line the match is excellent and is still very good even close to the light line cutoff, the slight differences observed here are likely due to the onset of bound mode resonances.

A physical interpretation for the leaky modes is that they represent electromagnetic fields that are only partially confined to the waveguide. For example, in the case of a simple three layer waveguide the guided mode was explained using the ray optics picture of a wave being totally internally reflected at both interfaces. However, if the angle of incidence is such that the wave is only partially reflected at the interfaces it is no longer fully bound but will still propagate some distance in the waveguide with energy being lost through transmission into the cladding layers. Radiative loss will result in an exponential decay of the fields in the waveguide along the propagation direction with a decay constant given by the imaginary component of the leaky mode solution. This description of partially confined energy also provides an interpretation for the divergent field profile. Considering the partially confined leaky mode shown in Fig. 4.5, it can be seen that the field amplitude in the cladding at a position (x_0, y) is proportional to the field emitted from the

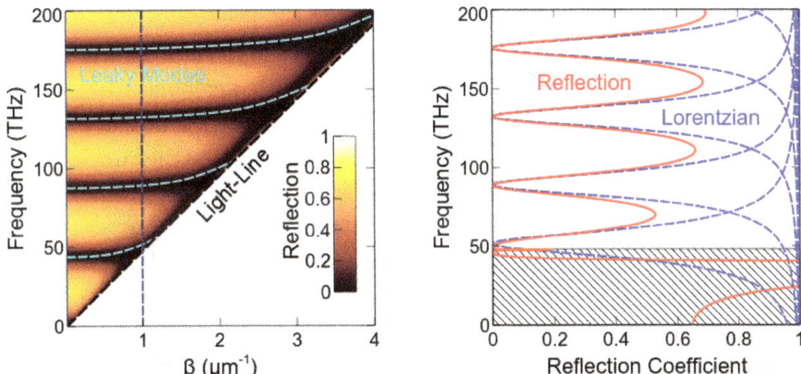

Figure 4.4: Discrete leaky mode representation of the radiation continuum for a simple silicon slab. The leaky modes are found at the minima of the reflection spectra (left) while the imaginary component accurately predicts the width of the resonance via a simple Lorentzian model (right) above the light line (shaded area).

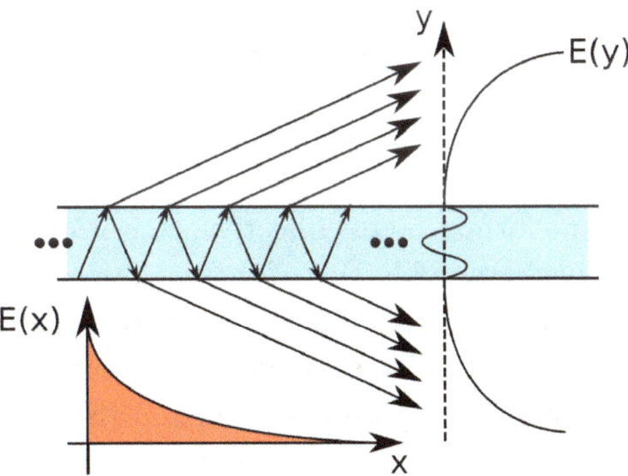

Figure 4.5: Schematic of a quasi-bound leaky mode. The partially confined light propagates in the positive x-direction loosing energy at with each reflection at the interfaces. After some distance this results in the y-dependence of the field E(y) appearing to exponentially increase away from the structure.

waveguide at a previous position $x = x_0 - \Delta x$. Assuming the field propagates in the positive x-direction this causes the field amplitude to increase exponentially into the cladding layers reproducing the behaviour predicted by the leaky modes. In practice the quasi-bound wave will have only been propagating for a finite distance and so the divergent profile will only appear over a short range close to the waveguide core.

An understanding of the leaky modes becomes particularly important when attempting to incouple radiation into a waveguide. A method known as prism coupling is often used to excite the waveguide modes. By utilising a high index material the incident wave can be momentum matched to the desired bound mode which is excited through evanescent fields. It is important to note, that, if light can couple in, it can also leak out again thus changing the nature of the mode from bound to leaky. Additionally the presence of the prism will perturb the modal band structure of the waveguide to a certain extent, dependent on the position and refractive index of the prism. As such it is important to calculate the leaky modes as well as the bound ones to fully understand the influence of the prism in such setups and avoid any unanticipated distortion of the band structure.

This concludes the general theory required to analyse the electromagnetic modes of multi-layer planar waveguides. In this work two specific types of waveguide, a plasmonic and a negative refractive index waveguide, are studied due to their ability to support modes with extremely low group velocities. This ability stems from the specific materials used in these types of waveguide which are, respectively, metals and artificial media known as metamaterials. In order to study these waveguides accurate models are needed to describe the electromagnetic response of these dispersive materials in terms of an effective permittivity and permeability. In the next section the optical properties of these materials and waveguides based on them are presented.

4.4 Metals and the Surface Plasmon Polariton

The optical properties of metals are primarily determined by the interaction of electro-magnetic radiation with the free electrons in the conduction band. This interaction can lead to the formation of collective excitations of electrons that oscillate with respect to the background lattice of the fixed positive ions [Ritchie (1957)]. The quanta of these excitations are known as plasmons. Under certain conditions electromagnetic waves can couple to plasmons at the surface of a metal to form a propagating wave called a surface plasmon polariton (SPP) [Maier (2007)]. SPPs have transverse field profiles that decay exponentially away from the surface into the surrounding metal and dielectric layers (see Fig. 4.6). Hence the electromag-

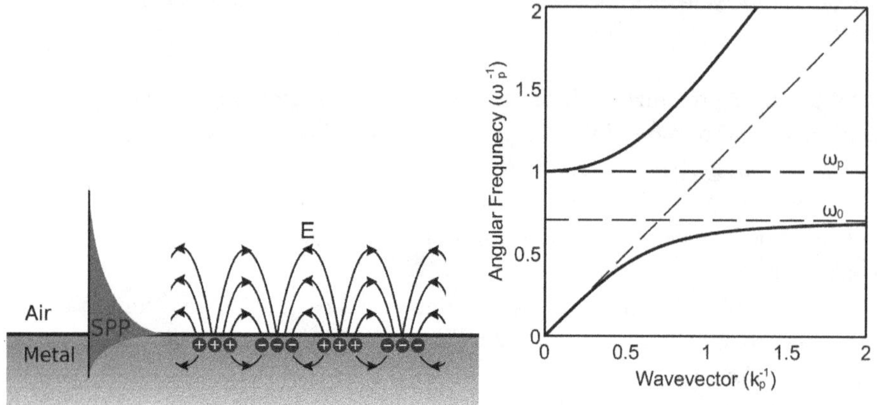

Figure 4.6: Schematic representation of the SPP field profile highlighting the coupling between the EM radiation and the charge oscillations. On the right the dispersion curves of the SPP and bulk plasmon are shown.

netic fields of SPPs are highly localised to the surface of the metal which makes them a promising option for the subwavelength confinement of optical waves, vital for applications such as integrated optical circuits. The strong confinement of energy also leads to an enhancement of the local electromagnetic field strength which is of interest for devices relying on strong light-matter interactions [Zayats *et al.* (2005b)].

4.4.1 *The Negative Permittivity of Metals*

At low frequencies the optical response of metals is mainly determined by the intraband excitation of electrons in the conduction band while, at higher frequencies, the presence of interband transitions also has to be accounted for. One of the simplest models used to describe the optical response resulting from intraband excitations is known as the Drude–Sommerfeld model which is based on the theory of a free electron gas. Here the presence of an external electric field leads to a displacement of the electrons relative to the background positive ions which generates an effective macroscopic polarisation field. The dependence of the polarisation on the external electric field can be derived from the equation of motion for electrons driven by a harmonically oscillating electric field [Maier (2007)].

The resulting frequency dependent permittivity is given as:

$$\varepsilon_{Drude}(\omega) = 1 - \frac{\omega_p^2}{\omega^2 + i\Gamma_D\omega}, \tag{4.39}$$

where $\omega_p = \sqrt{n_e^2/m^*\varepsilon_0}$ is the material dependent plasma frequency (the resonant frequency of the electron oscillation) with n_e being the number density of electrons, e is the fundamental electric charge and m^* is the effective mass of the electrons in the metal. A phenomenological damping term Γ_D is included representing the collision frequency of the electrons due to scattering with phonons and material impurities. The inclusion of this term results in the permittivity becoming complex valued, where the imaginary component describes the dissipation of energy due to the collisions of the electrons which is often termed the ohmic loss.

From Eq. (4.39) it can be seen that the permittivity becomes negative at frequencies below the plasma frequency. This is a necessary condition for the existence of surface waves and so the spectral range where SPPs can form is limited to $\omega < \omega_p$. As metals generally have large plasma frequencies they have historically been the main choice of material for plasmonics, however they do have several drawbacks which has lead to recent interest in alternative plasmonic materials [West *et al.* (2010)].

4.4.2 *Surface Plasmon at a Single Interface*

Although relatively simple, the Drude–Sommerfeld model reproduces many of the characteristics of light-metal interactions with reasonable accuracy, including the formation of SPPs at the interface between a metallic and a dielectric half-space.

From Maxwell's equations the dispersion equation of an SPP propagating along the interface between a metal with a negative permittivity ε_1 and a dielectric with positive permittivity ε_2 can be derived analytically as [Pitarke *et al.* (2007)]:

$$\beta = \frac{\omega}{c}\frac{\varepsilon_1\varepsilon_2}{\varepsilon_1 + \varepsilon_2}, \tag{4.40}$$

where β is the wavevector component in the direction of propagation. From Fig. 4.6 it can be seen that Eq. (4.39) results in the formation of two modes, represented by the two branches of the dispersion curve (here shown without loss, i.e. $\gamma = 0$). The lower curve, in red, is the SPP branch which can be seen to closely follow the light line of the dielectric at low frequencies before curving off and asymptotically approaching a frequency ω_0. This is known as the SPP resonance frequency and is found from the condition $\varepsilon_1 = -\varepsilon_2$, using Eq. (4.39) gives $\omega_0 = \omega_p/\sqrt{1 + \varepsilon_2}$. It is near this frequency that the SPP fields are most highly localised to the surface of the metal. As the group velocity is equal to the gradient of the dispersion curve it follows that the group velocity becomes very low in this region, approaching zero in the limit $k \to \infty$. The second branch appearing above the plasma frequency is sometimes called the bulk plasmon mode and describes the dispersion of electromagnetic waves propagating through the metal.

4.4.3 *Coupled SPPs in Gaps and on Thin Films*

When multiple thin metal films are brought close together, separated by thin layers of positive index dielectric, the interaction of SPPs on different interfaces leads to the formation of a new set of hybrid plasmon modes [Economou (1969)]. SPPs can form on the top and bottom interfaces of a metal layer surrounded by two dielectric claddings, this configuration is called an insulator-metal-insulator (IMI) waveguide. If the thickness of the metal is large compared to the skin depth of the fields then the two SPPs on either side of the layer will propagate independently and can both be described using the standard SPP dispersion Eq. (4.40). However, if the thickness of the metal is comparable or less than the skin depth, the two SPPs can couple and interact forming two new modes with different dispersion characteristics. These two modes are characterised by the profile of their electric fields where one is symmetric while the other is antisymmetric. As the antisymmetric mode has a lower field overlap with the metal it experiences a reduced loss in comparison to the symmetric mode. To reflect this these modes are sometimes called the long range SPP (LRSPP) and the short range SPP (SRSPP). By changing the dielectric cladding surrounding the metal film the dispersion properties of these two modes can be altered and under certain conditions the antisymmetric mode can exhibit negative or even zero group velocity at finite k values [Feigenbaum *et al.* (2008)]; an advantage over the case of a single interface where zero group velocity is only approached as $k \to \infty$.

A similar effect occurs if two metal layers are brought close together separated by a thin dielectric material, often called metal-insulator-metal (MIM) waveguide. Again when the gap is sufficiently thin the SPPs at the two interfaces can hybridise, resulting in the formation of two modes with symmetric and antisymmetric field profiles. However, in this case it is the symmetric mode that has a lower overlap with the metal and hence lower losses. This mode is also known as the gap-SPP. By varying the permittivity and thickness of the core layer, MIM waveguides have also been shown to exhibit negative and zero group velocities [Feigenbaum *et al.* (2008)]. It is this structure that forms the basic design of the stopped light plasmonic waveguide studied in this work. The specific geometry under consideration and modal characteristics are presented in the next chapter.

4.4.4 *Ohmic Loss and Alternative Plasmonic Materials*

One of the major obstacles to realising many plasmonic applications is the high material loss that greatly limits the propagation length of SPPs. For example two of the most commonly utilised metals, silver and gold, have collision frequencies of the order $10^{13} - 10^{13}$ s^{-1} resulting in lifetimes of tens of femtoseconds [Johnson and Christy (1972)]. If the SPPs propagate at the speed of light then the decay length is on the order of a few micrometers, dramatically reducing as the velocity decreases. There is great motivation to overcome this problem and several different solutions have been proposed. In some cases a modification of the geometry can greatly reduce the SPP loss, for example the long range SPP (LRSPP) present in thin metal films [Sarid (1981)] exhibits reduced losses as the field overlap with the metal is greatly reduced in comparison to SPPs at a single interface. However as the fields of the LRSPP are not highly localised to the metal surface energy can no longer be confined on subwavelength scales. In order to realise applications where subwavelength confinement is key it has been suggested that loss could be compensated by the supply of energy from a gain material in close proximity to the plasmonic structure. Recently a number of experimental investigations of have shown that partial and even full compensation of internal dissipative losses of SPPs can be achieved through the use of gain media [Seidel *et al.* (2005); Ambati *et al.* (2008); Grandidier *et al.* (2009b); Okamoto *et al.* (2004); Hill *et al.* (2009); Noginov *et al.* (2008b)].

Other groups have suggested alternative plasmonic materials that could potentially have intrinsically lower losses [West *et al.* (2010)] than currently used metals. These include alkali metals such as sodium and potassium which exhibit loss comparable to or even lower than silver, unfortunately these materials are highly reactive to air and water making them unsuitable for many applications. Another more promising set of materials are transparent conducting oxides (TCO) which, through doping, can exhibit a negative permittivity with a Drude like frequency response. In particular indium-tin-oxide (ITO) has been shown to act as a plasmonic material in the near infrared region. By varying concentrations of the two oxides (indium-oxide and tin-oxide) the plasma frequency can be altered, as confirmed in experimentally measured plasma frequencies ranging from 200–500 THz [Rhodes *et al.* (2006, 2008); Michelotti *et al.* (2009); Noginov *et al.* (2011)].

In the plasmonic waveguide presented in this work ITO is used due to its suitability for application in the NRI range. The design requirements leading to this choice are explained in the next chapter along with the specific Drude parameters used.

4.5   Modelling Effective Media: The Metamaterial Limit

The second waveguide investigated in this work is based on the incorporation of layers exhibiting a negative refractive index (NRI). This unusual property is not found in nature and has only recently been demonstrated using metamaterials [Smith *et al.* (2004b)]. NRI metamaterials have received considerable interest due to the exciting range of potential applications, such as perfect lenses [Pendry (2000)] with resolutions not limited by diffraction, as well as electromagnetic cloaks [Pendry *et al.* (2006); Schurig *et al.* (2006)], and slow light waveguides [Tsakmakidis *et al.* (2007)]. In this section the physical implications of NRI on wave propagation are discussed before presenting the design breakthroughs that led to the creation of NRI metamaterials. A more detailed exposition is presented in Chapter 5.

4.5.1   *Negative Refractive Index*

In the 1960s Victor Veselago [Veselago (1968)] considered the effect that simultaneously negative permittivity and permeability would have on the propagation of electromagnetic waves. He began by making a plane wave ansatz $(E_0 e^{-i(\omega t - \mathbf{k} \cdot \mathbf{r})})$ so that Maxwell's curl equations translate to:

$$\mathbf{k} \times \mathbf{E} = \omega \mu \mathbf{H}, \tag{4.41}$$

$$\mathbf{k} \times \mathbf{H} = -\omega \varepsilon \mathbf{E}. \tag{4.42}$$

From Eqs. (4.41) and (4.42) it can be seen that changing the sign of ε and μ will reverse the direction of the propagation vector $\mathbf{k}$, thus turning the right handed triplet of $\mathbf{E}$, $\mathbf{H}$, $\mathbf{k}$ into a left-handed one. As such, materials with negative ε and μ are often called left-handed media. Interestingly, even though the direction of the propagation vector is reversed, the flow of energy will remain the same as it only depends on the electric and magnetic field vectors. Consequently the energy in these media will propagate in the opposite direction to the phase fronts.

Veselago then considered the angle of transmission for a plane wave propagating from a positive index material into one with negative ε and μ. Maxwell's equations show that in the absence of free charge or current densities the field components parallel to an interface will be equal in both materials and that the perpendicular components are discontinuous experiencing a jump proportional to the ratio of layer permittivities (for the electric field) or permeabilities (for the magnetic field). So for an incident plane wave the electromagnetic boundary conditions lead to expressions for the angle of transmission as:

$$
\theta_t = \begin{cases} \tan^{-1}\left(\frac{\mu_1}{\mu_2}\tan\theta_i\right) & \text{for TE}, \\ \tan^{-1}\left(\frac{\varepsilon_1}{\varepsilon_2}\tan\theta_i\right) & \text{for TM}, \end{cases}
$$

where θ_i is the incidence angle while TE and TM denote transverse electric and transverse magnetic polarisation. From this expression it can be seen that waves propagating from a positive index material to one with negative permittivity and permeability will be transmitted with a negative angle. Through Snell's law this implies that a material with negative ε and μ will also have a negative refractive index (NRI). In practical materials where loss or gain may be present ($\varepsilon = \varepsilon' + i\varepsilon''$, $\mu = \mu' + i\mu''$) the condition for NRI is slightly modified to $\varepsilon'\mu'' + \varepsilon''\mu' < 0$ [Kinsler and McCall (2008)].

Unfortunately, while negative permittivities can be found in many metals, so far materials with simultaneously negative ε and μ have not been found in nature. As a result of this the study of NRI effects remained dormant until the late 1990s when the new field of metamaterials revealed that the permittivity and permeability of artificial media could be controlled via the design of subwavelength resonant elements [Pendry *et al.* (1996, 1999, 1998)]. Following these initial works the development of metamaterials with negative refractive index proceeded rapidly with the first successful demonstration being published only a few years later in 2000 by Smith *et al.* [Smith *et al.* (2000)]. To understand how a NRI can be realised, two important metamaterial designs shall now be highlighted.

4.5.2 *Electrical Response*

One of the first metamaterial designs was born from the desire to reduce the plasma frequency of metals into the infrared and microwave range. For most metals the plasma frequency occurs in the ultraviolet range and while SPPs can be excited at any frequency below ω_p their most interesting effects are limited to a small frequency range close to the plasmon resonance. This is because at lower frequencies the metal acts almost as a perfect conductor with very little field penetrating into the metal surface. As a result many of the interesting characteristics of SPPs, such as subwavelength confinement, are lost.

To overcome this limitation and extend the range of plasmonics effects, Pendry *et al.* [Pendry *et al.* (1996)] proposed that a 3D grid of thin metal wires could, under certain conditions, act as an effective material with a permittivity response well described by a Drude model with a reduced plasma frequency compared to the metal composing the wires. Two physical mechanisms lie behind this effect. First, as the metal is confined to thin wires the effective density of conduction electrons is reduced to $N_{eff} = N\pi r^2/a^2$, where N is the electron density in bulk metal, r is the wire radius and a is the lattice constant. The second effect is due to the self-inductance of the wires which opposes the driven motion of the electrons. This can be accounted for as an increase in the effective mass of the electrons to $m_{eff} = 2\mu_0 r^2 e^2 N \ln(a/r)$. Clearly this second effect becomes stronger as the radius of the wire decreases. Taking the effective medium approximation, the wire mesh can be assigned an effective permittivity following a Drude dependence [Pendry *et al.* (1996)]:

$$\varepsilon(\omega) = 1 - \frac{\omega_p^2}{\omega^2 + i\gamma_{eff}\omega},\tag{4.43}$$

where the effective plasma frequency is given by $\omega_p^2 = 2\pi c^2/\left(a^2 \ln(a/r)\right)$. In their paper, Pendry *et al.* calculated that for aluminium wires of radius 1 μm and spacing 5 mm the effective plasma frequency could be reduced by almost six orders of magnitude from the bulk plasma frequency of aluminium.

Another question was how would this affect the loss of the plasmon. By considering the electrical resistance of the wires it was shown that the damping term could be expressed as $\gamma_{eff} = \varepsilon_0 a^2 \omega_p^2/(\pi r^2 \sigma)$, where σ is the conductivity of the metal. For the same case of aluminium wires this gave a damping rate of $\sim 0.1\omega_p$, which is a similar ratio to that of the dielectric function of bulk aluminium. Thus, not only can the wire geometry reduce the plasma frequency it also results in lower dissipative loss of the SPP.

This effect was experimentally tested a couple of years later in a 3D periodic array of thin tungsten wires [Pendry *et al.* (1998)]. By comparing the measured transmission coefficient to that calculated from their analytic model, Pendry *et al.* were able to confirm the predicted low effective plasma frequency. Thus the wire mesh model provides a viable route to the creation of effective Drude-like media with controllable plasma frequencies.

4.5.3 *Magnetic Response*

The second requirement for NRI is the existence of a negative permeability. While metals can exhibit negative permittivities there are no known materials that have a negative permeability. In fact the magnetic response of most materials becomes very weak above GHz frequencies making it rare to find a material permeability that differs significantly from $\mu = 1$. The solution came from a new metamaterial design, this time based on metallic wire rings. When a time varying magnetic field penetrates these rings a current is generated through Faraday's law, which in turn creates a new magnetic field that can act in phase or out of phase with the external magnetic field. It was shown that if the magnetic response is strong enough this could lead to an effective magnetic permeability with a negative value. The strength of the response is relatively weak except near the resonance frequency of the rings. For simple rings or cylinders the corresponding wavelength is generally on the same scale as the dimensions of the ring, which is problematic as the validity of an effective medium only holds when the lattice constant $a \ll \lambda$. To meet this condition Pendry *et al.* [Pendry *et al.* (1999)] suggested a design based on split ring resonators (SRRs) which introduced capacitative elements that dramatically reduced the resonant frequency. Each SRR consists of two concentric split rings with their openings facing in opposite directions. The gaps prevent current from flowing around an individual ring but, by reducing the spacing between the rings, a large capacitance builds up which allows current to flow around the concentric ring pair. This structure can be viewed as an LC circuit with a resonant frequency $\omega_0 = 1/\sqrt{LC}$, where L and C represent the SRR inductance and capacitance. The effective permeability of a material consisting of a periodic array of SRRs can be described by a Lorentzian response:

$$\mu(\omega) = 1 - \frac{F\omega_p^2}{\omega^2 + i\gamma\omega - \omega_0^2}, \tag{4.44}$$

where $F = \pi r^2/a^2$ is the fraction of the unit cell area filled with the split rings, the resonant frequency is:

$$\omega_0 = \frac{3lc^2}{\pi r^2 \ln \frac{2t}{d}}, \tag{4.45}$$

and the damping coefficient is given by:

$$\gamma = \frac{2l\sigma}{r\mu_0}, \tag{4.46}$$

where r is the radius of the ring, a is the length of the unit cell, l is the distance between sheets of SRRs, t is the width of each ring, d is the width of the gap in between the concentric ring pairs, and σ is the resistance of the ring. Using this formula it was predicted that for copper rings with length scales in the mm range a small window of negative permeability could exist above the resonance frequency $\omega_0 = 2\pi \times 13.5$ GHz [Pendry *et al.* (1999)].

Experimental confirmation of this effect came just a few months later in the work of Smith *et al.* [Smith *et al.* (2000)]. Here SRRs were combined with uniformly placed metal wires in an attempt to create a left-handed medium. Transmission measurements performed on this structure were used to demonstrate the existence of a passband that would not be present unless both the permittivity and permeability were negative. Proof that this arrangement would result in negative refraction shortly followed in 2001 where Shelby *et al.* created a prism out of a SRR and wire mesh metamaterial [Shelby *et al.* (2001a)]. Using transmission measurements they were able to show that light passing through this prism was transmitted with negative angle. This work confirmed that the advent of metamaterials had opened a route to creating media with negative refractive index.

4.5.4 *Optical NRI Metamaterials*

The first NRI metamaterials were composed of SRRs combined with metallic wires in periodic arrays [Smith *et al.* (2000); Shelby *et al.* (2001a)]. As the geometry scales with the operating frequency these original demonstrations occurred in the microwave range where the structures could be fabricated relatively simply. Following these early proof of principle studies there came a strong push to create metamaterials with NRI at optical and telecommunications wavelengths. This was in large part due to the range of exciting new applications being proposed for NRI materials, such as optical cloaks [Pendry *et al.* (2006); Schurig *et al.* (2006)], perfect lenses [Pendry (2000)], with resolutions unlimited by diffraction as well as slow and stopped light waveguides [Tsakmakidis *et al.* (2007)].

In order to satisfy the effective medium approximation in the optical regime the metamaterial unit cell would have to be on the nanometer length scale. It was found that simply scaling down the design used by Smith and Shelby was not a viable option beyond a few THz. This is due to saturation of the SRR resonant frequency and a simultaneous decrease of the magnetic response at optical frequencies [Zhou *et al.* (2005)]. To overcome this new metamaterial designs were proposed, one of the most successful was the double fishnet metamaterial [Zhang *et al.* (2005); Dolling *et al.* (2006)]. Consisting of thin metal layers separated by a thin dielectric core (forming a MIM waveguide) with a periodic array of rectangular holes cutting through all three layers. By varying the dimensions of the holes the fishnet can be seen to resemble a grid of thick wires overlayed on a perpendicular grid of thinner wires. When light incident waves are polarised so that the electric field is in the direction of the thin wires, a current is generated in the thicker wires from the magnetic field. The metal-dielectric-metal acts as a capacitor and the circulating current results in a magnetic response. Using this design several groups have demonstrated a negative-refractive index response in or close to the visible spectrum [Zhang *et al.* (2005); Dolling *et al.* (2006); Shalaev *et al.* (2005)]. There are some drawbacks in the design that relate to the high metal losses in the optical region, unfortunately the use of metals is integral to the optical response can not be entirely eliminated. As such a figure of merit (FOM) is usually quoted for metamaterial designs describing the strength of the negative index response with respect to the amount of loss:

$$\text{FOM} = \frac{-\text{Re}(n)}{\text{Im}(n)}.$$
(4.47)

Measured values of the FOM are typically below 1 although values up to 3.5 have been demonstrated in a 3D metamaterial design consisting of stacked double fishnets [Valentine *et al.* (2008)]. Another issue with this design is that, as the fishnet is planar, it is highly anisotropic featuring, as mentioned previously, a polarisation dependent response. Subsequently experimental demonstrations of NRI in these structures have been performed with incident waves normal to the plane of the fishnet and polarised so that the electric field is in the direction of the thin wires. This may not be too problematic in certain situations, such as light propagating through waveguides, as here a NRI core could be constructed by stacking multiple fishnet layers in the plane orthogonal to the propagation wavevector, although in this case only one polarisation would experience NRI. A similar waveguide geometry, featuring an anisotropic metamaterial was studied in. Here the metamaterial consisted of alternating layers of positive and negative permittivity. It was shown that, if a TM mode is used, so that there is an component of the electric field orientated normal to this layered material, components the phase velocity and Poynting vector orientated along the propagation direction of the waveguide are anti-parallel [Alekseyev and Narimanov (2006)]. Thus features present in an isotropic NRI waveguide, such as forward and backward propagation and stopped light are retained [Alekseyev and Narimanov (2006)].

4.5.5 *NRI Waveguides*

Shortly after the early demonstrations of NRI metamaterials it was shown that surface polariton waves analogous to SPPs could form at the interface between a homogeneous NRI material and positive index half-space [Darmanyanab *et al.* (2003)]. The surface polaritons supported at this interface have similar field profiles to SPPs on metals but there are a number of differences. First, as both the permittivity and permeability are negative, surface polaritons of both TE and TM polarisation are supported. In total there are two TM polarised polaritons and a single TE polarised one. The TE surface wave exhibits negative group velocity while the two TM waves propagate with positive group velocities.

When a layer of NRI material surrounded by dielectric cladding is considered a larger range of modes are supported. This was first investigated by Shadrivov *et al.* [Shadrivov *et al.* (2003)] in 2003 who showed that NRI waveguides had a number of unique properties. The first was that the fundamental waveguide mode, that is the mode with no nodes in its transverse field profile which is always present in normal positive index waveguides, is not supported. Secondly many of the modes that are supported exhibit double degeneracy, meaning they have both forwards and backwards propagating branches. These two branches meet at the degeneracy point where the group velocity becomes zero. It was shown that this reduction in the group velocity was caused by the energy flow, quantified by the Poynting vector, being directly opposed in left and right handed media. By choice of the waveguide parameters the positive and negative flows can be made to cancel forming a closed-loop energy vortex associated with a zero group velocity. This property lead to NRI waveguides being investigated for slow light with novel applications being proposed such as the concept of a "trapped rainbow" introduced by Tsakmakidis *et al.* [Tsakmakidis *et al.* (2007)]. Here it was suggested that a broad range of frequencies could be stopped and trapped by an adiabatically tapered NRI waveguide. Different frequencies propagate along the waveguide until they reach a point of critical thickness where their associated energy flows cancel, resulting in zero group velocity. This enables frequencies to be stopped at different points along the waveguide, which could be useful for spectroscopic applications as well as optical signal processing.

4.6   Summary

Planar waveguides can be used to control the propagation of electromagnetic waves. The interaction of light with these structures is described in terms of the waveguide modes which can be split into different classifications based on their field profiles. Three important mode types have been introduced in this chapter along with a method for quickly and efficiently determining their characteristics in different waveguides. This technique is based on the TMM which allows the dispersion equation to be implicitly defined for waveguides consisting of arbitrary numbers of layers.

Next it was shown that the choice of materials can greatly influence the properties of the waveguide modes highlighting the important cases of plasmonic and NRI metamaterial waveguides. Response models describing the dispersive permittivity and permeability of these materials were presented which shall be used in the next chapter to calculate the modes of a plasmonic and a NRI waveguide. The unusual properties of these types of structure have been introduced and it was seen that both types can support slow light modes.

Chapter 5

Design Methods for Constructing Metamaterials

5.1 Introduction

In the previous chapter we examined in detail some of the intriguing characteristics of light propagation in bulk metamaterials and waveguides. There, we assumed that such materials were available and we were not concerned with how to actually construct them. However, as we remarked in the previous chapter, materials characterised by regions where their electromagnetic parameters are *simultaneously* negative do not occur naturally, and methods to construct them are not immediately apparent. Thus, though there are indeed naturally occurring media, well-known to scientists, that do possess *either* negative permittivity (such as plasmonic media) *or* negative permeability (such as gyrotropic media) in a given frequency region, double-negative metamaterials do not (under normal conditions) occur in nature and have to be "built" in the laboratory. It was this conception of practical recipes for the construction of man-made metamaterials, followed by their experimental realisation and demonstration of their unique properties, that led to a reappraisal of their usefulness and to a resurgence of interest by the scientific community in these exotic materials.

In this chapter we shall examine how one can construct effective materials whose interaction with light can accurately be described in terms of *negative* effective electromagnetic parameters, which can be engineered to exist in a desired frequency regime. We will start by studying the pioneering proposal towards obtaining materials with negative (effective) electric permittivity in the microwave regime. We will then see how one can also construct an engineered medium exhibiting negative magnetic permeability, again in the microwave regime – thus, opening the way to a man-made double-negative metamaterial, by suitably overlaying the negative ε and negative μ media. We will close this chapter by analysing in detail some novel configurations that can give rise to ultra-/zero-loss magnetic metamaterials or, indeed, engineered materials that can exhibit magnetic "gain" over a continuous range of frequencies *without* having to use gain media to overcome (dissipation, radiation, impedance- mismatch or surface-roughness) losses, solely by a judicious redesign, at the unit-cell level, of the previously studied magnetic metaparticles. These design

115

recipes are, thus, expected to open the way to a multitude of metamaterial-enabled applications, such as the "perfect" lens that we studied in Section 3.6, or slow-light metamaterial waveguides that we briefly considered in Section 3.8 – and to the more detailed study of which we shall focus in the next chapter.

5.2 Metamaterials with Negative Effective Permittivity in the Microwave Regime

It is well-known that metals at optical frequencies are characterised by an electric permittivity that varies with frequency according to the following, so called Drude, relation:

$$\varepsilon(\omega) = \varepsilon_0 \left[1 - \frac{\omega_p^2}{\omega(\omega + i\gamma)} \right], \tag{5.1}$$

where:

$$\omega_p^2 = \frac{Ne^2}{m\varepsilon_0} \tag{5.2}$$

is the plasma frequency, i.e. the frequency with which the collection of free electrons (plasma) oscillates in the presence of an external driving field, N, e and m being, respectively, the electronic density, charge and mass, and γ is the rate with which the amplitude of the plasma oscillation decreases. One can directly infer from Eq. (5.1) that, e.g., when $\gamma = 0$ and $\omega < \omega_p$ it is $\varepsilon < 0$, i.e. the medium is characterised by a negative electric permittivity. Typical values for ω_p are in the ultraviolet regime and, for example, for copper it is $\gamma \approx 4 \cdot 10^{13}$ rad/s. Unfortunately, for all frequencies ($\omega < \omega_p$) for which $\varepsilon < 0$, it is also $\omega \ll \gamma$, i.e. the dominant term in Eq. (5.1) is the imaginary part of the plasma electric permittivity, which is associated with losses (light absorption). Thus, if one is interested in obtaining a bulk structure characterised by a negative electric permittivity in, e.g., the microwave (GHz) regime – where the construction of an effective medium or device is, of course, considerably less demanding than the optical regime regime – simply deploying metals will result in an impractically lossy (mainly evanescent) propagation of light inside such a structure.

A method for overcoming this limitation was first proposed and analysed in detail by Pendry *et al.* [Pendry *et al.* (1998)], based on the observation that the plasma frequency, given by Eq. (5.2), depends critically on the density and mass of the collective electronic motion. They considered the structure illustrated in Fig. 5.1, wherein thin metallic wires (infinite in the vertical direction, z) of radius r are periodically arranged on a horizontal plane (xy). The unit cell of the periodic structure is a square whose sides have length equal to a. If an electric field $\mathbf{E} = E_0 e^{-i(\omega t - kz)} \mathbf{z_0}$ is incident on the structure (where $\mathbf{z_0}$ is the unit vector along z-axis), then the (free) electrons inside the wires will be forced to move in the direction of the incident field. If the wavelength of the incident field is considerably larger compared

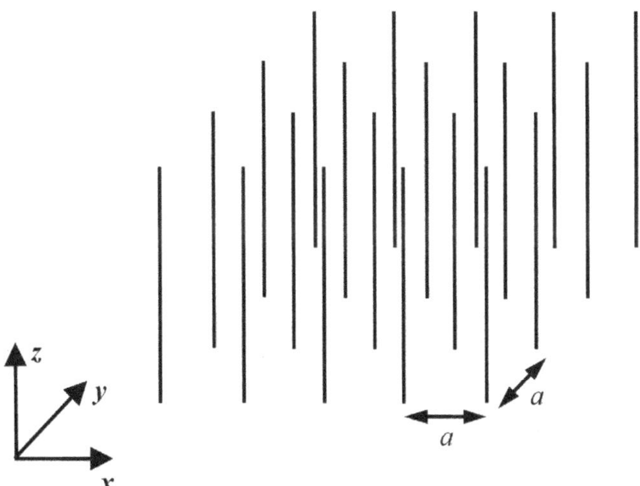

Figure 5.1: Schematic illustration of a period arrangement of infinitely long thin wires, used in the creation of an effective plasma medium at microwave frequencies.

to the side length of the unit cell, $\lambda \ll a$, then the whole structure will appear (to the incident electromagnetic field) as an effective medium whose electrons (confined in the wires) move in the $+\mathbf{z_0}$ direction.

The crucial observation here is that, since the electrons are confined to move only inside the thin wires, the effective electron density of the *whole* structure (effective medium) is:

$$N_{eff} = N\frac{\pi r^2}{a^2},\tag{5.3}$$

with N being the electron density inside each wire. Thus, for sufficiently thin wires the effective electron density, N_{eff}, of the engineered medium can become much smaller compared to N, thereby substantially decreasing the effective plasma frequency, ω_p, of the engineered medium according to Eq. (5.3).

For instance, for a wire radius $r = 1$ μm and wire spacing $a = 5$ mm, we find from Eq. (5.3) that $N_{eff} \approx 1.3 \cdot 10^{-7}N$, i.e. the effective electronic density of the new medium is reduced by seven orders of magnitude compared to that of the free electron gas inside an isolated wire.

Moreover, it also turns out that the effective mass, m_{eff}, of the electrons moving inside the engineered medium is considerably larger compared to the free electron mass, m. To prove this result, we start by assuming that the total current flowing in the $+z$ direction inside the whole engineered medium is I and that the associated current density is J, i.e.:

$$I = \int_0^\infty J(R)\,d(\pi R^2),\tag{5.4}$$

$$\mathbf{J} = \frac{I}{\pi R}\mathbf{z_0}.\tag{5.5}$$

We, then, for the sake of convenience, divide the xy plane into circles of radius R_c, centred at each wire and having area equal to that of the square unit cell, i.e. $R_c = a/\sqrt{\pi}$. Furthermore, we assume that the wires are sufficiently apart from each other ("isolated") so that the magnetic field inside each circle arises only from the current I that flows perpendicularly to the centre of the circle, and that the field at the circumference of each circle vanishes, i.e. $\mathbf{H}(R_c) = 0$. It, then, readily follows from Eq. (1.9) that the magnetic field intensity at a distance R from each wire is given by:

$$\mathbf{H} = \frac{I}{2\pi R}\left(1 - \frac{R^2}{R_c^2}\right)\boldsymbol{\phi_0}. \tag{5.6}$$

Realising that because of Eq. (1.10) the magnetic magnetic field $\mathbf{H}$ can be associated with a vector potential $\mathbf{A}$ according to the relation: $\mathbf{H} = \mu_0^{-1}\nabla \times \mathbf{A}$, yields the following expression for the vector potential yields the following expression for the vector potential $\mathbf{A}$:

$$\mathbf{A} = \frac{\mu_0 I}{2\pi}\left[\ln\left(\frac{R_c}{R}\right) + \frac{R^2 - R_c^2}{2R_c^2}\right]\mathbf{z_0}, \tag{5.7}$$

where, again, it has been assumed that $\mathbf{A}(R \geq R_c) = 0$.

For distances very close to the wires, i.e. for $R \to r \ll R_c$, Eq. (5.7) takes, to a great accuracy, the following form:

$$\mathbf{A} \simeq \frac{\mu_0 I}{2\pi}\ln\left(\frac{a}{R}\right)\mathbf{z_0}. \tag{5.8}$$

We may, now, recall that for any electromagnetic field we can write $\mathbf{B} = \nabla \times \mathbf{A}$ (as we mentioned above) and $\mathbf{E} = -\nabla\varphi - \partial\mathbf{A}/\partial t$, the latter relation stemming directly from Eq. (1.11). Thus, the equation of motion of a moving electron will be:

$$\frac{d(m\mathbf{v})}{dt} = q(\mathbf{E} + \mathbf{v} \times \mathbf{B}) \;\rightarrow\; \frac{d}{dt}(m\mathbf{v} + q\mathbf{A}) = -q\nabla(\varphi - v \cdot \mathbf{A}), \tag{5.9}$$

with $q = -e$ and v being the (absolute value of the) electronic charge, and the (magnitude of the) electron velocity, respectively. Since, both, $\mathbf{v}$ and $\mathbf{A}$ point along z, the RHS of the second part of Eq. (5.9) will vanish (for a given R), i.e. the, so called, "conjugate momentum" $(m\mathbf{v} + q\mathbf{A})$ of an electron will be *conserved along the z-direction*. As a result, an electron will be moving in the above engineered medium with an effective mass, $m_{eff} = qA(r)/v$. Realising that the current I can be simply re-expressed as: $I = -(\pi r^2)(Nev)$, we finally obtain using Eq. (5.8) the effective mass of a moving electron inside our effective medium:

$$m_{eff} = \frac{1}{2}\mu_0 N e^2 r^2 \ln\left(\frac{a}{R}\right). \tag{5.10}$$

Thus, for copper wires of radius $r = 1$ μm, being separated by $a = 5$ mm, we obtain $m_{eff} \simeq 1.3 \cdot 10^4$ m, i.e. the effective mass of an electron in our engineered medium is increased by more than four orders of magnitude. This, combined with the fact that (as we show before) the effective electron density is reduced by approximately seven orders of magnitude, leads to an effective plasma frequency that is in the microwave regime:

$$\omega_p^2 = \frac{N_{eff} e^2}{m_{eff} \varepsilon_0} = 5.1 \cdot 10^{10} \ [\text{rad/s}]^2 \ \rightarrow \ f_p = \frac{\omega_p}{2\pi} = 8.2 \ \text{GHz}. \tag{5.11}$$

It should be noted that, based on Eq. (5.11), the calculated wavelength, $\lambda_p = c/f_p$, which corresponds to our medium's effective plasma frequency turns out to be considerably larger compared to the periodicity of the structure ($\lambda_p \approx 7a$), justifying the description of the periodic structure as an effective medium. Therefore, with the herein presented methodology, we are indeed able to construct an engineered medium that can exhibit a *negative* electric permittivity in the *microwave* regime (with reasonably low losses and high field penetration inside the structure), thereby mimicking the interaction of light with real metals in the optical regime.

5.3 Metamaterials with Negative Effective Permeability in the Microwave Regime

In the previous section we examined how we can construct a metamaterial exhibiting negative electric permittivity (ε) in the microwave regime. However, harnessing the remarkable properties of left-handed metamaterials also requires, as we saw in the previous chapter, a design strategy for obtaining negative (effective) magnetic permeability (μ) at the same frequency region with the negative permittivity. Unfortunately, with the exception of some magnetic gyrotropic materials, media exhibiting negative (μ) do not occur naturally and should, thus, be built in the lab. In this section we shall see how one can construct such magnetic metamaterials in the microwave regime using entirely non-magnetic structured metallic elements [Pendry *et al.* (1999)], which act as the magnetic "molecules" of the engineered medium.

To this end, let us consider a three-dimensional periodic repetition of the external (larger) ring, shown in Fig. 5.2. The radius of the ring is r, and the whole arrangement is assumed to be immersed in air. This "split ring resonator" (SRR)

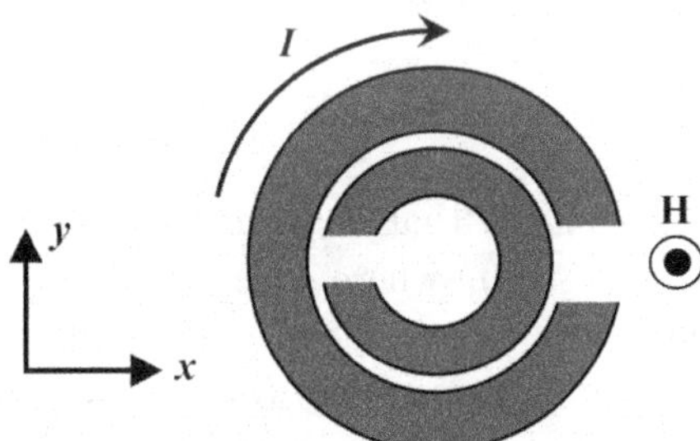

Figure 5.2: Schematic illustration of two concentric split-ring resonators (SRRs) for obtaining negative effective permeability in the microwave regime.

is equivalent to a simple RLC circuit, R being the resistance of the metallic ring, L its inductance and C (primarily) the capacitance between its unconnected ends. The rings residing on a given $z = zi$ plane have the same axis (i.e., they are "concentric") with the corresponding rings on the z-planes below and above them. The side of the square unit cell on an xy plane is equal to a.

Assuming that a magnetic field $\mathbf{H} = H_0 e^{-i(\omega t - \mathbf{k} \cdot \mathbf{z})} \mathbf{z_0}$ is incident on the structure, the induced (electromotive) source will (from Eq. (1.8)) be: $U = i\omega\mu_0\pi r^2 H_0$, generating an electric current I that circulates each ring (see Fig. 5.2). If the rings sitting on successive z-planes are closely together ("solenoid" approximation) there will be negligible "loss" of magnetic flux between the rings in each column, and therefore the magnetic flux will be: $\Phi = \mu_0\pi r^2 I/l$, l being the z-distance between corresponding SRRs lying on successive xy planes. Accordingly, the inductance L (in Henry) of each SRR will be: $L = \Phi/I = \mu_0\pi r^2/l$. One may, further, assume that the depolarizing magnetic flux lines generated by all rings are uniformly spread on a given xy plane, which results in a mutual inductance between two SRRs given simply by: $M = (\pi r^2/a^2)L = FL$, F being the fractional volume within a unit cell occupied by an SRR.

We may, now, apply Ohm's second law across a closed SRR "circuit" to obtain:

$$U = \left(R + \frac{i}{\omega C} - i\omega L + i\omega M \right) I, \tag{5.12}$$

where $R = 2\pi r\sigma$ is the (ohmic) resistance of each ring, σ being the resistance per unit length. Thus, the induced magnetic dipole moment per unit volume, M_d, will be:

$$M_d = I\frac{\pi r^2}{a^2 l}, \tag{5.13}$$

with the current I inferred from Eq. (5.12) to be:

$$I = \frac{H_0 l}{(1 - F) - 1/(\omega^2 LC) + iR/(\omega L)}. \tag{5.14}$$

As a result, the (relative) effective magnetic permeability associated with this medium will be (in the direction, z, that the incident magnetic field is polarised):

$$\mu = \frac{B/\mu_0}{B/\mu_0 - M_d} = 1 - \frac{F}{1 - 1/(\omega^2 LC) + iR/(\omega L)}. \tag{5.15}$$

From Eq. (5.15) we can see that μ assumes negative values in the range $1/\sqrt{LC} < \omega_p < 1/\sqrt{LC(1 - F)}$, where $\omega_{m0} = 1/\sqrt{LC}$ is the resonance frequency of the Lorentzian variation of the medium's magnetic permeability, and $\omega_{mp} = 1/\sqrt{LC(1 - F)}$ is the corresponding plasma frequency (where $\mathrm{Re}(\mu) = 0$). Crucially, we note that the resonant frequency (ω_{m0}) of the structure depends entirely on the rings' effective inductance (L) and capacitance (C), and can therefore be made considerably larger that the periodicity (a) of the structure, thereby fully justifying its description as an effective medium. Had we placed the SRRs on the other two planes (yz and xz) we would have, similarly, obtained negative effective permeabilities in the other two directions, x and y, as well, and the variation with frequency of these permeabilities would have been given by an expression similar to Eq. (5.15). Thus, with the present methodology we are able to construct a three-dimensional, isotropic metamaterials exhibiting negative effective permeability in a specified frequency region. The same effect can be obtained all the way from the radio up to the optical frequencies, simply by accordingly scaling the size of the SRR metaparticles.

5.4 Intrinsically Lossless Magnetic Metamaterials

As we have seen previously, it is by now well-established that not only do negative-μ and negative-index materials exist, but that they can even be constructed to exhibit broadband behaviour – in fact, they may even posses infinite bandwidth – [Eleftheriades (2007)], scaled down to optical frequencies and be built in three dimensions [Valentine *et al.* (2008)], as well as allow for efficient and fast tuning and switching [Chen *et al.* (2006b)]. Unfortunately, however, at present their performance is, as we remarked in Section 3.6, limited by the occurrence of losses (light absorption), which occasionally may reach levels of up to tens of dB/wavelength [Soukoulis *et al.* (2008)]. Clearly, if metamaterials are to find their way towards practical applications, this issue has to be adequately addressed.

To this end, a number of diverse strategies have been proposed [Shalaev (2007)]. A first approach relies on the use of gain media to compensate for the losses that originate from the presence of metallic elements in the "meta-molecules" [Kastel *et al.* (2007)]. This approach is very promising and it has, in fact, been theoretically shown that it can lead to zero-loss metamaterials [Webb and Thylen (2008)], even over a broad but finite bandwidth [Nistad and Skaar (2008)]. However, it may not always be possible to find a suitable gain medium to provide the necessary gain at the desired frequency regime. Another method for reducing losses in metamaterials relies on a judicious optimisation of the geometry of the "meta-molecules", whereby increasing the effective inductance to capacitance ratio, L/C, increases the quality factor, $Q = \frac{1}{2R}\sqrt{\frac{L}{C}}$, leading to reduction in the overall losses [Zhou *et al.* (2008)]. This approach results in metamaterial losses being substantially reduced, but normally by less than an order of magnitude, without being fully overcome, i.e. the resulting structures are never lossless. A third, recently proposed, interesting scheme makes use of negatively reflecting/refractive interfaces to reproduce the features of bulk negative-index metamaterials [Pendry (2008)]. Upon being negatively refracted at the engineered interfaces, light is allowed to propagate through lossless dielectric materials, such as air, thereby avoiding high-attenuation regions. This strategy relies on the use of nonlinear media and therefore requires intense incident light.

5.4.1 *Case of "Isolated" Unit Cells*

To arrive at a design for an *intrinsically lossless* resonant-type magnetic metaparticle let us start by noting that, in general, the magnetic metaparticles proposed so far correspond to equivalent electrical circuit models with only one mesh. For instance, as we show in the previous section, by determining the equivalent resistance (R), inductance (L) and capacitance (C) of a split-ring resonator (SRR) particle, one can accurately describe it in terms of a single RLC mesh, and then extract the effective magnetic permeability of a periodic array of such particles [Gorkunov *et al.* (2002); Chen *et al.* (2006a)]. Conversely, by connecting discrete R, L and C elements in a single loop, and for sufficiently large incident wavelengths to allow for being well within the "effective medium" regime, one can precisely reconstruct the magnetic behaviour of an SRR. Observing this common feature in the existing designs of magnetic meta-molecules, let us in the following analyse the quasi-static magnetic response of a pair of electrically connected RLC meshes, placed at the centre of a unit cell of volume $V = a^2 l$ which is periodically repeated in three dimensions. It is to be noted that electrical connection ("tight coupling") of RLC meshes in an infinite 1D array was recently considered in [Eleftheriades (2007)] in order to elucidate the broadband and low-loss behaviour of transmission-line metamaterials. It has to be emphasized that in the designs introduced herein, each pair of electrically connected RLC meshes resides at the centre of its own unit cell and is not connected to its neighbouring pairs. Furthermore, the wavelength of the incident light is considerably larger compared to the dimensions of the unit cell. For these reasons, the resulting structure is neither a backward transmission line [Eleftheriades (2007)] nor a magnetoinductive waveguide [Shamonina *et al.* (2007)], but a resonant-type effective magnetic medium.

Consider a plane wave with magnetic field component $\mathbf{H} = H_0^{i(\omega t - \mathbf{k} \cdot \mathbf{r})} \hat{\mathbf{e}}_0$, where $\mathbf{k}$ is the wavevector, $\mathbf{r}$ is the vector along the direction of the wave propagation, ω is the angular frequency, t is the time and $\hat{\mathbf{e}}_0$ the unit polarization vector. We assume that the magnetic field is incident on a pair of the electrically connected meshes (see Fig. 5.3) such that $\hat{\mathbf{e}}_0$ is parallel to $\hat{\mathbf{a}}$, where $\hat{\mathbf{a}}$ is the unit vector normal to the plane of the two meshes. From Faradays' law, the electromotive sources (voltages)

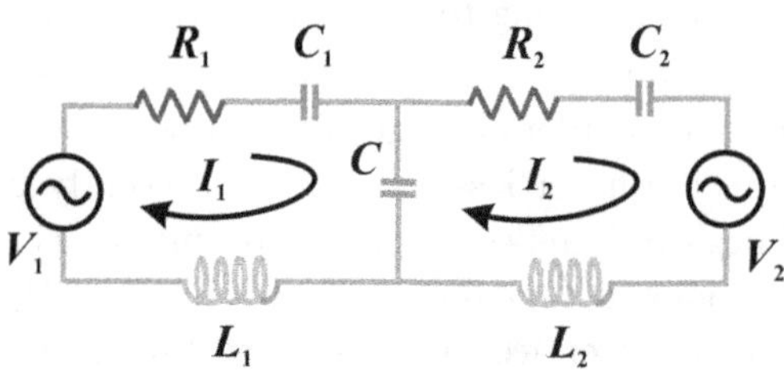

Figure 5.3: Schematic illustration of two electrically connected RLC meshes, each corresponding to a split-ring resonator (SRR). In the present case, both meshes reside on the same plane and are excited by the same incident magnetic field.

induced in the two closed meshes by the **H**-field will be $V_m = -i\omega\mu_0 S_m H_0$, $m = 1, 2$ with S_m being the surface of each mesh. From the well-known "mesh current method", one may determine the currents circulating in each loop in Fig. 5.3, as $I_m = \det(G_m)/\det(G)$, where:

$$G_1 = \begin{bmatrix} V_1 & i/\omega C_1 \\ V_2 & R_2 + i\omega L_2 - i/\omega C_2 - i/\omega C \end{bmatrix},$$

$$G_2 = \begin{bmatrix} R_1 + i\omega L_1 - i/\omega C_1 - i/\omega C & V_1 \\ -i/\omega C & V_2 \end{bmatrix},$$

$$G = \begin{bmatrix} R_1 + i\omega L_1 - i/\omega C_1 - i/\omega C & i/\omega C \\ i/\omega C & R_2 + i\omega L_2 - i/\omega C_2 - i/\omega C \end{bmatrix}, \quad (5.16)$$

with $R_m, L_m, C_m (m = 1, 2)$ being respectively the resistance, inductance and capacitance of the m-th mesh, C the capacitance of their common branch, and "det" designating the matrix determinant. Here, the resulting matrices are two-dimensional and the structure of the matrix G is, as we explain later in Section 5.4.6, typical of what is often called a "two-degrees-of-freedom" (2-DEG) system [Wahab (2008)]. Had we considered a cube with m electrically connected meshes, we would have obtained an M-DEG system described by m-dimensional matrices. Such systems are, indeed, well-known in diverse realms of science [Wahab (2008)], with some typical examples being highlighted in Section 5.4.6.

Having calculated the currents I_m circulating in each mesh, we can determine the total (emanating from both meshes) magnetic dipole moment per unit volume (magnetisation), M, as $M = (S_1 I_1 + S_2 I_2)/V$. Let us start by investigating the crucial for the points to be made later case where the meshes at the centres of neighbouring cubes are sufficiently apart from each other on the horizontal plane, i.e. the filling factor F — defined as the ratio between the total area of the two meshes and the area of the basis of a unit cell [Chen *et al.* (2006a)] — is sufficiently small, so that the cells can be regarded as weakly interacting ("isolated"). This is typical of low-density (dilute) gases, but note that contrary to the case of gases wherein the weak interaction of the molecules leads to the absence of any magnetic response and to $n \approx 1$, metamaterials can be designed to exhibit strong magnetic response [Pendry *et al.* (1999)], even for low densities, simply by increasing the Q-factor (e.g., by decreasing the value of the resistance R) of the resonant RLC meshes. We examine this crucial case of "isolated" cells in order to establish that the observed effects are intrinsic, i.e. that they occur at the unit-cell level and are not an artefact of the periodicity of the structure (see Section 5.4.3 for a detailed discussion of the latter situation). The usual case of "tightly coupled" meshes on the horizontal plane does not fundamentally differ in any of its conclusions with the case presented herein, as is explained in detail in Section 5.4.3. Note that in order to illuminate the subtle distinctions and facilitate a direct comparison between the two cases, in the numerical computations we used the same filling factors for both cases.

Under the aforementioned conditions, higher-order multipole terms are, indeed, negligible [Waterman and Pedersen (1986)] and the local fields in each cube are just $H_{loc} = H_0$ and $B_{loc} = B_0$, where B is the magnetic flux density, so that we may define $\mu_{r,eff} = 1 + M/H_0$. Note that in this case the effective (relative) permeability is directly proportional to the magnetization M. If the neighbouring cubes were considered "tightly coupled" then the expression for $\mu_{r,eff}$ would have become [Gorkunov *et al.* (2002)]

$$\mu_{r,eff} = \frac{1 + \frac{2}{3}\frac{M}{H_0}}{1 - \frac{1}{3}\frac{M}{H_0}}$$

(see Section 5.4.3). In that case, $\mu_{r,eff}$ is not any more simply proportional to neither the magnetisation M nor to the *total* active power

$$P = P_1 + P_2, \quad P_m = \mathrm{Re}\left[V_m I_m^*\right], \quad (m = 1, 2)$$

of the pair of meshes, and the reported effects become slightly more intricate, but do not fundamentally nor qualitatively differ from those presented below. It is useful to also note that, as we show in the previous section (Eq. (5.12)), in this case and for meshes that are stacked up closely together (solenoid approximation), the uniform depolarization magnetic field [Pendry *et al.* (1999)] only modifies the value of the inductances L_m in each cube [Chen *et al.* (2006a)], i.e. merely reducing the value of L_m in each mesh is sufficient to incorporate the effect of the depolarisation field in the analysis.

Figure 5.4 furnishes exemplary results for the real and imaginary parts, as well as for the corresponding figure of merit FOM = $-\mathrm{Re}[\mu]/\mathrm{Im}[\mu]$, of the (relative) effective permeability μ in structures where the area S_2 of the second mesh in each unit cell is much smaller compared to the area S_1 of the first mesh. Let us start with the case (shown in Figs. 5.4a,b) where the second mesh is also open-circuited. As expected, one observes a typical Lorentzian absorption pattern for the imaginary part of μ, as well as for the total (absorbed) active power P (see Section 5.4.2). The specific variation of the real part of μ with frequency stems from the fact that the engineered medium, since it is physically realizable, is causal with $\mathrm{Re}[\mu]$ and $\mathrm{Im}[\mu]$ strictly adhering to Kramers–Kronig relations. Figure 5.4b presents the FOM for this case, which can assume appreciable values only in regions where $\mathrm{Re}[\mu]$ is either positive or very close to zero. Figures 5.4c,d report the corresponding results for the case where the second mesh is included in each cube, but has area $S_2 \ll S_1$ (with $S_2 \to 0$). Note the abrupt dip in the spectra of, both, the imaginary part of μ and the total active power P (the latter shown in Section 5.4.2), accompanied by a sudden small jump in the FOM, shown in Fig. 5.4d. This behaviour is, indeed, expected since it is well-known that macroscopic coupled resonators (in this case, electrical resonators) can mimic the electromagnetically induced transparency effect that occurs in atomic gases [Hau *et al.* (1999); Liu *et al.* (2001)] leading to a, so

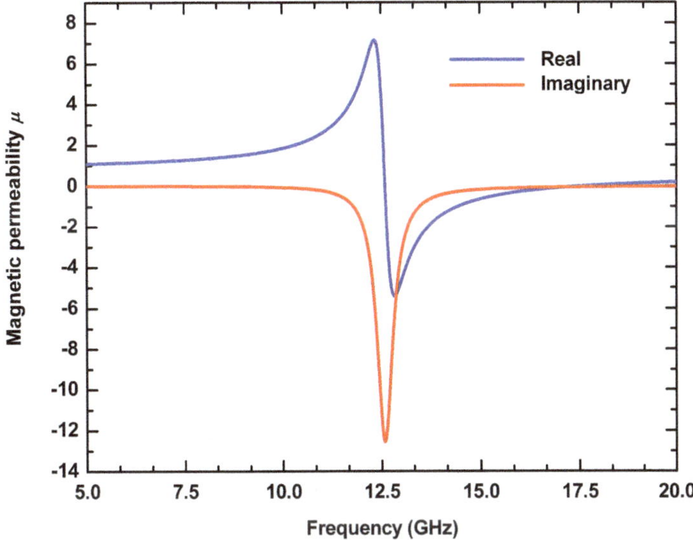

(a) Real and imaginary part of effective μ for the case where the second mesh is open circuited and the value of the coupling capacitance is $C = 0.1 \cdot 10^{-12}$ F.

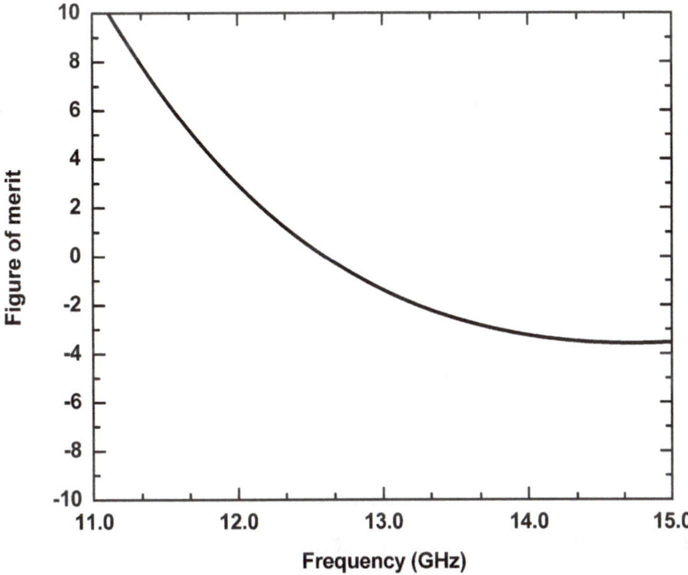

(b) Corresponding figure of merit for the configuration with the parameters of (a).

Figure 5.4: Calculated effective (relative) magnetic permeability (μ) and figure of merit (FOM = $-\text{Re}[\mu]/\text{Im}[\mu]$) for the case where both loops lie on the yz plane. A perpendicularly incident magnetic field $H_0 = 1$ A/m induces two electromotive voltage sources, V_1 and V_2 (see Fig. 5.3). In all cases, the values of the lumped elements were: $R_1 = 50$ Ohms, $R_2 = 0.1$ Ohms, $C_1 = C_2 = 10^{-14}$ F, $L_1 = L_2 = 1.6 \cdot 10^{-8}$ H. The surface of the first loop is $S_1 = \pi(2 \cdot 10^{-3})^2$ m^2 and both loops are located at the centres of unit cells of volume $(5 \cdot 10^{-3}) \times (5 \cdot 10^{-3}) \times (1 \cdot 10^{-3})$ m^3.

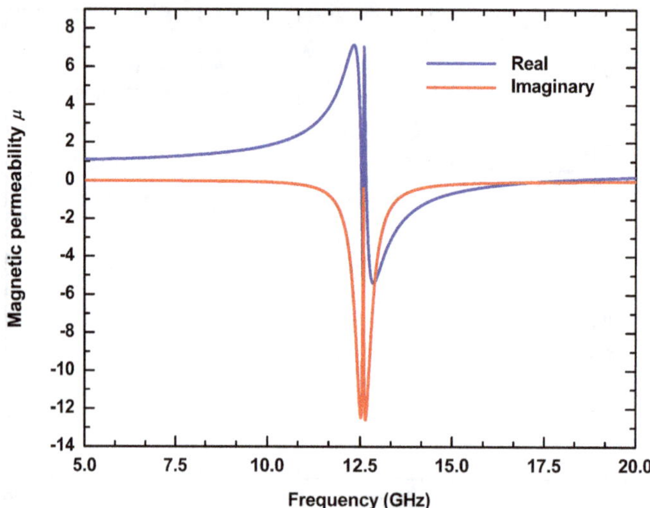

(c) Real and imaginary part of μ for the case where the second mesh is included, but has surface $S_2 \to 0$. The value of the coupling capacitance is, now, $C = 1 \cdot 10^{-12}$ F.

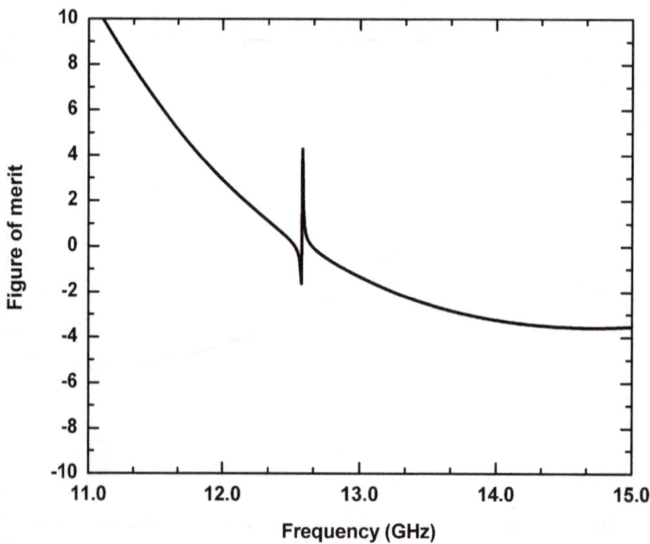

(d) Corresponding figure of merit for the configuration with the parameters of (c).

Figure 5.4: (*Continued*)

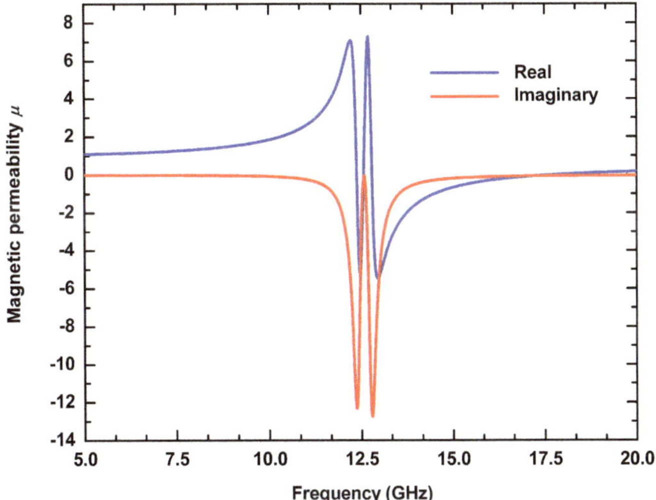

(e) Same as in (c), but now with $C = 0.3 \cdot 10^{-12}$ F.

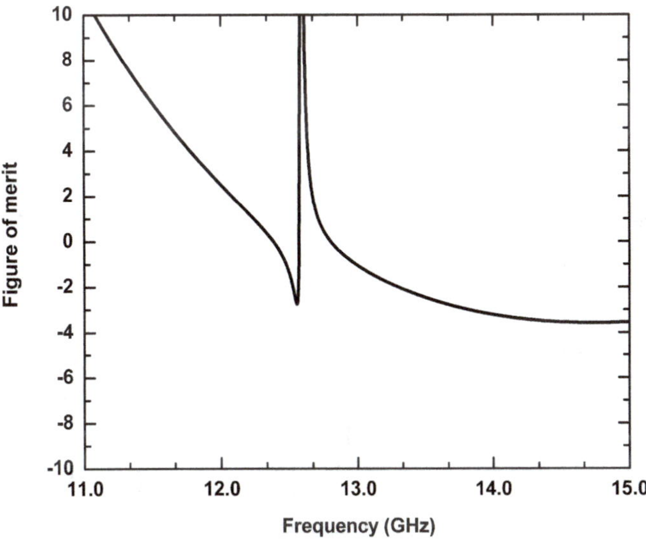

(f) Corresponding figure of merit for the configuration with the parameters of (e).

Figure 5.4: (*Continued*)

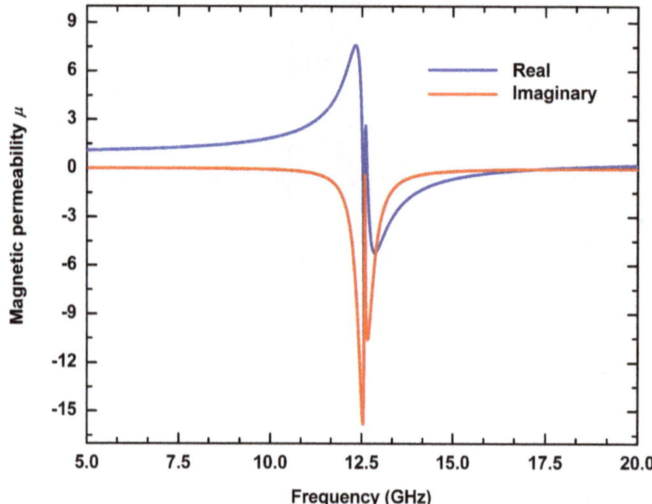

(g) Same as in (c), but now with $S_2 = S_1/10$ and $C = 1 \cdot 10^{-12}$ F.

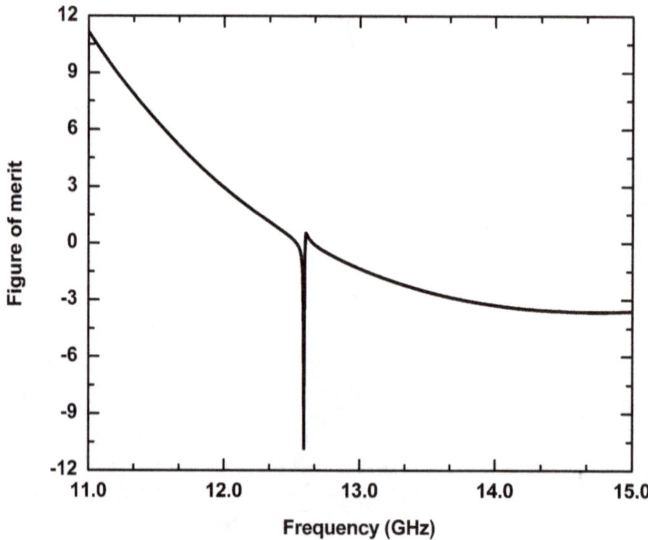

(h) Corresponding figure of merit for the configuration with the parameters of (g).

Figure 5.4: (*Continued*)

called, "coupled resonator induced transparency" [Alzar *et al.* (2002); Boyd and Gauthier (2006); Smith *et al.* (2004a); Totsuka *et al.* (2007)]. The associated very strong dispersion in a close region around the dip can be used for decelerating light [Hau *et al.* (1999)], although over only a very narrow bandwidth. Figures 5.4e,f present similar results to the previous case, but for a coupling capacitance reduced tenfold. One can clearly observe the evolution of the spectra of P and the imaginary part of μ according to the well-known Autler–Townes effect [Alzar *et al.* (2002)]. The calculated FOM in this case becomes large, assuming a maximum value of approximately 60, but mainly in regions where $\text{Re}[\mu] > 0$. These low-loss regions of μ (with $0 < \text{Re}[\mu] < 1$) could be useful for good-quality diamagnetic metamaterials, as well as in the design of "invisibility" cloaks [Pendry *et al.* (2006); Schurig *et al.* (2006)]. For instance, at $f = 12.58$ GHz, one obtains $\mu \approx 0.9346 - i0.0355$, with a FOM ≈ 26. To complete the presentation of the results for this general case, Figs. 5.4g,h report the corresponding calculations to Figs. 5.4c,d, but now with $S_2 = S_1/10$. As shown in Section 5.4.2, this is the first instance where P_2 becomes nonzero, and it does so in an intriguing way: there is a narrow region wherein P_2 actually becomes negative, which suggests that in this region the second mesh does not absorb but, instead, remits active power to the first mesh. We will return to this important observation later when we discuss the design of an intrinsically lossless magnetic metamaterial, but at the moment let us note that in this region the total power P absorbed by the pair of meshes is reduced, owing to the negative contribution from P_2, and consequently the FOM exhibits an increased jump — within the $\text{Re}[\mu] < 0$ region — compared with that of Fig. 5.4d.

Let us, now, turn our attention to the important case where the areas of the two meshes inside each unit cell are equal, i.e. $S_2 = S_1$. Figures 5.5a–e report the real/imaginary parts of μ, the corresponding FOM and the absorbed powers P_1, P_2 and $P = P_1 + P_2$. The exchange of active power between the two meshes (electrical oscillators) is readily inferred from Figs. 5.5c–e. There are, indeed, regions wherein P_1 or P_2 become negative, but *never* simultaneously (which would violate the conservation of energy). In the region where $P_1 < 0$ the second mesh receives active power from the first mesh, with the opposite being true for the region where $P_2 < 0$. The fact that P_1 and P_2 are never simultaneously negative is evident from the fact that $P > 0$ throughout, as required from the passivity of the structure. Interestingly, one can observe a region where the total absorbed power P reduces abruptly towards zero, i.e. it loses its Lorentzian shape. This is also reflected in the imaginary part of μ, shown in Fig 5.5a, and as a result the FOM at precisely this region increases dramatically to around 987; for instance, at $f = 13.196$ GHz, we obtain $\mu \approx -4.4575 - i0.0045165$, i.e. we achieve a substantially negative $\text{Re}[\mu]$ with simultaneously excellent FOM.

Until now we have assumed that the two electrically connected meshes lie on the same plane inside each unit cell, but this restriction is not necessary. Consider the case where the first first mesh is placed on the xz plane and the second mesh

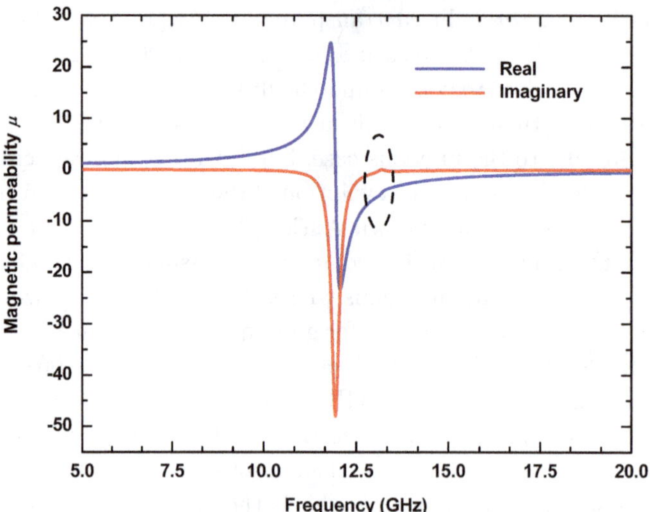

(a) Variation of real and imaginary part of effective μ with frequency.

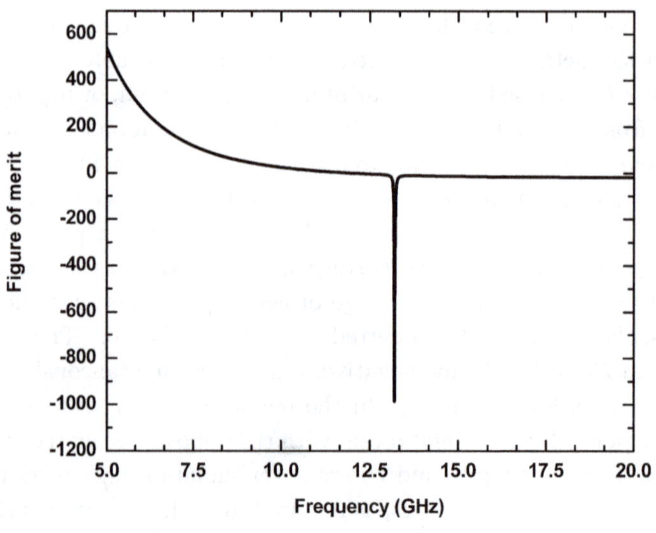

(b) Corresponding figure of merit.

Figure 5.5: In all cases the electromagnetic and geometric parameters are those of Fig. 5.4, but now with $S_2 = S_1$ and $C = 0.1 \cdot 10^{-12}$ F.

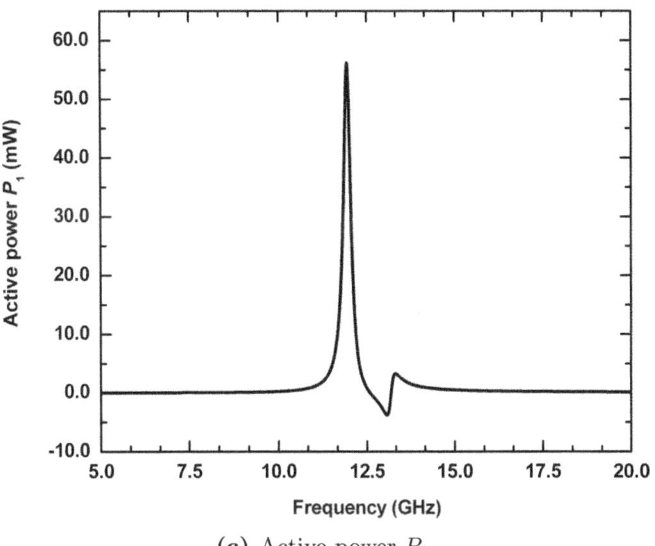

(c) Active power P_1.

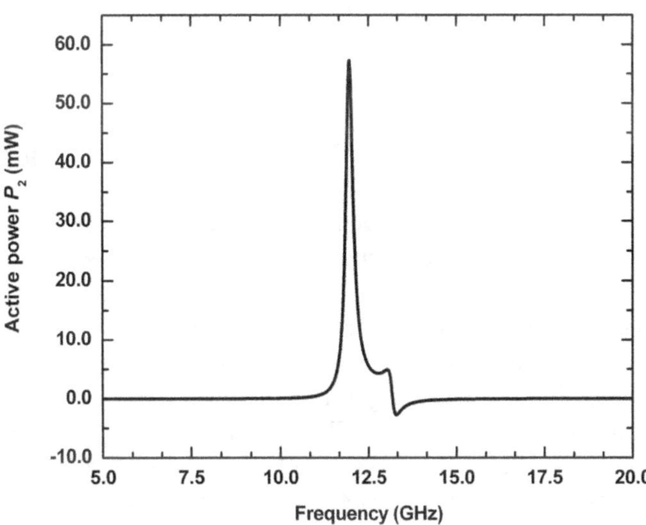

(d) Active power P_2.

Figure 5.5: (*Continued*)

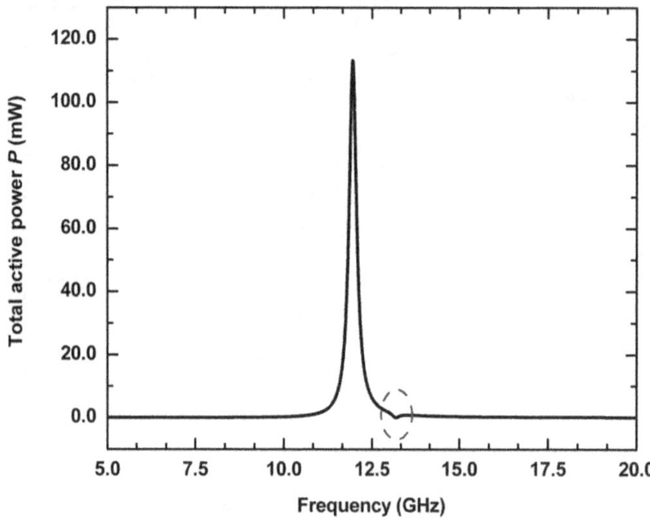

(e) Total active power $P = P_1 + P_2$.

Figure 5.5: (*Continued*)

on the yz plane, as illustrated in Fig. 5.6. Let us, further, assume that two uniform harmonic magnetic fields of equal amplitude, $H_x = H_2 = \mathbf{H} \cdot \hat{\mathbf{x}}_0 = H_0 e^{i(\mathbf{k} \cdot \mathbf{r_1} - \omega t)}$ and $H_y = H_1 = \mathbf{H} \cdot \hat{\mathbf{x}}_0 = H_0 e^{i(\mathbf{k} \cdot \mathbf{r_2} - \omega t)}$, are simultaneously incident on the structure, generating the electromotive voltages V_1 and V_2 shown in Fig. 5.3. In this particular arrangement, the magnetic field H_1 generates electric currents circulating in *both* meshes. As a result, the magnetic field $H_y = H_1$ generates magnetic moments on both xz and yz planes, i.e. it is responsible for the generation of μ_{yy} and μ_{xy}. The corresponding is also true for the magnetic field $H_x = H_2$. The "total" $\mu_x = \mu_2$ that the x-component of the magnetic flux density, B_x, experiences — $B_x = \mu_{xx} H_x + \mu_{xy} H_y = (\mu_{xx} + \mu_{xy}) H_0 = \mu_x H_0$ — can, now, be calculated by finding the "total" current I_2 shown in Fig. 5.3 and following the methodology for the calculation of $\mu_{r,eff}$ that was outlined in the previous examples. Figures 5.7a,b report the results of such a calculation of $\mu_x = \mu_2$ and $\mu_y = \mu_1$ for the case where the coupling capacitance, C, of the two meshes is $C = 0.4$ pF. Note that the imaginary parts of μ_1 and μ_2 follow closely the variation with frequency of the active powers P_1 and P_2, respectively, which were previously studied in Figs. 5.5c–e, and as such there are now regions where either μ_1 (μ_y) or μ_2 (μ_x) become zero or even positive. Furthermore, owing to the passivity of the structure, μ_1 and μ_2 are never simultaneously positive; in fact, $\text{Im}[\mu_1] + \text{Im}[\mu_2]$ is exactly equal to the imaginary part of μ that was shown in Fig. 5.5a.

Thus, with the present anisotropic design we are, in principle, able to harness zero-loss negative-μ magnetic metamaterials — in fact, even with artificial magnetic "gain" — over a continuous range of frequencies along one direction, at the cost of having increased magnetic losses in the other direction. For instance, assuming

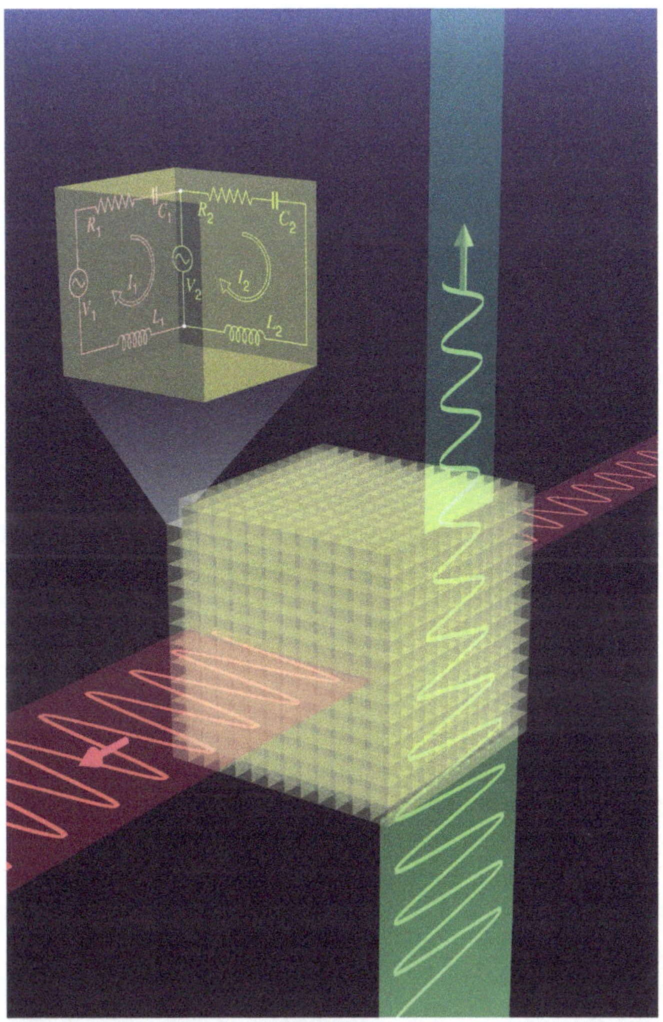

Figure 5.6: To arrive at a design that can allow for harnessing intrinsically lossless metamaterial magnetism, we place – in each unit cell – the first mesh on the xz plane and the second mesh on the yz plane (see top left inset). We then use two separate light beams of the same incident amplitude: one (shown in green) whose magnetic field oscillates perpendicularly to the second mesh, i.e. parallel to the x-axis, and a second beam (shown in red) whose magnetic field oscillates perpendicularly to the first mesh, i.e. parallel to the y-axis. Our analysis reveals that there are certain frequency regions wherein the imaginary part of the structure's effective magnetic permeability, $\text{Im}[\mu]$, becomes positive (corresponding to magnetic "gain") along the y (x) direction and negative along the x (y) direction. The magnetic losses in the direction where $\text{Im}[\mu] < 0$ are larger compared to the "gain" in the direction where $\text{Im}[\mu] > 0$, so that the conservation of the magnetic energy is satisfied.

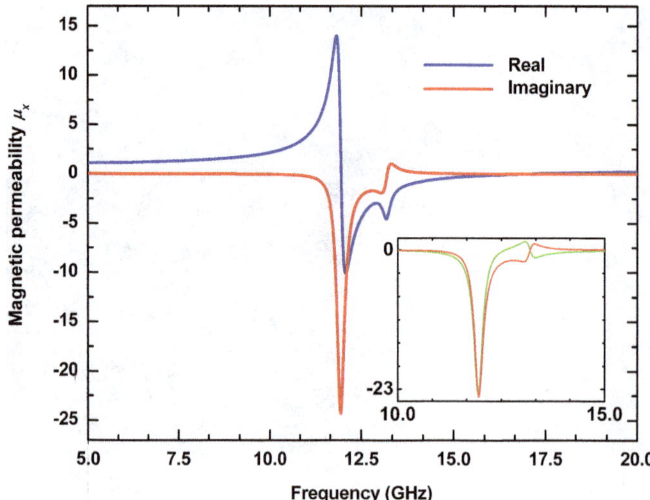

(a) Variation with frequency of the real and imaginary part of the effective relative permeability along the x-axis (μ_x). The inset illustrates the variation with frequency of the imaginary part of μ_x (red) and μ_y (green), from where one can observe that in the region where μ_y (μ_x) > 0 it is, also, μ_x (μ_y) < 0.

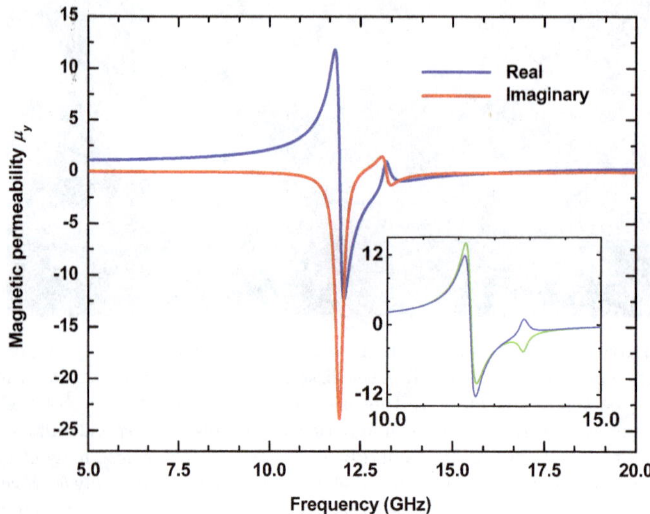

(b) Variation with frequency of the real and imaginary part of the effective relative permeability along the y direction (μ_y). The inset illustrates the variation with frequency of the real part of μ_x (green) and μ_y (blue).

Figure 5.7: The electromagnetic and geometric parameters are those of Fig. 5.3, but now the first mesh lies on the xz plane, while the second mesh on the yz mesh.

that the imaginary part of μ_1 is positive in a frequency region, one will observe an *increase* (artificial "gain") in the magnetic density B_y in the direction of its propagation, owing to $\text{Im}[\mu_1] > 0$; at the same time, however, the decrease of the magnetic density B_x along the direction to which it propagates, owing to $\text{Im}[\mu_2] < 0$ in the same frequency region, will be *larger* compared with the increase that B_y experienced, so that — at every frequency point — there will be no violation of the conservation of the magnetic energy.

In summary, in this section we have presented a design paradigm that allows for ultra-low or (uniaxially) zero-loss magnetic metamaterials. The proposed blueprint is based on a 2-DEG topology and relies critically on the presence of electrically connected RLC meshes that exchange active power. In certain frequency regions, one mesh is able to naturally "pump" energy to the other mesh, which leads to the imaginary part of μ in a specified direction becoming exactly zero or even positive. The scheme is, in principle, scalable and realizable at any frequency regime, and does not require nonlinear or gain media to compensate for the losses. This design paradigm may conceivably lead to a new generation of ultralow — or zero-loss metamaterials that could be exploited in the design of "perfect" lenses, "invisibility" cloaks, slow-light waveguides, optical nanocircuits and magnetic resonance imaging systems. In the following subsections we present a number of elaborating remarks on the aforementioned scheme, particularly pertaining to the calculations of the active powers and to the case of "tightly coupled" neighbouring unit cells.

5.4.2 *Calculations of Active Powers in the Equivalent Electrical Circuits*

Consider two circuits, C_1 and C_2, electrically connected with two conducting wires, as shown in Fig. 5.8. Circuit C_1 contains sources and passive elements (resistors, inductors, capacitors), while circuit C_2 contains only passive elements. Let us assume that the time-domain (designated with low case symbols) voltage, ν, and the current, i, shown in Fig. 5.8, are sinusoidal functions of time, i.e.: $\nu(t) = V_0 \cos(\omega t)$, and $i(t) = I_0 \cos(\omega t - \varphi)$, where φ is the phase difference between the current and the voltage. For the sign conventions shown in Fig. 5.8, a positive time-domain (instant) power, $p(t) = \nu(t)i(t) > 0$, means that electrical power flows from circuit C_1 towards circuit C_2. Conversely, if $p(t) < 0$ then power flows from circuit C_2 to circuit C_1.

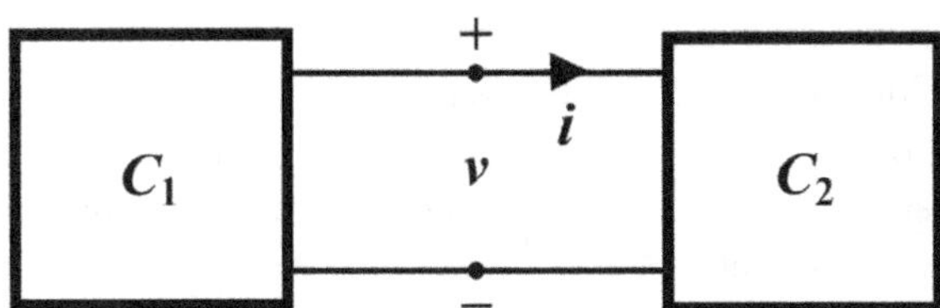

Figure 5.8: Schematic illustration of two electrically connected circuits.

It is instructive to write the expression for the instant power p in the form:

$$p(t) = VI \cos\varphi\left(1 + \cos(2\omega t)\right) + VI \sin\varphi \sin(2\omega t), \qquad (5.17)$$

where $V = V_0/\sqrt{2}$ and $I = I_0/\sqrt{2}$. Since we have assumed that the circuit C_2 contains only passive elements, the phase difference φ between the current and the voltage will always be $-\pi/2 < \varphi < \pi/2$, and therefore the first term on the RHS of Eq. (5.17) will be positive, corresponding to power flowing from circuit C_1 to circuit C_2, where it is consumed (dissipated). For this reason, this power, $p_R(t) = VI \cos\varphi\left(1 + \cos(2\omega t)\right)$, is called *real* or *active* power. The active power varies between 0 and $2VI \cos\varphi$ in one period, and a metric that is frequently used for it, is its time-averaged value: $\overline{P} = VI \cos\varphi$, which customarily is again referred to as *active* power. By contrast, the second term in the RHS of Eq. (5.17), $VI \sin\varphi \sin(2\omega t)$, changes sign twice during the period $T = 2\pi/\omega$. During the first half-period, p_x flows from C_1 to C_2, so that the overall effect is zero. Because of the fact that p_x does not (on average) produce any work, it is usually referred to as *reactive* power. Since the time-averaged value of p_x is zero, a metric that is frequently used for its quantification is its amplitude, $Q = VI \sin\varphi$. Generally, the reactive power refers to the part of the instant power p that is send by a source to the inductors and capacitors of the circuit, stored there temporarily, and then send back to the source.

In many situations it is useful to consider the transformations of the voltage V and current I in the frequency domain, $\tilde{V} = Ve^{i\theta_1}$ and $\tilde{I} = Ie^{i\theta_2}$, θ_1 and θ_2 being the angles of the rotating vectors $\tilde{V}$ and $\tilde{I}$ with the real axis, and to work with the so called complex power, S:

$$S = \tilde{V}\tilde{I}^* = VIe^{i(\theta_1 - \theta_2)} = VIe^{i\varphi}. \qquad (5.18)$$

It can, then, be readily observed that the reactive power is simply given by: $Q = \text{Im}[S]$, while the (time-averaged) active power by: $\overline{P} = \text{Re}[S]$, which is the relation that we used in the calculations of the active powers referred to in the main text.

It should be noted that when the circuit C_2 does not contain sources, the active power $\overline{P}_2$ in C_2 is strictly positive, i.e. C_2 absorbs (consumes) real power, which is converted into heat at the resistances. In contrast, if C_2 also includes sources of electrical energy, then it is possible that $\overline{P}_2$ may become negative (for the sign convention shown in Fig. 5.8) in a frequency region, corresponding to active power flowing *away* from C_2, i.e. in that region C_2 acts as a *source* of real power: it does not absorb real power, but remits it to the neighbouring meshes that it is electrically connected with. However, the total active power, $P = \sum_i P_i$, $i = 1, 2, ..., n$, n being the total number of electrically connected meshes, must still be positive at *every frequency point*, since overall we have a net consumption of real power. These features are, indeed, precisely what we observe in the analytic calculations (shown below) of the active powers in the meshes of the 2-DEG system that we described in the main body of this work.

In the following Figure 5.9 we present the results of the calculations of the active powers for the various mesh configurations that were analysed in the main text. Note, in each case, the precise correspondence (in their variation with frequency) between the total active power $P = P_1 + P_2$ and the imaginary part of the effective magnetic permeability (μ) that was presented in Section 5.4.1.

5.4.3 *Case of "Tightly Coupled" Unit Cells*

In Section 5.4.1 we studied the quasi-static magnetic response of "isolated" pairs of electrically connected meshes, in order to establish the fact that the observed regions of positive imaginary parts of the effective magnetic permeabilities do not arise owing to the periodicity of the structure [O'Brien and Pendry (2002); Markos and Soukoulis (2003); Koschny *et al.* (2003); Depine and Lakhtakia (2004); Efros (2004); Koschny *et al.* (2004)] but that they are an intrinsic feature of the 2-DEG system, occurring within each individual, "isolated" unit cell. There, we showed that such regions are ultimately the result of the exchange of active power within each pair of meshes.

Having established that the aforementioned observed effects are intrinsic, let us in the following examine the corresponding situations with those of the main text, but now for the case of strongly interacting ("tightly coupled") cells. For an incident magnetic field of amplitude H_0, inducing a magnetisation M per unit cell, a pertinent analysis reveals that, in this case, the effective magnetic permeability may be given by:

$$\mu_{r,eff} = \frac{1 + \frac{2}{3}\frac{M}{H_0}}{1 - \frac{1}{3}\frac{M}{H_0}} \tag{5.19}$$

with care exercised to accordingly modify the total impedance, Z, in each mesh [Gorkunov *et al.* (2002); Chen *et al.* (2006a)], (see also later in this subsection). It was further shown in [Chen *et al.* (2006a)] that for meshes stacked closely together, one can accurately consider each pile to be a "solenoid". In that approach, the depolarisation magnetic field results in an additional (mutual) inductance, M, which is simply subtracted from the (lumped) inductance L of each mesh.

To facilitate direct comparison with the corresponding results in the main text, here, we retain the same value (16 nH) for the inductance of each mesh as in the main body of the article. This is, of course, equivalent to assuming that the actual inductance, L, in each mesh is slightly larger, so that when M is subtracted from L we end up with a total inductance, in each mesh, equal to 16 nH.

All the results concerning the active powers in each mesh are precisely the same as the corresponding ones presented in the previous section or in the main article. Accordingly, all the dips or peaks in the imaginary part of the effective permeabilities associated with the exchange of active power between the meshes continue to occur at the same frequency ($\approx$ 12.5 GHz) as in the case of "isolated" cells. However, since the effects of neighbouring cells are now incorporated in the analysis, one

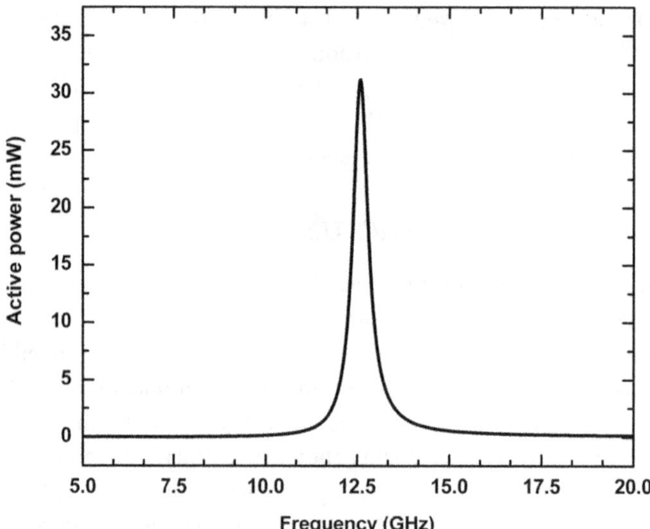

(a) Total (time-averaged) active power for the configuration with the parameters of Fig. 5.4a.

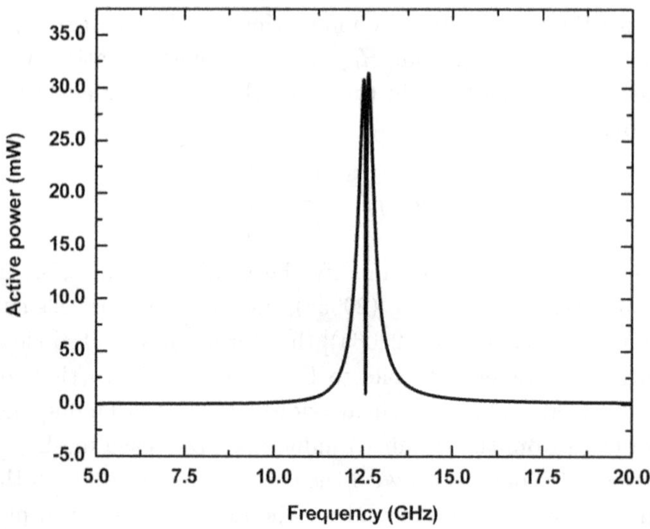

(b) Same as (a), but now with the parameters of Fig. 5.4c.

Figure 5.9: Results of the calculations of the active powers for the various mesh configurations that were analysed in the main text.

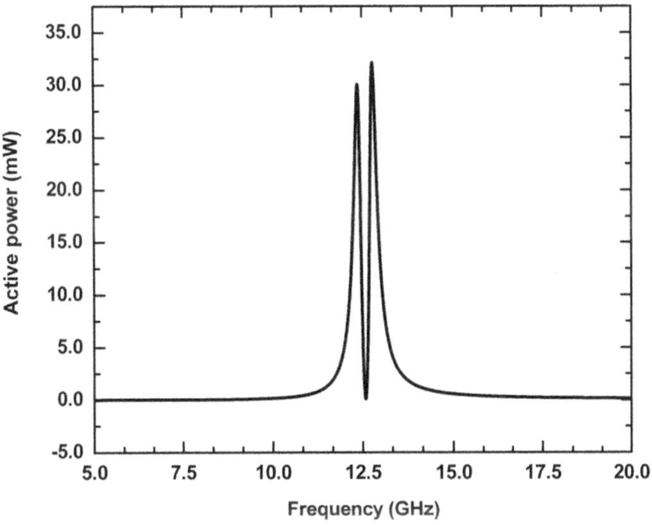

(c) Same as (a), but now with the parameters of Fig. 5.4e.

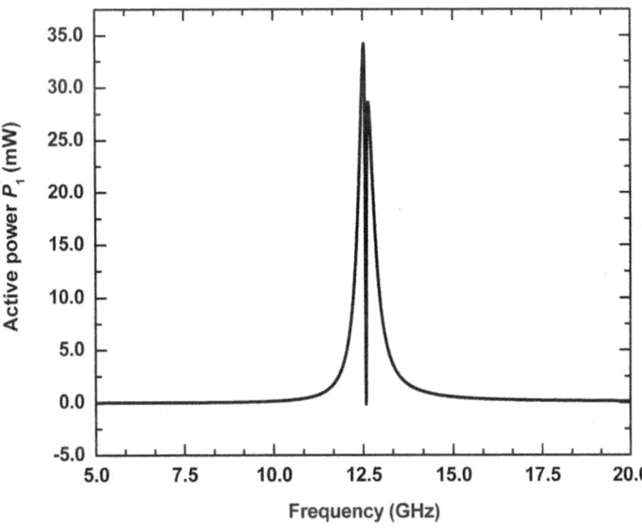

(d) Active power of the first mesh, with the parameters of Fig. 5.4g.

Figure 5.9: (*Continued*)

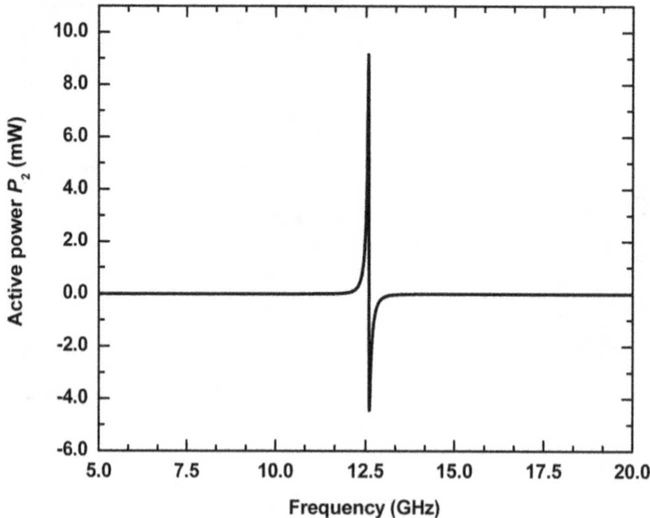

(e) Active power of the 2nd mesh, with the parameters of Fig. 5.4g.

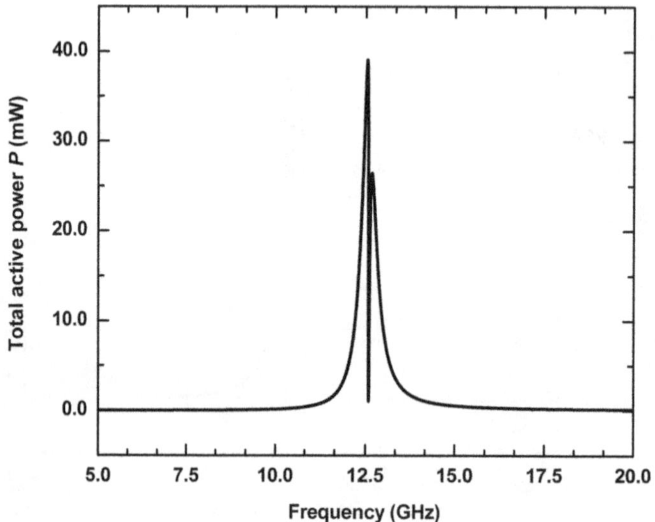

(f) Total (time-averaged) active power for the configuration with the parameters of Fig. 5.4g.

Figure 5.9: (*Continued*)

does expect to observe a redshift in the collective resonant response of the composite medium. For instance, it is well-known from the classical Lorentz theory for the description of dielectric molecules that the use of the Clausius–Mossotti relation (which is very similar to Eq. (5.19) above) results in a redshift of the resonant frequency from ω_0^2, when the density of the molecules is reasonably low ("isolated" unit cells, as in a gas), to $\omega_0^2 - Ne^2/3\varepsilon_0 m$, where N is the density of the molecules, e the electronic charge, m its mass, and ε_0 the free-space permittivity. This is also what we observe in the results presented below.

Inspection of Figs. 5.10–5.12 reveals that the variation with frequency of the calculated relative effective permeabilities are, as expected, very similar to the ones that was presented in Figs. 5.4, 5.5 and 5.7. In particular, compared with the case of "isolated" unit cells, there is a redshift in the resonant response of the system (e.g., compare Fig. 5.4a with Fig. 5.10a), as highlighted above. The location of the sudden dips and rises, however, which is due to exchange of active power between the electrically connected meshes in each unit cell, remains at around 12.5 GHz, similarly to Figs. 5.4, 5.5 and 5.7, as one would expect. This is because the influence of the neighbouring cells to the currents circulating the pair of meshes in each cell arises primarily through the mutual inductance M, which is subtracted from the (herein assumed slightly increased) inductance L of each mesh, so that $L - M = 16$ nH, equal to the value of L that was used in the main text. As a result, the currents circulating each pair of meshes inside the "tightly" coupled cells are exactly equal to those of the case of "isolated" cells. Hence, here, the variations of the active powers with frequency, which are responsible for stipulating the location of the dips and rises, are the same with the variations presented in the main part of the article.

It is interesting to note from Fig. 5.12c that using Eq. (5.19) to describe the quasi-static response of a medium with "tightly"' coupled unit cells, in which the meshes are located at different planes (anisotropic medium), results in a region where the summation of the imaginary parts of the effective permeabilities along the directions perpendicular to these planes becomes positive, implying the occurrence of "net" gain in that region. As explained in Section 5.4.5, such an outcome, highlighted by a dashed red line in Fig. 5.12c, emergences ultimately owing to the violation of the assumptions used in the derivation of Eq. (5.19).

5.4.4 *Remarks on the Working Principle of the Magnetically Lossless 2-DEG Configuration*

The presented 2-DEG structure allows for the realisation of ultralow- or zero-loss magnetic metamaterials over a continuous range of frequencies. As was highlighted in the main text, a number of approaches have previously been proposed for compensating losses [Shalaev (2007); Kastel *et al.* (2007)] or even creating stable gain [Boardman *et al.* (2007)] in metamaterials. Those approaches usually relied on providing optical gain, e.g. in the form of optical parametric amplification [Popov and Shalaev (2006)] or electromagnetically induced chirality [Kastel *et al.* (2007)], or

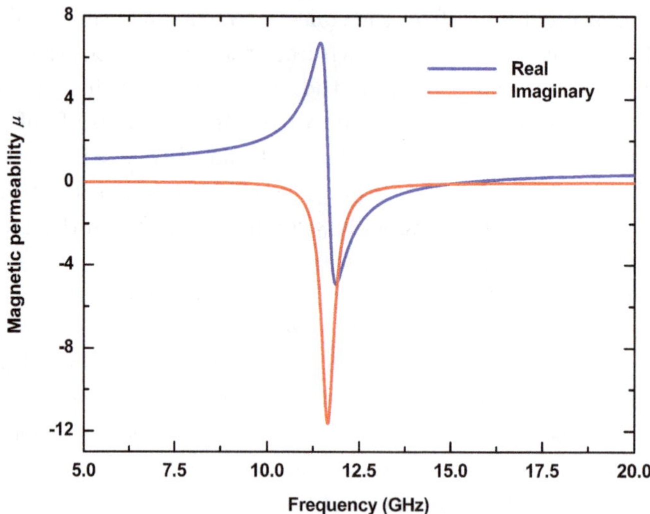

(a) Relative effective permeability when the second mesh is open-circuited.

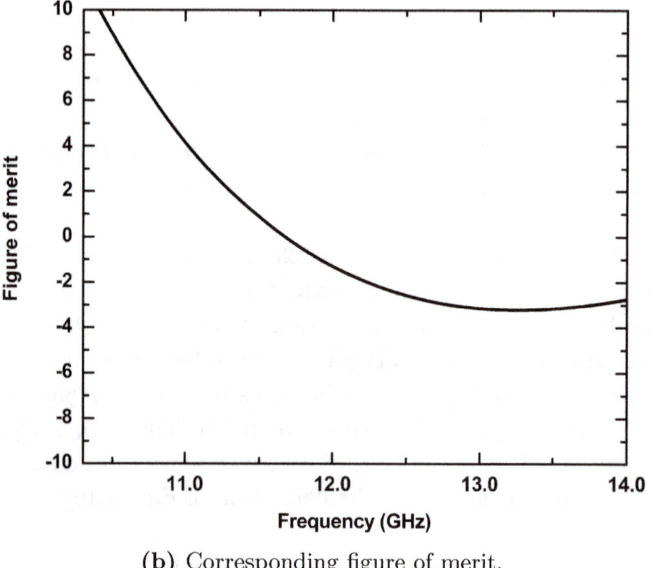

(b) Corresponding figure of merit.

Figure 5.10: In all cases, the electromagnetic and geometric parameters are those of Figs. 5.4.

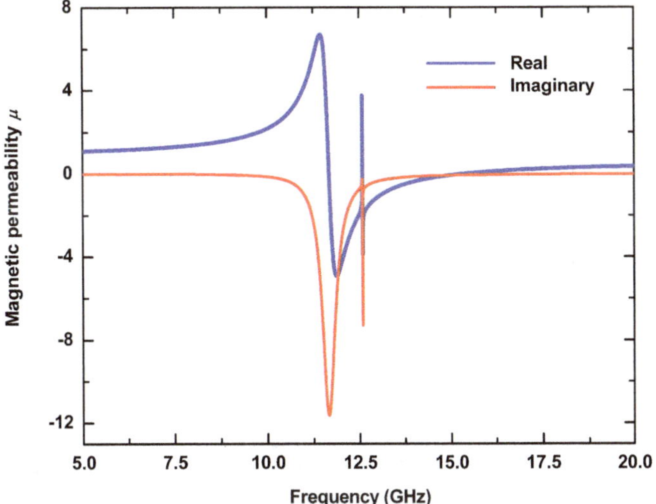

(c) Relative effective permeability for $S_2 = 0$ and $C = 1$ pF.

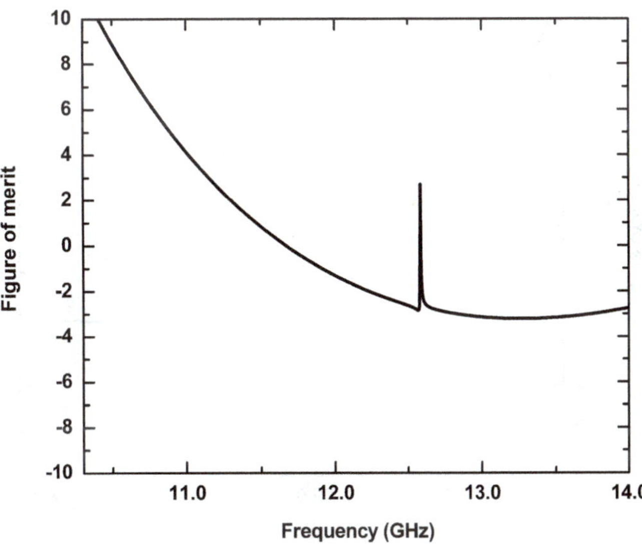

(d) Corresponding figure of merit.

Figure 5.10: (*Continued*)

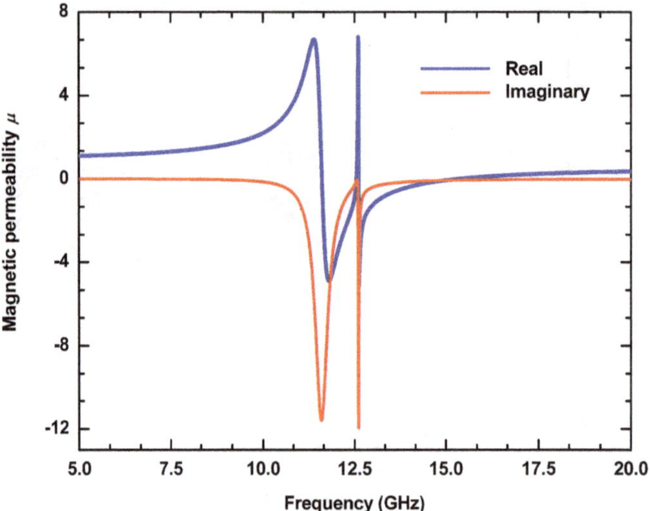

(e) Relative effective permeability for $S_2 = 0$ and $C = 0.3$ pF.

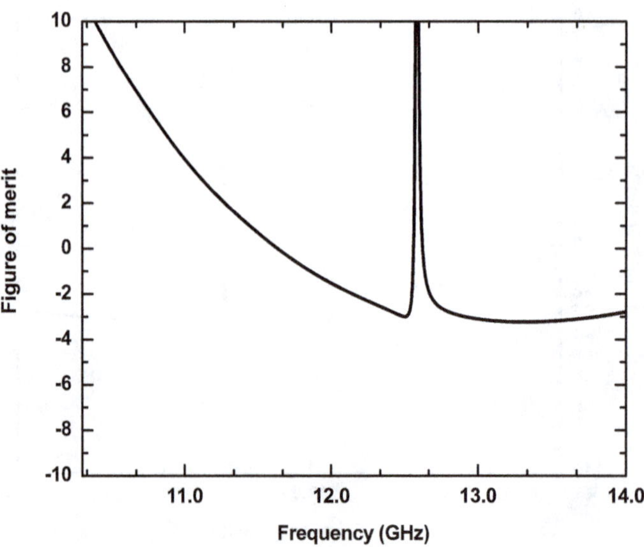

(f) Corresponding figure of merit.

Figure 5.10: (*Continued*)

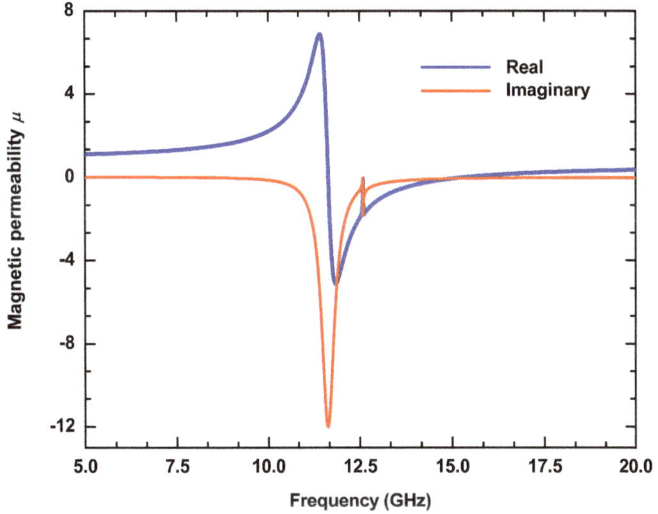

(**g**) Relative effective permeability for $S_2 = S_1/10$ and $C = 1$ pF.

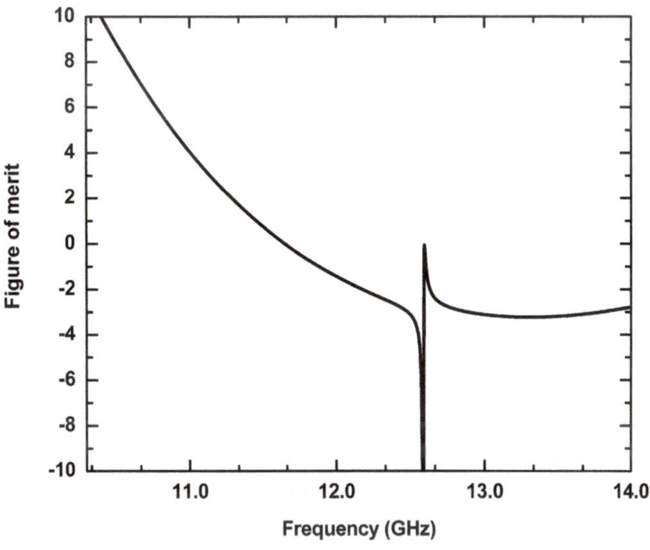

(**h**) Corresponding figure of merit.

Figure 5.10: (*Continued*)

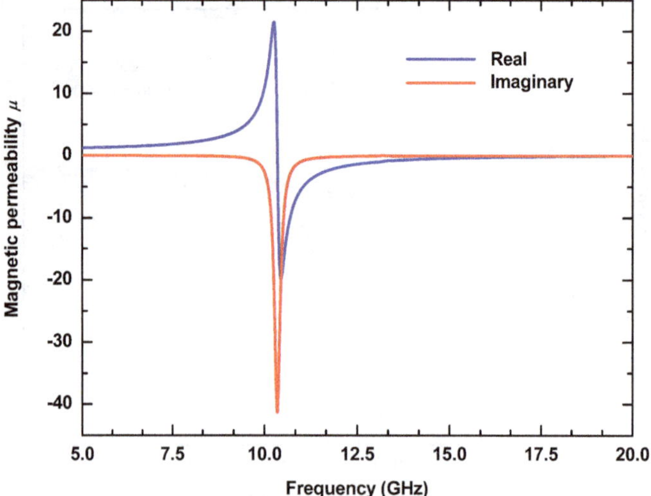

(**a**) Relative effective permeability for $S_2 = S_1$ and $C = 1$ pF.

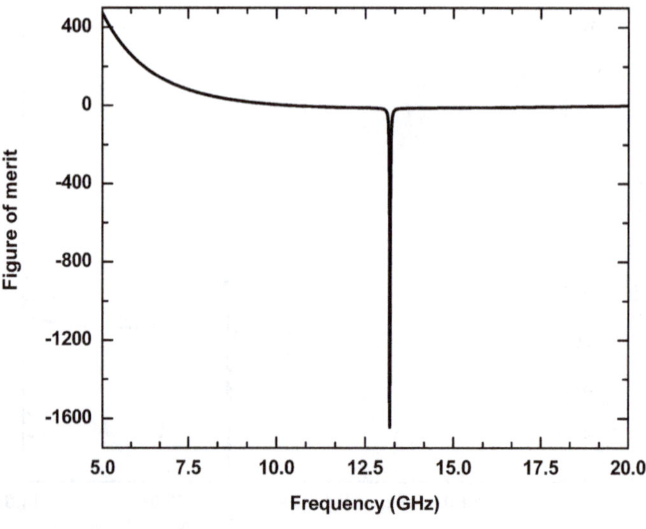

(**b**) Corresponding figure of merit.

Figure 5.11: In all cases, the electromagnetic and geometric parameters are those of Fig. 5.5.

on the use of active, negative-resistance, diode elements, such as Gunn or resonant tunnel diodes [Boardman *et al.* (2007)], which enable "cancelling" the metaparticles' ohmic losses and even leading to magnetic metamaterials with gain. However, though promising, such approaches normally require high field intensities and/or are accompanied by nonlinearities in the response of the engineered medium, which may not always be desirable. Moreover, their scaling from radio to visible frequencies (or vice versa) can be challenging. Clearly, a much more desirable and convenient

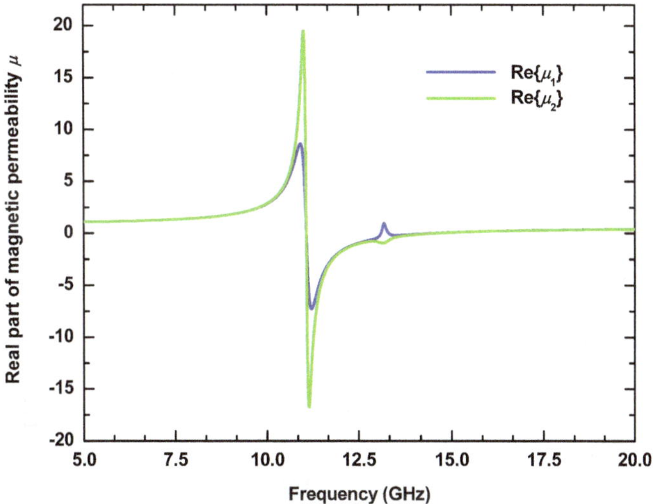

(a) Real parts of the relative effective permeabilities along the x-(μ_x) and the y-directions (μ_y).

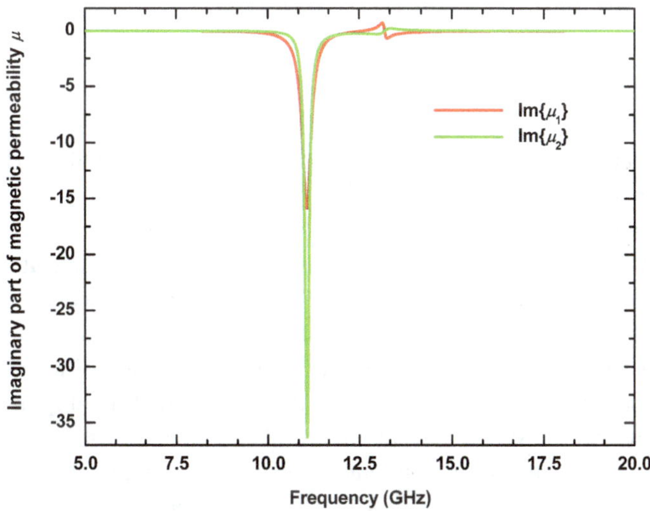

(b) Corresponding imaginary parts.

Figure 5.12: The electromagnetic and geometric parameters and arrangements are those of Fig. 5.7a.

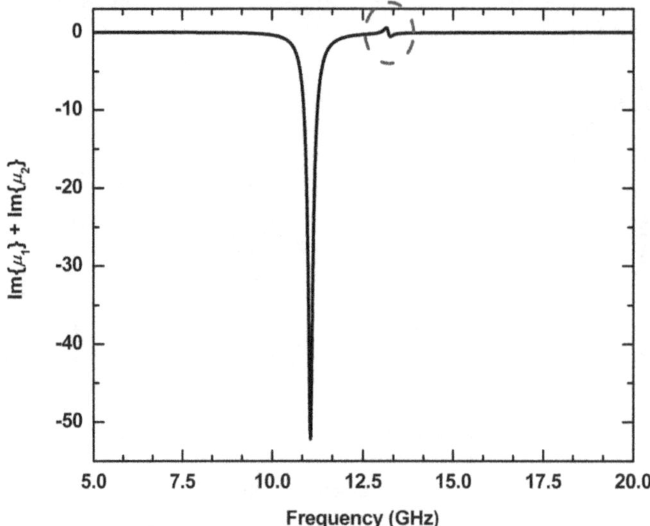

(**c**) Sum of the imaginary parts of μ_x and μ_y.

Figure 5.12: (*Continued*)

approach would be to judiciously redesign the structure of the metaparticles at the unit-cell level, such that it could open a "window" for harnessing perfectly lossless artificial magnetism over a continuous range of frequencies. This is the route that was followed here.

The approach presented herein is based solely on the exchange of active powers between the two electrically connected meshes inside each unit cell. This results in frequency regions wherein the sign of the active powers becomes negative (the meaning of which was explained in Section 5.4.2) and the imaginary parts of the susceptibilities associated with the two meshes become positive (for the convention $\exp^{i(\omega t - \mathbf{k} \cdot \mathbf{z})}$ spatiotemporal dependence that was assumed herein), but not at the same frequency region. When an RLC mesh related to a permeability with positive imaginary part is placed on a different plane plane (say xz) than the mesh it is electrically connected to, this results in lossless artificial magnetism for a plane wave that has its magnetic field component polarized perpendicularly to that plane (H_y) and propagating along any of the orthogonal axes (x or z) of that plane.

The scheme works by invoking (at the unit-cell level) a mechanism that manages to *naturally* "pump" magnetic energy in a specified direction – along which we obtain lossless metamaterial magnetism. As expected, the mesh that absorbs (receives) the electrical active power from its pair is associated with increased magnetic losses (in the same frequency region over which its pair is magnetically active), so that the conservation of energy is preserved. The scheme results in a light beam propagating without magnetic losses along a specified direction inside the metamaterial, whilst another beam, propagating perpendicularly to the first, naturally "supplies" the required "gain" to the first beam and is itself experiencing increased magnetic losses. Note the subtle, but important, conceptual difference between the present approach and those that rely on provision of optical or electrical gain: the latter approaches provide gain to cancel (or even reverse) the magnetic losses that are already present in the metamaterial. By contrast, the approach presented herein relies on "enforcing" one of the two meshes *not to absorb* the incident magnetic energy at all, but remit it to the mesh it is electrically connected with. As a result, one of the two meshes naturally becomes a "source" of energy, which is the reason behind the occurrence of positive imaginary part for the magnetic susceptibility associated with that mesh.

From the examples presented in the main text it should also be clear that the present scheme does not require large intensities for the incident fields — in fact, it works equally well even with low field intensities. Indeed, the results shown in Section 5.4.1 (Fig. 5.4) were obtained assuming incident magnetic fields having intensity of just 1 A/m. Obviously, similar results can be obtained with even smaller magnetic intensities, so long as they suffice to excite a useful or interesting collective magnetic response of the effective medium.

Furthermore, it turns out that the underlying mechanism (i.e., the exchange of active power between the meshes) that is responsible for enabling lossless artificial magnetism is very robust against the presence of ohmic losses, insofar as the *difference* in (*not the actual values of*) the ohmic resistances of the meshes is substantial — in the examples of the main text it was: $R_1 = 50 \; \Omega$ and $R_2 = 0.1 \; \Omega$. Indeed, from Fig. 5.13 one observes that even when the resistance in each mesh increases by an order of magnitude ($R_1 = 500 \; \Omega$, $R_2 = 1 \; \Omega$) there are, as before, regions wherein the imaginary parts of the effective permeabilities become positive. The same holds true for any increase in R_1 and R_2, to the extend that we do not enter the "overdamped" region wherein there is no effective magnetic oscillation (response) in the effective medium at all. Ultimately, this robustness against losses is, as highlighted before, owing to the fact that the underline mechanism relies critically on an *imbalance* in the values of the resistances of the pair of meshes, and not on the actual values of the resistances themselves. As long as such an imbalance is present (and provided that we are not in the "overdamped" regime) there will always be power flowing *away* from a mesh to its pair, resulting in algebraically negative active power for the mesh that remits this power and in, correspondingly, positive imaginary part of the associated magnetic susceptibility/permeability.

This suggests the remarkable (and counterintuitive) prospect of actually deploying "poor" conductors to create a perfectly lossless magnetic metamaterial in a specified direction (or directions, see Section 5.4.6). For instance, let us assume that an 1-DEG system, corresponding to the equivalent electrical circuit of a split ring resonator (SRR), uses a "good" conductors of resistance 0.1–1 Ω, which is typical for conductors in the GHz regime [Pendry *et al.* (1999)]. As we show in Section 5.2, a periodic arrangement of the single RLC mesh (or SRR) in this 1-DEG system

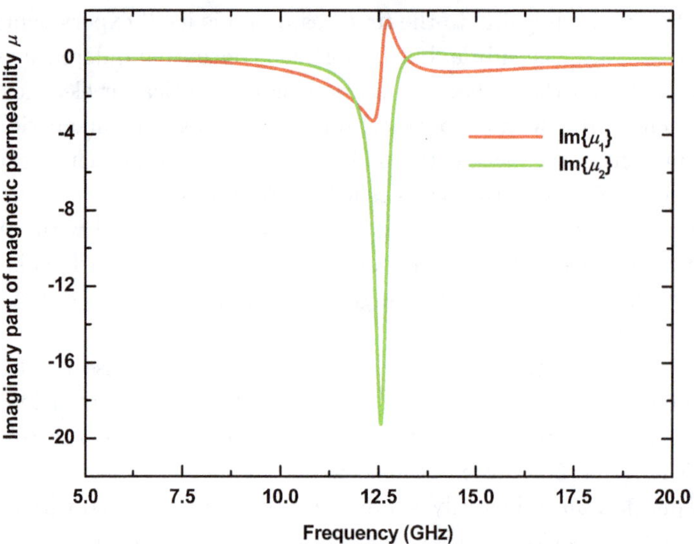

(a) Case of low-density (weakly interacting) unit cells.

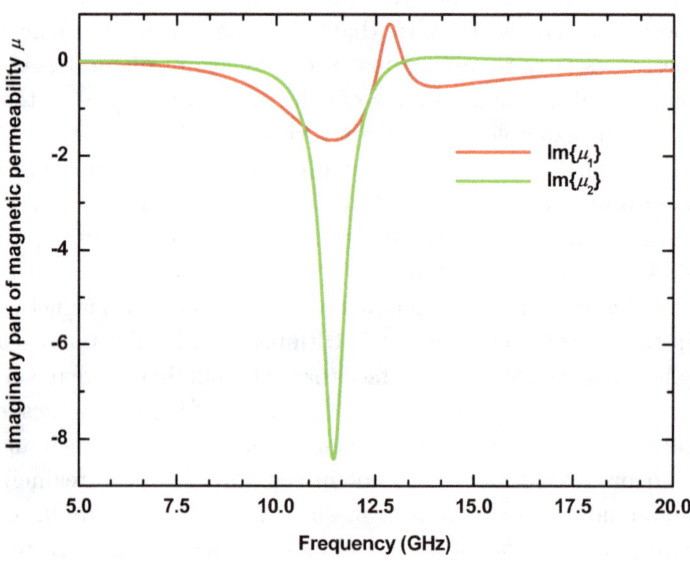

(b) Case of "tightly coupled" unit cells.

Figure 5.13: Imaginary parts of μ_x and μ_x for the electromagnetic and geometric parameters used in Fig. 5.4, but now with $R_1 = 500\ \Omega$ and $R_2 = 1\ \Omega$.

will, in the quasi-static regime, result in an effective medium exhibiting artificial magnetism [Pendry *et al.* (1999)], but limited by the presence of the ohmic losses. Instead of attempting to further reduce the losses present in this medium (by, e.g., using even better conductors for the single meshes), our analysis shows that one may (or, in fact, should) deploy a "poor" conductor of resistance 50–500 Ω to create a second RLC mesh on a different plane than the first. The imbalance in the values of the meshes' resistances residing at the two planes will give rise to power being exchanged between the two meshes which, according to what was explained above, will cause positive imaginary parts for the associated permeabilities over certain frequency regions, and therefore in perfectly lossless magnetism in the corresponding directions — in fact, even with the presence of gain.

The approach introduced here for creating lossless magnetic metamaterials can be most conveniently realized experimentally in the radio and microwave frequencies by using discrete lumped resistors, inductors and capacitors, or SRRs placed at different planes and made of different conductors. The scheme is also scalable down to optical frequencies, where magnetic metamaterials made of arrays of SRRs have already been demonstrated. A further method for the construction of the herein proposed structures at optical frequencies could be the use of discrete nanoresistors, nanoinductors and nanocapacitors [Engheta *et al.* (2005)], which have already been studied and were shown to hold promise in connection with the creation of optical nanocircuits [Engheta (2007)] and nanoantennas [Alu and Engheta (2002)].

Finally, it should be noted that the electrical connection of the meshes is a crucial aspect of the proposed mechanism for overcoming metamaterial losses, not only because it allows electrical power to flow and be exchanged more easily between the meshes, but also because it allows each magnetic field component to generate a magnetic moment at, both, the plane to which it is perpendicular *and* at a plane to which it is parallel. For instance, the H_y-field component generates currents circulating both meshes residing at the xz and yz planes. As a result, the H_y-field component induces a magnetic moment not only along the y-direction (perpendicularly to the xz plane), but also along the x-direction (perpendicularly to the yz plane, to which the H_y-field component is parallel). The corresponding is, of course, also true for the H_x-field component. This superposition of the magnetic moments generated, at both planes, by both **H**-field components is an essential feature of the present design. Note also that in order for the "superposition principle" to be in force, our present scheme not only is it not nonlinear, but it actually requires linearity and low or moderate field intensities to function according to its conception.

5.4.5 *Issues with the Homogenisation Method for the Case of "Tightly Coupled" Unit Cells*

As was noted in Fig. 5.12c, when the density of the unit cells is high ("tightly coupled" unit cells) application of Eq. (5.19) leads to a certain frequency region wherein wherein $\text{Im}[\mu_1] + \text{Im}[\mu_2] > 0$, implying the occurrence of "net" gain in that frequency region, i.e. that the magnetic gain that a beam experiences along, e.g., the x-direction is larger than the dissipation that a second beam experiences along, e.g., the y-direction. Such an outcome cannot be physically justified, since in our scheme we do not use an active medium or electronic element to provide gain to the structure. It is interesting to note that similar observations, i.e. occurrence of positive imaginary parts (corresponding to "gain") in the spectra of effective permittivities and permeabilities, have in the past also been reported in a number of theoretical [O'Brien and Pendry (2002); Markos and Soukoulis (2003); Koschny *et al.* (2003); Depine and Lakhtakia (2004); Efros (2004); Koschny *et al.* (2004)], but also experimental [Liu *et al.* (2007); Rockstuhl *et al.* (2008)] studies. However, in those cases such outcomes were not occurring at the unit-cell basis and were entirely an artifact of the periodicity of the structure, i.e. they were only appearing owing to the inappropriateness of describing a periodic medium in terms of a homogeneous effective medium for wavelengths that are not sufficiently larger compared to the periodicity of the engineered structure. As was noted in [Koschny *et al.* (2004)], in such a pseudo-effective medium, spatial dispersion (i.e., dependence of the effective electromagnetic parameters not only on the frequency, but also on the wavevector), as well as resonant-antiresonant coupling between the electric and magnetic responses of the medium, should not be ignored. Those effects, though substantially diminished, still persisted even for relatively large wavelengths [Koschny *et al.* (2004)].

As was explained in Section 5.4.1, as well as in the previous Section 5.4.4, in the designs presented herein the presence of positive imaginary parts in the spectra of the effective permeabilities is physically fully justified on the basis of the exchange of active power between the electrically connected meshes. The possibility that the observed effects could arise due to the periodicity of the structure was further eliminated by studying the magnetic moments of the individual unit cells and noting that that they possessed positive imaginary parts. No periodic boundary conditions were used at the edges of the cells, and the analytic, exact circuit calculations in each unit cell revealed the presence of active powers with negative algebraic sign. The herein reported positive imaginary parts in the spectra of the calculated effective permeabilities will clearly continue to occur even in the deep-subwavelength ("true" effective medium) regime, as long as frequency regions wherein $P_i < 0$, $i = 1, 2$ are suitably engineered. For the same reason, the effects reported here are, also, not related to a resonant-antiresonant coupling between electric and magnetic dipoles induced in the structure.

It should be noted that in our structures frequency regions wherein $\mathrm{Im}[\mu_1] + \mathrm{Im}[\mu_1] > 0$ *never* occur for the case of low-density, weakly interacting ("isolated") unit cells. This is simply because in this case the effective (relative) permeability of the structure along a particular direction is directly proportional to the magnetisation M ($\mu_{r,eff} = 1 + M/H_0$) and therefore, ultimately, directly proportional to the active power of the corresponding mesh, which takes on negative values in a certain frequency region. This observation is further verified in Fig. 5.14, which shows the frequency variation of the sum $\mathrm{Im}[\mu_1] + \mathrm{Im}[\mu_2]$ for the case of "isolated" unit cells and for the structure that we studied previously in Fig. 5.13a. Evidently, the aforesaid sum is negative throughout, in accord with what one would expect for a structure to which there is no net supply of gain.

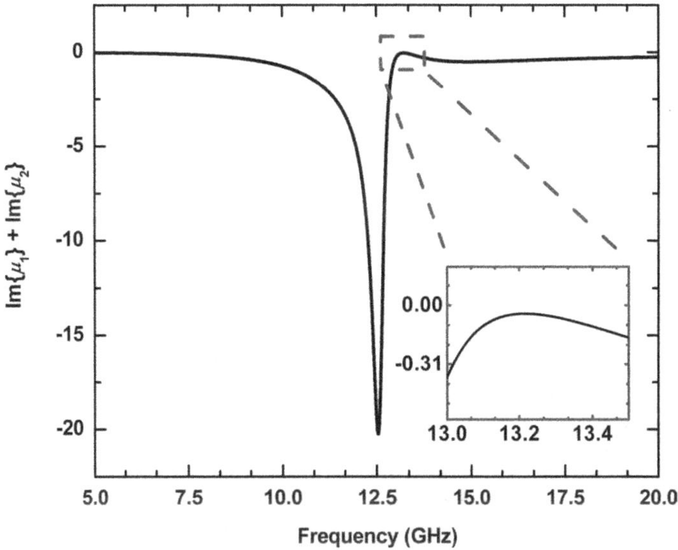

(a) Case of low-density (weakly interacting) unit cells.

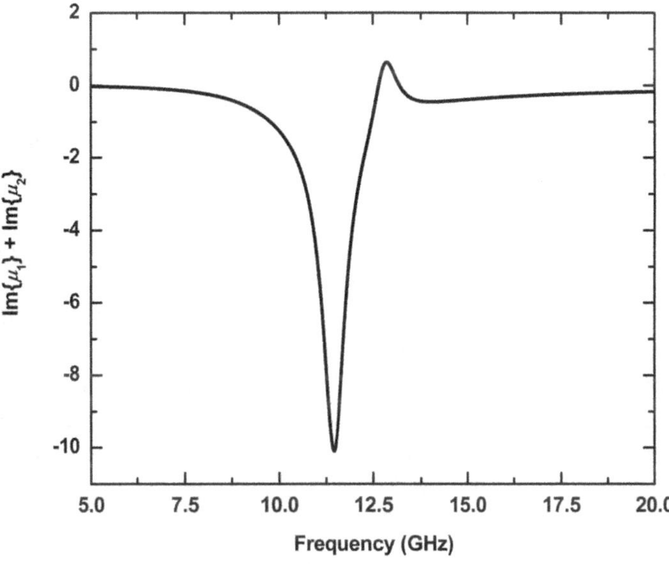

(b) Case of "tightly coupled" unit cells.

Figure 5.14: Sum of the imaginary parts of μ_x and μ_y for the electromagnetic and geometric parameters used in Fig. 5.4, but now with $R_1 = 500\ \Omega$ and $R_2 = 1\ \Omega$.

Figure 5.14b reports the variation of the same quantity with frequency, but now for the case of strongly interacting cells. Here, because of the damping of the magnetic resonance caused by the increased values of resistances in each mesh, the region wherein $\mathrm{Im}[\mu_1] + \mathrm{Im}[\mu_2] > 0$ is more clearly seen. Before we proceed into furnishing a plausible explanation for this observation, it is important at this point to reiterate that the mechanism for overcoming losses and producing magnetic gain in our structure has a *local* origin, i.e. it occurs because we are able to design the meshes in *each* unit cell so that they do not consume active power in certain frequency regions, but remit it to each other. As a result, *each* individual unit cell does not absorb but amplifies the corresponding magnetic flux density component. The working principle of our scheme does not rely at all on (destructive) interference amongst the cells to eliminate losses, i.e. it does not have a "global" origin, but a much stronger "local" one, occurring at the unit-cell level. It follows that regardless of how strongly or weakly do the unit cells interact with each other or the precise form of their interaction, the collective (effective) medium must necessarily also exhibit zero absorption of the magnetic flux density component in a specified direction (and, also, $\text{Im}[\mu_1] + \text{Im}[\mu_2] > 0$, so that the conservation of the magnetic energy is honoured). As a result, the reason for the presence of frequency region wherein $\text{Im}[\mu_1] + \text{Im}[\mu_2] > 0$ should be traced to the limitations of Eq. (5.19) in describing such a medium.

Such an explanation, indeed, becomes plausible when one considers the precise methodology that should be followed in assigning bulk electromagnetic parameters in a medium. The local quasi-static electromagnetic field components will be owing to, both, the incident field and the field *scattered* and/or induced by the neighbouring metaparticles. For the methodology to be self-consistent, one should start by computing the electric and magnetic multipoles induced by the incident field [Waterman and Pedersen (1986)]. In doing so, one expands the electric and magnetic dipoles in terms of the incident electromagnetic field components and their derivatives, and the electric quadrupole tensor in terms of incident electric field. One then proceeds by determining the multipolar coefficients in the previous expansion by deploying, e.g., time-dependent quantum perturbation theory. In our case, the determination of the scattered fields should also take into account the electrical connectivity of the equivalent meshes which, as we show, results in (active and reactive) power being exchanged between them. However, the most important point for our discussion here is that, unless the unit cells are very weakly interacting, the higher-order multipolar terms cannot be ignored, even in the long-wavelength, quasi-static regime [Barron (2004)]. Moreover, it turns out that these terms further depend on the origin of our coordinate system, a point that requires careful treatment before meaningful effective-medium parameters can be assigned. Such a methodology should –according to what was explained above– result in the summation $\text{Im}[\mu_1] + \text{Im}[\mu_2] < 0$ at every frequency point, as is the case with the "isolated" cells. Here, however, it is the prospect of overcoming losses in metamaterial that is investigated, leaving the detailed development of a more appropriate homogenisation methodology to be the subject of a future work.

5.4.6 *Systems with M Degrees of Freedom*

The notion of, so-called, systems having M degrees of freedom is a one that is very frequently encountered in diverse realm of science, from civil or mechanical engineering to quantum mechanics. In these situations, the properties (e.g., movement or oscillation) of a multi-degrees-of-freedom system in the real or in the phase space are described with the aid of M independent parameters, $u_1 - u_m$, known as *generalised Lagrange coordinates* of the system. The number of these parameters (or "coordinates") depends upon the particular form or structure of the system, on the way it is excited, as well as on the required accuracy. Generally, increasing the number of the independent parameters in the description of a system also increases the accuracy of the obtained results. As a result, there is only a limited number of cases where, e.g., an infinite-degrees-of-freedom system (also know as a *continuous* system), such as a transmission line, is described in terms of an 1-DEG equivalent system. By contrast, with the aid of a relatively small number of independent parameters, the description of a continuous system in terms of an M-DEG system can normally be accommodated with sufficient accuracy.

Figure 5.15 schematically illustrates an example of a 3-DEG system, frequently encountered in the realms of, e.g., civil or mechanical engineering. It shows the three main modes of oscillation of 3-DEG oscillator composed of three masses, m_1–m_3, which are elastically attached into a vertical pole. The latter may, e.g., be simulating a concrete rod in a building or a shaft joining parts of a machine. It turns out [Wahab (2008)] that the 3×3 matrix describing the movement of this system is remarkably similar in its structure to the matrix resulting from the application of the "mesh current method" to an electrical circuit of three coupled RLC meshes. This is simply owing to the well-known analogy between mechanical and electrical oscillators, which is also applicable in the case of infinite-degree-of-freedom (continuous) systems.

The system that we introduced in this work for overcoming losses in metamaterials consisted of two electrically connected meshes; hence, it was described in terms of two independent parameters (I_1 and I_2) and the resulting matrix was 2×2, i.e. the system was a 2-DEG one. This arrangement enabled lossless propagation for

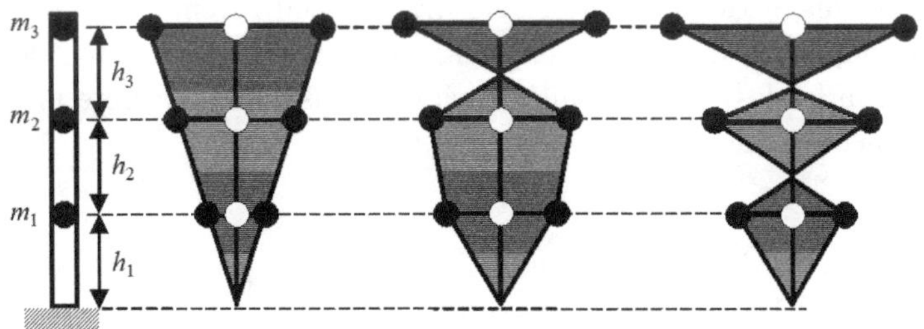

Figure 5.15: Main modes of oscillation of a 3-DEG vertical rod oscillator.

one component of the magnetic flux density (propagating along any of the allowed two directions, orthogonal to the direction of this component) at a given frequency region — actually, as we show in Figs. 5.7, 5.12 and 5.13, the propagation could, indeed, be made lossless for *two* magnetic flux density components, but this occurred at different frequency regions for each component.

It is possible, however, to envisage a system with a larger number of meshes being electrically connected, e.g., a system wherein each mesh resides at a separate plane in a cubical cell, i.e. a 6-DEG system. Depending on the design, the structure may, of course, posses even more degrees of freedom (e.g., a 12-DEG system) if more than one meshes are electrically connected in each side of the unit cell. We speculate that with a judicious choice of the electrical (R, L, C) parameters, such a *discrete* (i.e., non-continuous) M-DEG system may enable lossless propagation for two out of the three orthogonal component of the magnetic flux density, in the *same* frequency region. This can occur as long as there is sufficient imbalance in the values of the resistances of the meshes to allow for active power to flow away from the meshes residing at two orthogonal planes (e.g., the xz– and yz–planes), towards the meshes residing at the third plane or planes (e.g., the xy–plane). In this manner, the meshes at the xz– and yz–planes will, at the same frequency region, act as "sources" of electrical power, i.e. they will not absorb the incident magnetic energy but they will remit it to the meshes residing at the xy–plane. It follows that the imaginary parts of the magnetic susceptibilities and permeabilities associated with the meshes at the xz– and yz–planes will be positive in the aforementioned frequency region (or regions), i.e. the structure will be magnetically active for the magnetic flux density components perpendicular to the xz– and yz–planes.

Finally, it should be emphasised that the present scheme requires that the meshes in each unit cell are *not* electrically connected to the meshes of their neighbouring cells, i.e. the system should be discrete at the unit-cell level. If the meshes of neighbouring unit cells are electrically connected then, in the long-wavelength regime, the system becomes a continuous one (such as, e.g., a backward transmission line). These systems have been well-studied in the recent past [Eleftheriades and Balmain (2005)] and, though they allow for the attainment of broadband and relatively low-loss metamaterials, they do not exhibit regions of positive imaginary parts for their effective electromagnetic parameters. As was detailed herein, the latter feat is ultimately accomplished due to the exchange of active power amongst the meshes, which results in lossless magnetism in each of the isolated, discrete unit-cells that "built" the effective medium.

Chapter 6

"Trapped Rainbow": Stopping of Light in Metamaterials

6.1 Introduction

Light usually propagates inside transparent materials with well-known ways that science students are taught. However, as we show in the previous chapter, researchers have recently been examining the possibility of taking a normal transparent material and inserting tiny metallic inclusions (meta-molecules/particles) of various shapes and arrangements. As light passes through these structures, oscillating electric currents are set up that generate electromagnetic field moments (see, e.g., Eq. (5.13)), which modify the way the light travels through the material. The effects can be dramatic leading to light propagation or bending in very unusual ways (e.g. negative refraction, Section 3.4). As we saw in Section 3.6, lenses that break traditional diffraction limits are one possibility [Pendry (2000)]; an "invisibility cloak" another [Pendry *et al.* (2006); Schurig *et al.* (2006)].

Significantly less research has focused on the potential of such structures for slowing, trapping and releasing *broadband* light signals. In this chapter, we shall show that an axially varying heterostructure with a core of negative refractive index metamaterial can be used to efficiently and coherently bring *broadband* light to a *complete* standstill. We will see that in stark contrast to previously proposed schemes for decelerating and storing light [Hau *et al.* (1999); Kash *et al.* (1999); Bigelow *et al.* (2003a,b); Liu *et al.* (2001); Gehrig *et al.* (2006); Vlasov *et al.* (2005); Gersen *et al.* (2005); Okawachi *et al.* (2005); Stockman (2004); Karalis *et al.* (2005)], the present one simultaneously allows for high in-coupling efficiencies and broadband, room-temperature, operation. We shall see that at a critical point the effective thickness of the waveguide reduces to *zero*, preventing the lightwave to propagate further. At this point, a light ray is found to be permanently trapped, its trajectory forming a double light-cone that we call an "optical clepsydra". Each frequency component of a wave packet is stopped at a different guide thickness, leading to the spatial separation of the packet's spectrum and the formation of a "trapped rainbow". The results in this chapter, therefore, bridge the gap between two important contemporary realms of science, that is metamaterials and slow light, and may open a host of combined investigations. Such macroscopic control of photons may, also, conceivably find applications in optical data processing and storage and in the realisation of quantum optical memories.

Section 6.2 will present the main theory behind the slowing and stopping of light in metamaterial waveguides, with the remaining Sections 6.3–6.6 presenting and elaborating remarks on the derivations of the main results described in Section 6.2. The case of lossy structures is analyzed in detail later on, in Chapter 7.

6.2   Broadband Stopping of Light in Metamaterial Waveguides

For decades scientists were arguing [Blumenthal *et al.* (1994)] "optical data cannot be stored statically and must be processed and switched on the fly". The reason for this conclusion was that stopping and storing an optical signal by dramatically reducing the speed of light itself was thought to be infeasible. Undeniably, the absence of any form of interaction between photons and other elementary particles, as well as their enormous speed, makes confining them to a finite volume by reducing their velocity down to zero excessively difficult. However, in recent years a series of major scientific breakthroughs contributed towards annulling such assertion and proved conclusively that it is, indeed, possible to bring light to a complete standstill. Amongst others, electromagnetically induced transparency (EIT) [Hau *et al.* (1999); Liu *et al.* (2001)], quantum-dot semiconductor optical amplifiers (QD-SOAs) [Gehrig *et al.* (2006)], photonic crystals (PhCs) [Vlasov *et al.* (2005); Gersen *et al.* (2005)], coherent population oscillations (CPOs) [Bigelow *et al.* (2003a,b)], stimulated Brillouin scattering (SBS) [Okawachi *et al.* (2005)] and surface plasmon polaritons (SPPs) [Stockman (2004); Karalis *et al.* (2005)] in metallodielectric waveguides have been proposed as means of producing "slow light". However, so far most of these methods bear inherent limitations that may hinder their practical deployment. For instance, EIT uses ultracold atomic gases and not solid state materials, QD-SOAs usually allow for only modest delays but for potentially ultra-broadband light pulses, CPOs and SBS are very narrowband owing to the narrow transparency window of the former and the narrow Brillouin gain bandwidth (around 30 MHz in standard single-mode optical fibres) of the latter, SPPs are very sensitive to surface roughness and are relatively difficult to excite, while PhCs are normally highly multimodal [Gersen *et al.* (2005)]; this, combined with the strong impedance mismatch in the "slow-light regime" causes launching the incoming light energy to a single, slow mode alone overly difficult [Vlasov and McNab (2006)].

As we saw in Chapters 3 and 5, during the same period and in parallel with the above advances a different, wide-ranging realm of contemporary research has also been developing. It largely followed from a sequence of works by Pendry, Smith and co-workers, wherein they proposed practical means for realizing negative refractive index metamaterials [Veselago (1967, 1968)] and meticulously demonstrated their operation [Pendry *et al.* (1998, 1999); Smith *et al.* (2000); Shelby *et al.* (2001b,a)]. These materials offer a completely new perspective to the optical world and provide additional "degrees of freedom" in the design of photonic devices [Pendry *et al.* (2006)], thereby allowing unprecedented control over the flow of light.

A "perfect" lens [Pendry (2000)], highly anisotropic electromagnetic "cloaks" that render objects invisible to incident radiation [Pendry *et al.* (2006); Schurig *et al.* (2006)], as well as focusing of electron de Broglie waves by sharp p-n junctions in graphene [Cheianov *et al.* (2007)] are a few characteristic examples. Moreover, there is by now compelling evidence that, via building on familiar transmission-line concepts borrowed from microwave analysis, such materials can be designed to exhibit broadband negative-index behaviour and relative robustness to losses, all through the radio [Wiltshire *et al.* (2001)] up to the visible [Shalaev (2007)] regime.

In this work we bring together the realms of metamaterials and slow light as a result of studying the stimulating physics associated with wave propagation in slowly, spatially varying negative-index heterostructures. To gain an insight into the physics of the problem at hand, let us for a moment imagine a ray of light zigzag propagating along a waveguide with a left-handed core. The ray experiences negative Goos-Hänchen lateral displacements [Berman (2005)] (see Eq. (3.15)) each time it strikes the interfaces of the core with the right-handed claddings. Accordingly, the cross points of the incident and reflected rays will sit *inside* the LH core and the effective thickness of the guide will be *smaller* than its natural thickness. It is reasonable to expect that by gradually reducing the core physical thickness, the dotted border lines indicating the effective thickness of the guide will eventually meet, i.e. the effective thickness of the guide, t_{eff}, will vanish. Obviously, beyond that point the ray will not be able to propagate further down, and will effectively be trapped inside the left-handed heterostructure (LHH). For nonradiative trapping of the ray, we hope that the point where the effective thickness vanishes will not occur below the lower cutoff frequency of the guide. If this condition is fulfilled, then one of the frequencies of a guided wave packet will be completely arrested. The arresting points for the other frequency components will be continuously spaced out. Thereby, one will in principle have a means of stopping light in its track and storing it indefinitely, without radiation, over a whole range of frequencies. Such an approach will also be totally immune to atomic decoherence effects [Milonni (2005)] and will not require atomic media nor group index resonances.

In the following, we show that efficient, broadband, nonradiative storage of light can be realized inside an LHH. To this endeavour, we consider an axially nonuniform, linearly tapered, planar waveguide with a core of lossless (see Section 5.4), isotropic and homogeneous negative-index metamaterial, bounded asymmetrically by two positive-index media, as illustrated in Fig. 6.1. The simultaneous negativeness of ε and μ in the left-handed core allows for the existence of efficiently excitable (oscillatory) waveguide modes (see Section 3.8) [Tsakmakidis *et al.* (2006a)], which are not supported by single-negative (SNG) plasmonic structures. We arrange the variation of the core half thickness a with distance z to be appropriately slow [Snyder and Love (1983)] (adiabatic), $da/dz < 0.05 \min \left(a k_0 (n_{eff,2} - n_{eff,3})/(2\pi) \right)$, with $a k_0$ being the reduced guide thickness, k_0 the free space wavenumber and $n_{eff,2}$, $n_{eff,3}$ the effective indices of the 2nd- and 3rd-order oscillatory modes, so that the power of a local mode is conserved along the structure.

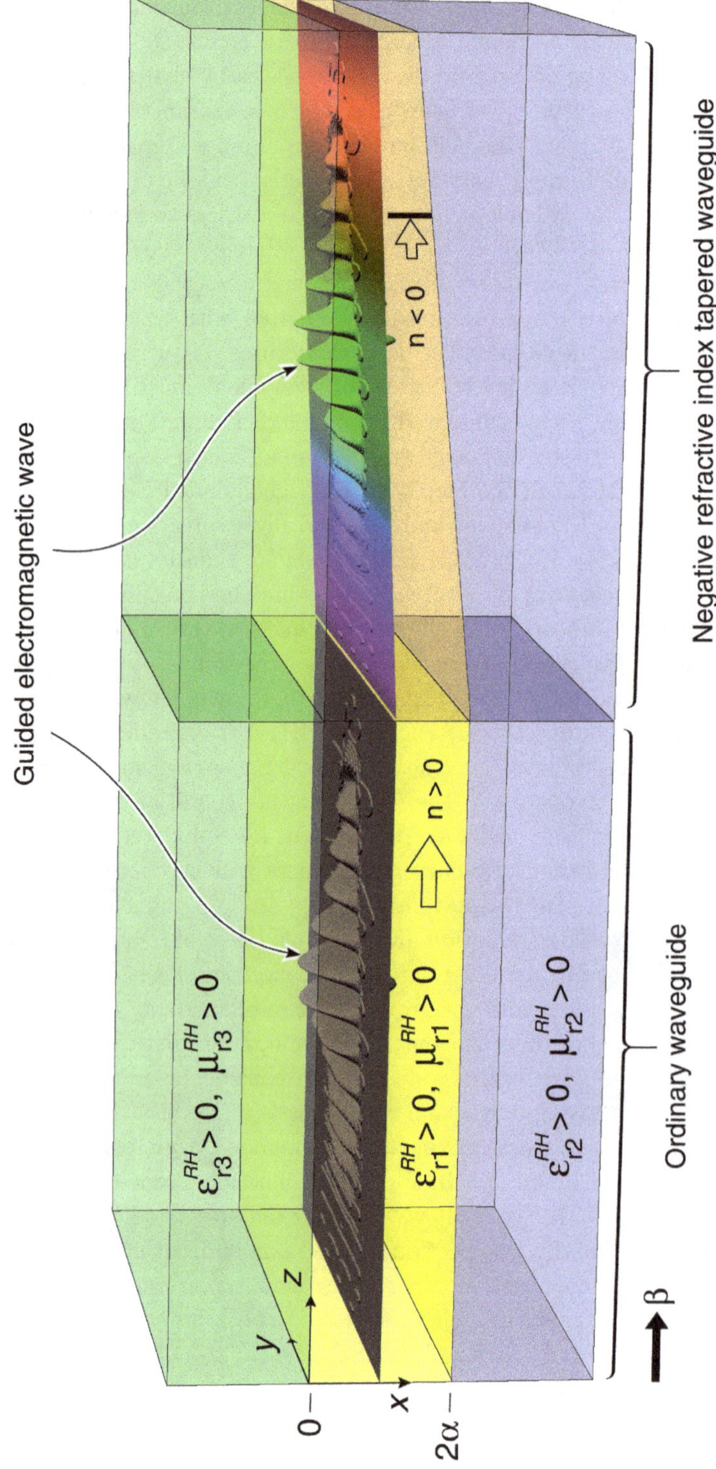

Figure 6.1: Trapped rainbow. Different frequency components of a guided wavepacket stop at correspondingly different thicknesses inside a tapered LHH.

explanations for Fig. 6.1

The upper (green) and bottom (cyan) dielectric layers, generally, vary between the right- and left-handed (RH and LH) waveguides. The thick empty arrows reveal the direction of power flow propagation (P_{tot}^{+z}), while the thin black arrows show the direction of phase propagation (β). A guided wave packet is efficiently injected from the ordinary waveguide to the LHH, inside which propagates smoothly owing to the slow (adiabatic) reduction in the thickness of the core. Inside the RHH P_{tot}^{+z} and β are parallel, but become oppositely directed in the LHH. The guided wave is altogether halted within a continuous range of "critical" thicknesses, as designated by the short vertical line inside the LH core. The smallest (red) frequency components of the wave are stopped at the smallest core thicknesses, while the largest (blue) components stop at correspondingly larger core thicknesses. Thereby, the spectrum of the oscillatory field will be spatially decomposed into its frequency constituents, similar to the decomposition of sunlight when it illuminates water droplets and the appearance of a rainbow. The inset shows an example of the dependence of the "critical" thicknesses of the LHH on the spectrum of the wave packet. The optogeometric parameters used in this example are those of Fig. 6.2.

One may then analytically show (see Section 6.3) that the electromagnetic fields at a position z_t are given by:

$$\mathbf{G}(x, y, z_t, t) = F(z_t)\mathbf{g}(x, y, \beta(z_t)) \exp\left(i \sum_{z=0}^{z=z_t} \beta(z_t)\Delta z - i\omega t\right). \tag{6.1}$$

Here, $\mathbf{G} = \mathbf{E}$ or $\mathbf{H}$, $\mathbf{g}$ is the solution to the E- or H-field vector wave equation for $z = z_t$, β is the longitudinal propagation constant, ω is the wave angular frequency and F is an appropriate factor, which carries all the information about the mode energy conservation and is calculated as:

$$F = 2\sigma_\varepsilon U \left[\frac{\omega\varepsilon_0|\varepsilon_{r1}|P_{tot}^{con}}{a\beta(W_3^2 + \sigma_\varepsilon^2 U^2)\Theta}\right]^{1/2}, \tag{6.2}$$

where $\sigma_\varepsilon = \varepsilon_{r3}/|\varepsilon_{r1}|$, $U = a\kappa$, $W_3 = a\gamma_3$ are the reduced transverse and layer-3 decay constants, respectively (κ and γ_3), $P_{tot} = \sum_{i=1}^{3} P_i$ is the conserved total time-averaged power flow with P_i ($i = 1, 2, 3$) being the time-averaged power flow in layer i, propagating in the increasing z direction, and:

$$\Theta = 2 + \frac{\rho_\varepsilon}{W_2}\frac{U^2 - W_2^2}{W_2^2 - \rho_\varepsilon^2 U^2} + \frac{\sigma_\varepsilon}{W_3}\frac{U^2 - W_3^2}{W_3^2 - \sigma_\varepsilon^2 U^2}, \tag{6.3}$$

$\rho_\varepsilon = \varepsilon_{r2}/|\varepsilon_{r1}|$ and $W_2 = a\gamma_2$ being the reduced decay constant in the lower layer (γ_2 is the unnormalised decay constant for layer 2).

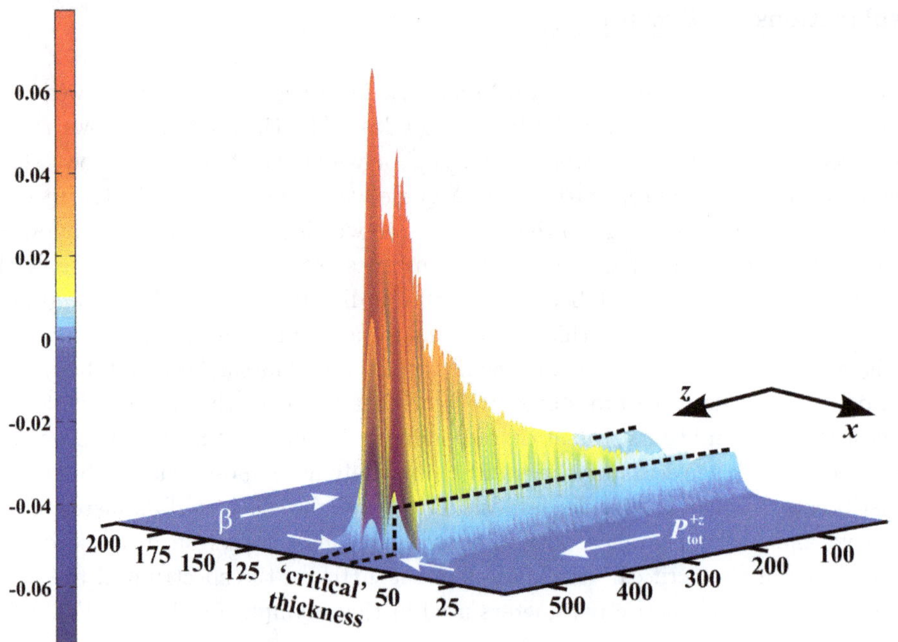

Figure 6.2: Wave propagation in an axially varying left-handed heterostructure.

explanations for Fig. 6.2

The optical parameters of the structure are $\varepsilon_{r1}^{\text{LHH}} = -5$, $\varepsilon_{r2}^{\text{LHH}} = 2.56$, $\varepsilon_{r3}^{\text{LHH}} = 2.25$, $\mu_{r1}^{\text{LHH}} = -5$, $\mu_{r2}^{\text{LHH}} = \mu_{r3}^{\text{LHH}} = 1$. Shown is a snapshot of the propagation of the monochromatic ($f = 1$ THz) p-polarised magnetic-field component, which enters the LHH from the wide-thickness end ($ak_0 = 1.7$) and stops at the pre-arranged "critical" thickness of $ak_0 \approx 0.55$. The light signal carries a total power $P_{tot}^{con} = 82.6$ μW/m^2/sec, which to an excellent approximation is assumed to be unaffected by the negligibly small reflections from the approaching media interfaces. The light signal propagates inside the LHH with its group $\mathbf{v}_g$ and phase $\mathbf{v}_{ph}$ velocities being antiparallel. The group velocity is parallel to the *total* power flow P_{tot}^{+z}, while the phase velocity is parallel to the longitudinal propagation constant β. As the lightwave propagates, its amplitude progressively increases; at the critical core thickness it has increased by a factor of 4 compared to the amplitude of the field at the wide-thickness end. Moreover, we find that the exponentially decaying extension of the H_y-field field inside the lower dielectric layer increases, while the wavelength gradually reduces and the field becomes spatially compressed.

One can judiciously choose the optical parameters of the left-handed heterostructure (LHH) to facilitate total suppression of all surface polariton (SP) modes, i.e. the minimum thickness of the core layer can be chosen above the upper cutoff of the parasitic SP mode, so that only oscillatory guided modes may exist [Tsakmakidis *et al.* (2007, 2006c, 2009)]. For such a structure, Fig. 6.2 furnishes an example of *ab initio* calculations, following the previously outlined methodology, of the propagating, monochromatic, p-polarised magnetic-field component at a time instant $t = 0$. We find that the magnitude of the total time-averaged power flow propagating in the positive $+z$ direction $P_{tot}^{+z} = \frac{1}{2} \int_{-\infty}^{\infty} \mathrm{Re}(\mathbf{E} \times \mathbf{H}^*)_z \, dx$, gradually drops off until it totally peters out, even though P_{tot}^{con} is conserved along the nonuniform waveguide.

Accordingly, one discovers that whilst the guided oscillatory fields propagate along the structure, with their phase ($\mathbf{v}_{ph}$) and group ($\mathbf{v}_g$) velocities being antiparallel, the group and energy ($\mathbf{v}_E$) velocities [Loudon (1970)] progressively decrease, eventually becoming zero at a "critical", pre-determined, guide thickness. At this point, the fields are slowly spatially compressed and amassed [Tsakmakidis *et al.* (2007)], with their total amplitude increasing by a factor of 4 for the present set of optogeometric parameters. The exponentially decaying extension of the field inside the lower positive-index layer (medium 3) also progressively increases, as anticipated. Note that in a regular dielectric waveguide the monochromatic oscillatory field would have been reflected and radiated off at the guide cut off thickness. In that case, an increase in the total amplitude of the field would have been the result of interference between the two lightwaves propagating in opposite directions. In the present LHH case, however, the wave travels solely in the positive z direction and completely stops upon reaching the "critical" guide thickness; hence, neither reflection nor radiation or interference occurs.

The trapping of light is more intuitively recognized by tracing the trajectory of a light ray inside the core for guide thicknesses nearly equal to the critical one mentioned above. Recall that the variation of the core refractive index with distance z is much smaller than the ray half period z_{hp}, so that locally the guide appears practically uniform. Let us assume that the light ray arrives at the 1-3 media interface of the LHH with an angle θ (Figs. 6.3c–e). Following reflection from this interface, the ray experiences a Goos-Hänchen phase shift [Berman (2005)] $\delta_{p13} = -2\tan^{-1}\left(W_3/(\sigma_\varepsilon U)\right)$, which is equal in magnitude but opposite in sign to the shift of the corresponding positive-index case. Detailed calculations then reveal that the distance x_{p13} between the cross point A of the two rays (Figs. 6.3c–e) and the 1-3 media interface is (see Section 6.3) $x_{p13} = (a\sigma_\varepsilon V_3^2)/(W_3(W_3^2 + \sigma_\varepsilon^2 U^2))$, where $V_3^2 = U^2 + W_3^2$. Likewise, the distance x_{p12} between point B and the 1-2 interface is $x_{p12} = (a\rho_\varepsilon V_2^2)/(W_2(W_2^2 + \rho_\varepsilon^2 U^2))$, where $V_2^2 = U^2 + W_2^2$. One is, thus, led to discern (Fig. 6.3c, blue line) that the light ray is effectively altogether confined within the middle region of thickness t_{eff} wherein it repeatedly bounces off points A and B.

explanations for Fig. 6.3

(a) Variations of normalised effective guide thickness t_{eff}/a (solid blue line), conserved $-P_{tot}^{con}$ (dotted green line) and forward P_{tot}^{+z} (solid red line) total time-averaged power flow with the reduced guide thickness ak_0. The forward component (red) of the conserved power flow (green) gradually decreases in magnitude, until it becomes exactly zero at the critical core thickness. At this point, the effective thickness of the LHH also vanishes. For larger guide thicknesses, t_{eff}/a tends asymptotically to the value of 2. The insets associate characteristic regions of P_{tot}^{+z} with the ray analysis results shown in (c)–(e).

In (b)–(e), a ray of light (black), which here signifies power propagation, hits the media interfaces with an angle θ while propagating down the waveguide, and experiences a Goos-Hänchen (GH) lateral displacement. The black dotted arrows denote the evanescent field power flow from the optically denser core to the rarer claddings. In a regular dielectric waveguide (a), the GH phase shifts are positive, the core appears to extend (dotted orange lines) inside the cladding layers and the effective guide thickness is always larger than the physical core thickness $2a$.

In the slowly varying LHH case (c)–(e), the thickness of the core remains practically constant over many ray periods owing to the slow variation criterion. Here, the ray experiences a negative, i.e. antiparallel to P_{tot}^{+z}, later displacement originating from the reversed power flow from the core to the claddings and the associated negative GH phase shift.

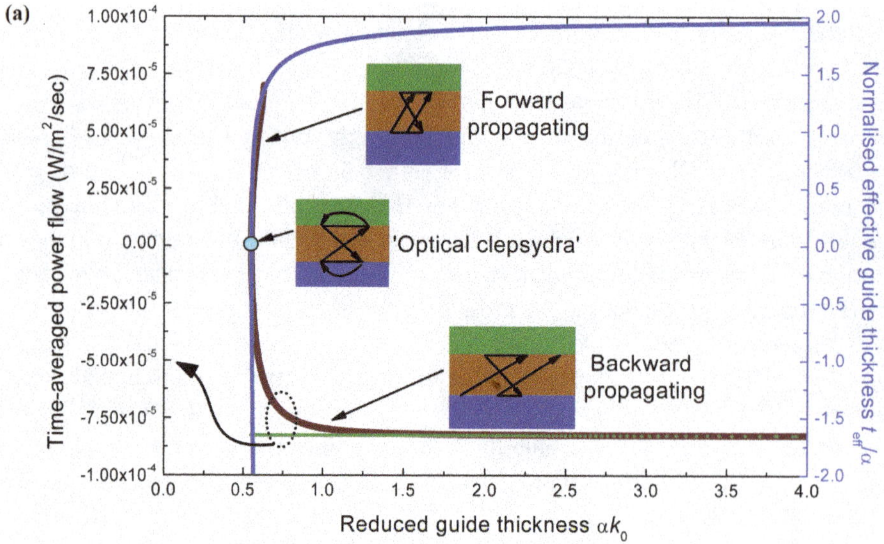

Figure 6.3: (a) Ray analysis reveals that the effective LH guide thickness is smaller than the physical thickness and can become zero or even negative.

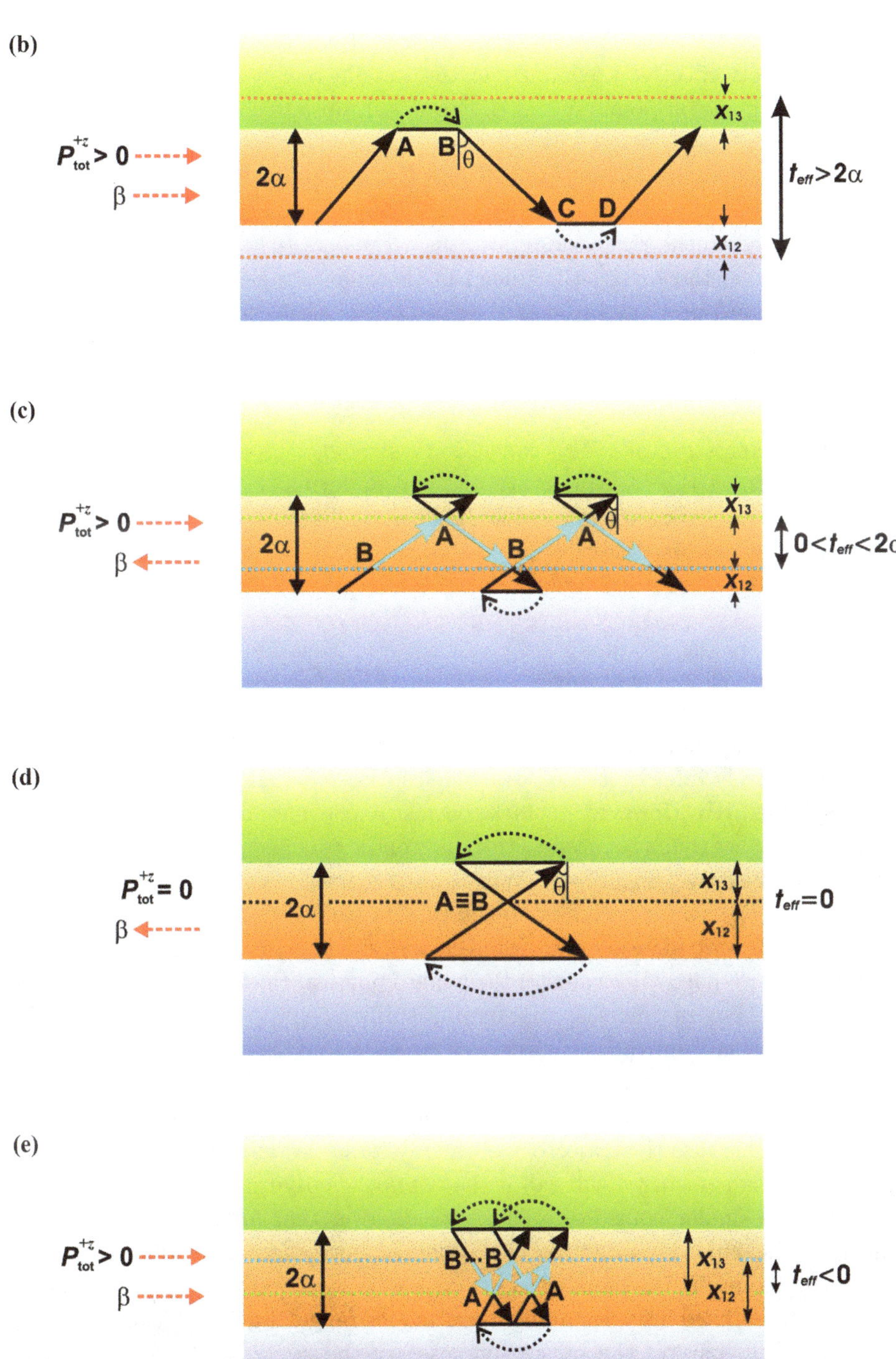

Figure 6.3: (b)–(e) (*Continued*)

In (b), the shifts are relatively small, such that $x_{12} + x_{13} < 2a$. Accordingly, the ray (blue) is effectively confined in the middle region of thickness t_{eff}, by repeatedly bouncing off points A and B. For an appropriate choice of optogeometric parameters (c), the two phase shifts can become such that $x_{12} + x_{13} = 2a$. exactly. In this case, the effective thickness of the guide (t_{eff}) vanishes and the ray becomes permanently trapped, forming a double light cone ("optical clepsydra"). For even larger, compared to the ray period, later displacements (d) the effective thickness becomes negative; the ray is still guided in the middle region of thickness $|t_{eff}|$, but now is forward-propagating, i.e. the direction of phase propagation is parallel to the direction of power flow.

We argue that $t_{eff} = 2a - x_{p12} - x_{p13}$ is the effective thickness of the left-handed waveguiding heterostructure. This conclusion is strongly supported by the preceding remarks concerning the ray trajectory and can be formally established by noting that the total time-averaged power flow P_{tot}^{+z} can be directly linked to t_{eff} through the following relation (see Section 6.4):

$$P_{tot}^{+z} = \frac{1}{4} E_x^{max} H_y^{max} t_{eff}, \tag{6.4}$$

where E_x^{max} and H_y^{max} are the maximum values of the E_x- and H_y-field components in the guide, respectively. The similarities between Eq. (6.4) and its counterpart in the case of conventional, right-handed heterostructures (RHH) [Tamir (1979)], allow one to conclusively infer that t_{eff}, as defined in Eq. (6.4) is, indeed, the effective thickness of the LHH. However, in stark contrast to conventional waveguides, t_{eff} is here *always* smaller than the physical thickness of the core and can become zero (Fig. 6.3d) or even negative Fig. 6.3e). When t_{eff} becomes negative, we deduce from Eq. (6.4) that P_{tot}^{+z} and β become parallel; thereby, the corresponding guided mode will be a forward-propagating one. Interestingly, based on Fig. 6.3a and Eq. (6.4) we infer that for a particular value of the guide's physical thickness $2a$, the effective thickness vanishes. In this case, the lateral shifts x_{p12}, x_{p13} experienced by the ray upon reflection from the two interfaces are such that $x_{p12} + x_{p13} = 2a$ exactly. For the aforementioned "critical" physical thickness, the light ray is permanently trapped inside the LHH, being unable to propagate further down. From Fig. 6.3d we see that in this case the trajectory of the ray forms a double light-cone. In view of its characteristic form we will call it the "optical clepsydra".

Following a similar course of analysis we discover that for different excitation frequencies the guided oscillatory fields stop at correspondingly different guide thicknesses. Accordingly, a guided electromagnetic wave packet can be altogether trapped within a fixed area, spanning a continuous range of guide thicknesses. The leading (trailing) part of the pulse, composed of the *smallest* (*highest*) frequencies (Fig. 6.1), stops at the smallest (highest) guide thicknesses. Thereby, in the small-intensity, linear case wherein the propagating spectral power densities of a "white" wave packet do not couple [Snyder and Love (1983)] and the guided field is a linear, weighted sum of its single frequency constituents, the "red" and "blue" components of the field will be spatially separated (Fig. 6.1), similar to the separation of the colours of the visible spectrum and the appearance of a rainbow when sunlight illuminates a transparent prism or falling water droplets. For this reason, we shall henceforth call the stopping and storing of light in such LHHs the "trapped rainbow" effect.

A critical characteristic of the axially varying LHH is that further away from the point where the trapping of the light beam is arranged to occur, i.e. for larger guide thicknesses, it is possible to achieve complete impedance matching with a dielectric waveguide. We recall that for waveguide structures, the characteristic impedance is defined as $Z_0^{PV} = V_0 V_0^* / (2P_{tot}^{+z})$, where V_0 is a "voltage" defined as the line integral of the electric field along some path, which starts from below the lower interface and ends amply above the upper one [Helszajn (2000)]. Figure 6.4 illustrates the variation with reduced guide thickness of the analytically calculated (see Supplementary Information) ordinary and LH waveguide characteristic impedances. At a point sufficiently far from the "critical" thickness of the LHH, its characteristic impedance becomes equal to that of a regular waveguide. At this point ($\approx 12.76 a k_0$ in our case) the two structures also have equal thicknesses. Moreover, the spatial distribution of the guided field at the wide end of the LHH (Fig. 6.4 inset, red line) closely matches the field distribution of a single-mode optical waveguide. Accordingly, a lightwave launched from a dielectric guide to a wide-thickness LHH will experience minimal reflection, mainly owing to minute mode-mismatch, which can be further adjusted and optimised at will.

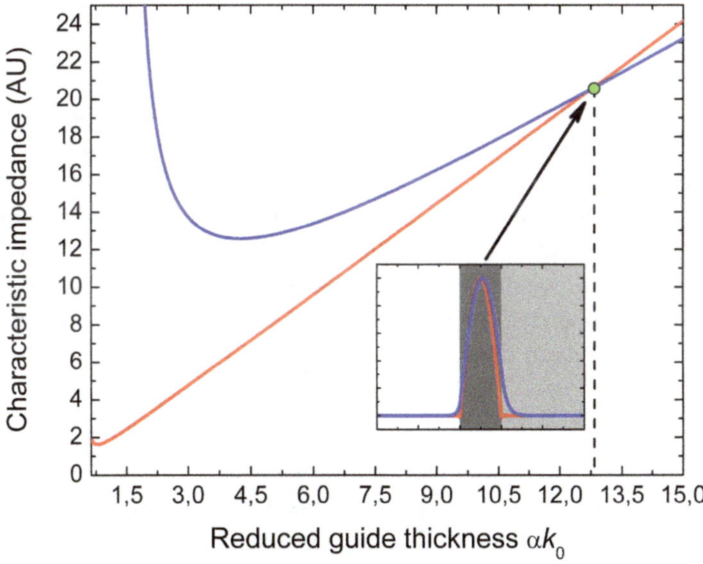

Figure 6.4: Honed conditions for waveguide coupling: simultaneous impedance, thickness and mode matching in adjoining RHH and LHHs.

explanations for Fig. 6.4

The optical parameters of the LHH are similar to those in Fig. 6.2. For the RHH we have $\varepsilon_{r1}^{\text{RHH}} = 1.5625$, $\varepsilon_{r2}^{\text{RHH}} = 1.44$, $\varepsilon_{r3}^{\text{RHH}} = 1.21$, $\mu_{r1}^{\text{RHH}} = \mu_{r2}^{\text{RHH}} = \mu_{r3}^{\text{RHH}} = 1$. We note that the characteristic impedance of the dielectric waveguide (blue) exhibits a deep at $ak_0 \approx 4.53$, after which it grows monotonically. A similar trend is found for the characteristic impedance of the LHH (red). Here, the minimum value occurs at a guide thickness $ak_0 \approx 0.89$ and at the "critical" LHH thickness it diverges. The two curves cross at $ak_0 \approx 12.77$. The inset shows the profile of the fundamental (blue) and second-order oscillatory (red) mode of the dielectric waveguide and LHH, respectively, at the cross point. The darker the shaded region in the inset, the higher is the magnitude of the refractive index.

In summary, we have shown how guided electromagnetic fields can efficiently be brought to a complete standstill whilst travelling inside axially varying LH waveguiding heterostructures. By nature, the scheme invokes solid-state materials and, as such, is not subject to low-temperature or atomic coherence limitations. Moreover, it inherently allows for high in-coupling efficiencies and broadband function, since the deceleration of light does not rely on refractive index resonances. This "trapped rainbow" method for storing photons opens the way to a multitude of hybrid, optoelectronic devices to be used in "quantum information" processing, communication networks and signal processors, and conceivably heralds a new realm of combined metamaterials and slow light research.

6.3 Derivation of Spatiotemporal Field-Component Equations Used in the Adiabatic Variation

For the applicability of the adiabatic approximation one needs to ensure that the variation of core half-thickness a with propagation distance z is properly slow. Starting from Maxwell's equations and by deploying coupled-mode theory, we can formally show (Chapters 19 and 28 of [Snyder and Love (1983)]) that the requirement for slow core thickness variation is fulfilled when the length of each tapered waveguide segment is large compared with the largest distance over which the guided fields can change appreciably owing to phase differences between the supported local modes. This leads to the following "axial variation" criterion:

$$\frac{da}{dz} \ll \frac{a(\beta_2 - \beta_3)}{2\pi}, \tag{6.5}$$

where β_2 and β_3 are the scalar propagation constants of the 2nd- ($m = 1$) and 3rd-order ($m = 2$) backward waveguide modes TM_{m+1}^{b} of the LHH [Tsakmakidis *et al.* (2006a)]. This criterion can also take the form:

$$\frac{da}{dz} \ll \frac{ak_0 \Delta n_{eff}^{2-3}}{2\pi}, \tag{6.6}$$

which is used in the main text of the latter.

The electromagnetic fields $\mathbf{G} = \mathbf{E}$ or $\mathbf{H}$ at a distance $z = z_t$ inside an axially varying LHH, which satisfies the above criterion, are given by:

$$\mathbf{G}(x, y, z_t, t) = F(z_t)\mathbf{g}(x, y, \beta(z_t)) \exp\left(i \sum_{z=0}^{z=z_t} \beta(z_t)\Delta z - i\omega t\right). \tag{6.7}$$

Here the parameter F, which is the positive constant used in the solution ansatz to the wave equation (see Eq. (6.1)), is chosen in such a way that $\partial P_{tot}^{con}\partial z = 0$, where $P_{tot}^{con} = \sum_{i=1}^{3} |P_i^{+z}|$ is the conserved total time-averaged power flow in the LHH [Tsakmakidis *et al.* (2006a, 2007)] and P_i^{+z} the time-averaged power flow in the i-layer ($i = 1, 2, 3$), propagating in the increasing $+z$ direction. In order to enforce the conservation of P_{tot}^{con} in the analytic computations, we normalise the fields in such a way that, at each tapered waveguide segment, they carry a total (conserved) time-averaged power flow equal to P_{tot}^{con}. To this end, we start by calculating $P_i^{+z} = \frac{1}{2} \int_{A_i} \mathrm{Re}(\mathbf{E} \times \mathbf{H}^*)_z \, dA$, in each waveguide layer and, after some algebraic manipulations [Tsakmakidis *et al.* (2006a, 2007)], we arrive at:

$$P_{tot}^{con} = \frac{F^2}{4\omega\varepsilon_0} \frac{a\beta}{|\varepsilon_{r1}|} \frac{W_3^2 + \sigma_\varepsilon^2 U^2}{\sigma_\varepsilon^2 U^2} \left[2 + \frac{\rho_\varepsilon}{W_2} \frac{U^2 - W_2^2}{W_2^2 + \rho_\varepsilon^2 U^2} + \frac{\sigma_\varepsilon}{W_3} \frac{U^2 - W_3^2}{W_3^2 + \sigma_\varepsilon^2} \right]. \tag{6.8}$$

From Eq. (6.8), it follows that by requiring at each segment of the tapered waveguide:

$$F = 2\sigma_\varepsilon U \left[\frac{\omega\varepsilon_0 |\varepsilon_{r1}| P_{tot}^{con}}{a\beta(W_3^2 + \sigma_\varepsilon^2 U^2)\Theta} \right]^{1/2}, \tag{6.9}$$

with Θ defined in Eq. (6.3), we ensure that the guided electromagnetic field carries a constant (conserved) total power flow, equal to P_{tot}^{con}, throughout the LHH. Note that the fields are normalised with respect to P_{tot}^{con}, not $P_{tot}^{+z} = \frac{1}{2} \int_{-\infty}^{\infty} \mathrm{Re}(\mathbf{E} \times \mathbf{H}^*)_z \, dx$, since the latter one does not remain constant along the LHH, as in regular dielectric guides but, instead, it continuously decreases until it becomes zero at the "critical" guide thickness. Normalising the fields with P_{tot}^{+z} instead of P_{tot}^{con} would have caused their unphysical divergence at the point where they are stopped.

6.4 Derivation of the Expressions for Light-Ray Goos–Hänchen Spatial Displacements

Let us assume that a ray of p-polarised light impinges upon the 1-3 media interface with angle θ (see Fig. 6.3c). Following a course of analysis similar to that followed for dielectric waveguides, one may show that the associated Goos–Hänchen phase shift will be [Snyder and Love (1983); Berman (2005)]:

$$\delta_{p13} = -2\tan^{-1}\left(\frac{W_3}{\sigma_\varepsilon U}\right). \tag{6.10}$$

For the sake of convenience in the subsequent algebraic manipulations, let us rewrite Eq. (6.10) in the following form:

$$\delta_{p13} = -2\tan^{-1}\left(f(\theta)\right) = -2\tan^{-1}\left(\frac{(\sin^2\theta - \sigma_\varepsilon\sigma_\mu)^{1/2}}{\sigma_\varepsilon\cos\theta}\right), \tag{6.11}$$

from whence we obtain:

$$\frac{df(\theta)}{d\theta} = \frac{(1 - \sigma_\varepsilon\sigma_\mu)\sin\theta}{\sigma_\varepsilon\cos^2\theta(\sin^2\theta - \sigma_\varepsilon\sigma_\mu)^{1/2}}, \tag{6.12}$$

$$\frac{d}{d\theta}\tan^{-1}\left(f(\theta)\right) = \frac{\sigma_\varepsilon(1 - \sigma_\varepsilon\sigma_\mu)\sin\theta}{(\sin^2\theta - \sigma_\varepsilon\sigma_\mu)^{1/2}\left((\sigma_\varepsilon^2 - 1)\cos^2\theta + 1 - \sigma_\varepsilon\sigma_\mu\right)}, \tag{6.13}$$

The inverted "penetration" distance x_{p13} (see Figs. 6.3c–e), can now be calculated by means of the following relationship:

$$x_{p13} = \frac{1}{k_0 n_1 \sin\theta}\frac{d}{d\theta}\tan^{-1}\left(f(\theta)\right), \tag{6.14}$$

and we successively have:

$$\begin{aligned}
x_{p13} &= \frac{\sigma_\varepsilon}{k_0(n_1^2\sin^2\theta - n_3^2)^{1/2}}\frac{1 - \sigma_\varepsilon\sigma_\mu}{(\sigma_\varepsilon^2 - 1)\cos^2\theta + (1 - \sigma_\varepsilon\sigma_\mu)} \\[2ex]
&= \frac{\sigma_\varepsilon}{\gamma_3}\frac{n_1^2 - n_3^2}{\sigma_\varepsilon^2\frac{\kappa^2}{k_0^2} - \frac{\kappa^2}{k_0^2} + n_1^2 - n_3^2} \\[2ex]
&= \frac{\sigma_\varepsilon}{\gamma_3}\frac{n_1^2 - n_3^2}{\sigma_\varepsilon^2\frac{\kappa^2(n_1^2 - n_3^2)}{\gamma_3^2 + \kappa^2} - \frac{\kappa^2(n_1^2 - n_3^2)}{\gamma_3^2 + \kappa^2} + n_1^2 - n_3^2} \\[2ex]
&= \frac{\sigma_\varepsilon}{\gamma_3}\frac{\gamma_3^2 + \kappa^2}{\gamma_3^2 + \sigma_\varepsilon^2\kappa^2},
\end{aligned} \tag{6.15}$$

where we used the identity $k_0^2 = (\gamma_3^2 + \kappa^2)/(n_1^2 - n_3^2)$.

From Eq. (6.15), it directly follows that:

$$x_{p13} = \frac{a\sigma_\varepsilon}{W_3}\frac{V_3^2}{W_3^2 + \sigma_\varepsilon^2 U^2}, \tag{6.16}$$

which is the relation used in Section 6.2.

In a similar vein, one can prove that:

$$x_{p12} = \frac{a\rho_\varepsilon}{W_2}\frac{V_2^2}{W_2^2 + \rho_\varepsilon^2 U^2}. \tag{6.17}$$

6.5 Derivation of the Relation between the Total Time-Averaged Power Flow and the Effective Thickness of the Waveguide

The solution ansatz to the wave equation for the p-polarised oscillatory waveguide modes supported by the LHH has the following form:

$$H_y = \begin{cases} F \exp(\gamma_3 x), & x \le 0 \\ G \cos(\kappa x) + K \sin(\kappa x), & 0 \le x \le 2a \\ L \exp\left(-(x - 2a)\gamma_2\right), & x \ge 2a \end{cases} \tag{6.18}$$

where

$$G = F,$$

$$K = -\frac{\gamma_3}{\sigma_\varepsilon \kappa},$$

$$L = \left(\cos(2a\kappa) - \frac{\gamma_3}{\sigma_\varepsilon \kappa} \sin(2a\kappa)\right) F,$$

and $E_z = -\frac{j}{\omega\varepsilon}\frac{\partial H_y}{\partial x}$, $E_x = \frac{\beta}{\omega\varepsilon} H_y$.

From Eq. (6.18) we deduce that the maximum value for the H_y field component is

$$H_y^{\max} = F\frac{\left(W_3^2 + \sigma_\varepsilon^2 U^2\right)^{1/2}}{\sigma_\varepsilon U},$$

and occurs within the middle layer-1 at point

$$x^{\max} = \frac{a}{U}\tan^{-1}\left(-\frac{W_3}{\sigma_\varepsilon U}\right).$$

After some algebraic manipulations, we can analytically calculate the total time-averaged power flow in the increasing $+z$ direction, P_{tot}^{+z}, as [Tsakmakidis *et al.* (2006a, 2007)]:

$$P_{tot}^{+z}\big|_{\text{LHH}} = \frac{F^2}{4\omega\varepsilon_0}\frac{a\beta}{|\varepsilon_{r1}|}\frac{W_3^2 + \sigma_\varepsilon^2 U^2}{\sigma_\varepsilon^2 U^2}\left[\underbrace{\frac{\rho_\varepsilon}{W_2}\frac{V_2^2}{W_2^2 + \rho_\varepsilon^2 U^2}}_{x_{p12}/a} + \underbrace{\frac{\sigma_\varepsilon}{W_3}\frac{V_3^2}{W_3^2 + \sigma_\varepsilon^2}}_{x_{p13}/a} - 2\right], \tag{6.19}$$

from whence we immediately obtain:

$$P_{tot}^{+z} = \frac{1}{4}E_x^{max} H_y^{max} t_{eff}, \tag{6.20}$$

where t_{eff} is defined in the main text.

Note that, owing to the negativeness of the permittivity ε_1 in the core of the LHH, the term $E_x^{max} H_y^{max}$ in the right-hand side of Eq. (6.20) is always negative; hence, P_{tot}^{+z} and t_{eff} are oppositely signed. It is should be herein noted that a negative P_{tot}^{+z} corresponds to a negative phase velocity mode (P_{tot}^{+z} antiparallel to the mode longitudinal propagation constant β) and a positive P_{tot}^{+z} to a positive phase velocity mode (P_{tot}^{+z} and β are parallel).

6.6 Characteristic Impedance of Left- and Right-Handed Waveguides

In both cases we begin by calculating the "voltage" V_0 across the waveguide by means of the relation:

$$\int_{-\infty}^{\infty} E_x \, dx = \frac{\beta}{\omega} \int_{-\infty}^{\infty} \frac{H_y}{\varepsilon} \, dx. \tag{6.21}$$

We may then obtain the following expressions for the voltages V_i ($i = 1, 2, 3$) crosswise each i-layer of the LHH:

$$V_1^{\mathrm{LHH}} = -\frac{F}{\omega\varepsilon_0} \frac{a\beta}{|\varepsilon_{r1}|} \frac{\sqrt{W_3^2 + \sigma_\varepsilon^2 U^2}}{\sigma_\varepsilon U^2} \left[\pm \frac{W_2}{\sqrt{W_2^2 + \rho_\varepsilon^2 U^2}} - \frac{W_3}{\sqrt{W_3^2 + \sigma_\varepsilon^2 U^2}} \right], \tag{6.22}$$

$$V_2^{\mathrm{LHH}} = \mp \frac{F}{\omega\varepsilon_0} \frac{a\beta}{|\varepsilon_{r1}|} \frac{1}{\sigma_\varepsilon W_2} \frac{\sqrt{W_3^2 + \sigma_\varepsilon^2 U^2}}{\sqrt{W_2^2 + \rho_\varepsilon^2 U^2}}, \tag{6.23}$$

$$V_3^{\mathrm{LHH}} = \frac{F}{\omega\varepsilon_0} \frac{a\beta}{|\varepsilon_{r1}|} \frac{1}{\sigma_\varepsilon W_3}, \tag{6.24}$$

from whence we find:

$$V_0^{\mathrm{LHH}} = \sum_{i=1}^{i=3} V_i^{\mathrm{LHH}}$$

$$= \frac{F}{\omega\varepsilon_0} \frac{a\beta}{|\varepsilon_{r1}|} \frac{\sqrt{W_3^2 + \sigma_\varepsilon^2 U^2}}{\sigma_\varepsilon U^2} \left[\mp \frac{V_2^2}{W_2\sqrt{W_2^2 + \rho_\varepsilon^2 U^2}} + \frac{V_3^2}{W_3\sqrt{W_3^2 + \sigma_\varepsilon^2 U^2}} \right], \tag{6.25}$$

where the "+" (plus) sign is used for $U \in \left[(m - 1/4)\pi, (m + 1/4)\pi\right]$ and the "−" (minus) sign for $U \in \left[(m + 1/4)\pi, (m + 3/4)\pi\right]$ with $m \in \mathbb{N}$, $U > 0$.

In a similar vein, using Eq. (6.18) and the parameter definitions that follow it, we obtain the following expressions for the V_i ($i = 1, 2, 3$) voltages of the RHH:

$$V_1^{\mathrm{LHH}} = \frac{F}{\omega\varepsilon_0} \frac{a\beta}{\varepsilon_{r1}} \frac{\sqrt{W_3^2 + \sigma_\varepsilon^2 U^2}}{\sigma_\varepsilon U^2} \left[\pm \frac{W_2}{\sqrt{W_2^2 + \rho_\varepsilon^2 U^2}} + \frac{W_3}{\sqrt{W_3^2 + \sigma_\varepsilon^2 U^2}} \right], \tag{6.26}$$

$$V_2^{\mathrm{LHH}} = \pm \frac{F}{\omega\varepsilon_0} \frac{a\beta}{\varepsilon_{r1}} \frac{1}{\sigma_\varepsilon W_2} \frac{\sqrt{W_3^2 + \sigma_\varepsilon^2 U^2}}{\sqrt{W_2^2 + \rho_\varepsilon^2 U^2}}, \tag{6.27}$$

$$V_3^{\mathrm{LHH}} = \frac{F}{\omega\varepsilon_0} \frac{a\beta}{\varepsilon_{r1}} \frac{1}{\sigma_\varepsilon W_3}, \tag{6.28}$$

from whence we find:

$$V_0^{\text{LHH}} = \sum_{i=1}^{i=3} V_i^{\text{RHH}}$$

$$= \frac{F}{\omega \varepsilon_0} \frac{a\beta}{\varepsilon_{r1}} \frac{\sqrt{W_3^2 + \sigma_\varepsilon^2 U^2}}{\sigma_\varepsilon U^2} \left[\pm \frac{V_2^2}{W_2 \sqrt{W_2^2 + \rho_\varepsilon^2 U^2}} + \frac{V_3^2}{W_3 \sqrt{W_3^2 + \sigma_\varepsilon^2 U^2}} \right],$$

$$(6.29)$$

where the "+" (plus) sign signs follows the same rule as in the case of the LHH.

Moreover, one can show that the time-averaged power flow propagating in the increasing $+z$ direction inside the RHH is given by [Tsakmakidis *et al.* (2006a)]:

$$P_{tot}^{+z}|_{\text{RHH}} = \frac{F^2}{4\omega\varepsilon_0} \frac{a\beta}{\varepsilon_{r1}} \frac{W_3^2 + \sigma_\varepsilon^2 U^2}{\sigma_\varepsilon^2 U^2} \left[2 + \frac{\rho_\varepsilon}{W_2} \frac{V_2^2}{W_2^2 + \rho_\varepsilon^2 U^2} + \frac{\sigma_\varepsilon}{W_3} \frac{V_3^2}{W_3^2 + \sigma_\varepsilon^2} \right]. \quad (6.30)$$

By means of the power-voltage definition of the waveguide characteristic impedance [Helszajn (2000)]:

$$Z_0^{\text{PV}} = \frac{|V_0|^2}{2 P_{tot}^{+z}}, \quad (6.31)$$

using Eqs. (6.19) and (6.29) for the LHH, or Eqs. (6.29) and (6.30) for the RHH, one can now directly calculate the impedance for each waveguide.

We note from Eq. (6.31) that the characteristic impedance of the LHH diverges at the "critical" guide thickness, as anticipated, since in this case the heterostructure is in the "stopped light regime" and, hence, the corresponding light signal can not penetrate it and propagate inside.

Chapter 7

Passive Stopped-Light Waveguides

7.1 Introduction

As we saw in the previous chapters, negative refractive index (NRI) and plasmonic metal-insulator-metal (MIM) waveguides allow for unprecedented level of control over the propagation of electromagnetic waves, with a potential for realising a number of slow-light applications, such as optical buffers and "trapped rainbow" storage of light [Tsakmakidis *et al.* (2007)]. However dissipative loss inherent to the metallic components of these waveguides may impede this potential [Reza *et al.* (2008)].

In this chapter, we show in some detail that for complex-k solutions, a backbending occurs in the modal dispersion curve around the point where the group velocity would be zero in a loss-free waveguide. Based upon this theoretical finding it has, until recently, been assumed that dissipative losses would therefore prevent the stopping of light in realistic waveguides [Reza *et al.* (2008)].

To overcome this problem, it has been suggested that gain media could be incorporated into the waveguide design to reduce the modal loss [Genov *et al.* (2007)], as also shown later on in Chapter 9. TMM calculations have shown that if the gain can fully compensate the dissipative loss experienced by the waveguide mode then backbending is eliminated from the modal dispersion and extremely low group velocities are again possible [Lu *et al.* (2010)]. The use of optical gain materials has been shown to be effective for compensating losses in a number of plasmonic and metamaterial [Xiao *et al.* (2010a)] systems even resulting in the demonstration of surface plasmon lasers [Berini and Leon (2012)].

However, as shall be shown in this chapter, this is not necessary for the realisation of stopped light. This misunderstanding arises from the fact that, typically, the influence of dissipation on the waveguide modes is considered theoretically only for real-frequency (ω), complex-wavevector (k) solutions to the dispersion equation but an alternative set of solutions can be found with complex-ω and real-k. The dispersion curves of this alternate set differ greatly in the slow light regime in that they do not exhibit backbending when loss is present, thus retaining points where the group velocity goes to zero. Although this behaviour of the complex-ω modes has been noted in the literature [Yao *et al.* (2009); Lu *et al.* (2010)] an in-depth

175

study into their applicability to slow and stopped light waveguides has, until now, been lacking.

In this chapter, the ability to stop light in *dissipative* (lossy) NRI and plasmonic waveguides is presented in detail. In Section 7.2 the modal characteristics of both types of waveguides are considered in the absence of loss in order to highlight the dependence of the group velocity on geometric parameters. Following this introduction, the influence of dissipation in the slow-light regime is studied using, both, complex-k and complex-ω solutions to the dispersion equation. Section 7.4 presents a scheme for exciting the modes of the plasmonic waveguide directly from air based on the principles of prism coupling. This scheme is then demonstrated in Section 7.5 using full-wave finite-difference time-domain simulations. We also go through an essentially new technique for computing the centre of energy of an excited wavepacket in a lossy waveguide, showning that extremely low speeds, matching those predicted by the complex-ω solutions, can be attained. Spatial broadening and shifts in the central wavevector of the wavepacket resulting from both real and imaginary components of the modal dispersion are discussed in Section 7.5 before looking at methods to control the radiative loss of this leaky mode in Section 7.6. Finally, the main conclusions of this chapter are summarised in Section 7.8.

7.2 Modal Characteristics of the Plasmonic and Metamaterial Waveguides

The NRI metamaterial waveguide studied in this work consists of a single NRI layer of thickness $d = 200$ nm, surrounded by semi-infinite half spaces of air. A double Drude model [Ziolkowski (2003); Zhao *et al.* (2009); Cummer (2003)] is used to describe the frequency response of the effective permittivity and permeability of the NRI medium so that the refractive index $n_{NRI} = \varepsilon_{NRI} = \mu_{NRI} = \varepsilon_\infty - \omega_p^2/(\omega^2 - i\Gamma_D\omega)$, where $\varepsilon_\infty = 1$, $\omega_p = 2\pi \times 893.8 \times 10^{12}$ rad/s and $\Gamma_D = 0$ or 10^{12} rad/s depending on whether or not dissipative loss is being included. Modal properties are determined using the TMM presented in the previous chapter. In this section the NRI metamaterial is considered to be loss-free i.e. $\Gamma_D = 0$ rad/s.

Figure 7.1 compares the first six TM modes[1] of the NRI waveguide in terms of their spectral dispersion and field profiles. The order of each mode is determined by the number of nodes in the field profile. As noted in the previous chapter the two unusual properties of this type of waveguide are the absence of the fundamental TM_0 mode and that higher order modes have both a forwards and backwards going component [Shadrivov *et al.* (2003)]. At the point on the dispersion curve where the forwards and backwards branches meet, the group velocity goes to zero. Investigations into slow and stopped light in this waveguide shall focus on the specific case of the TM_2 mode although the results are in principle applicable to all the higher order modes.

---

[1]As the permittivity and permeability have the same response the dispersion and field profiles of the TE modes will be identical except for the substitution of E for H.

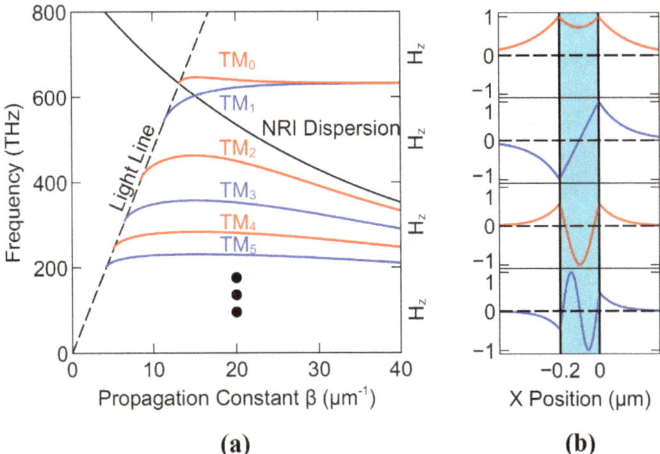

Figure 7.1: **(a)** Dispersion curves for the first six TM modes of the NRI waveguide, even modes are shown in red while the odd modes are blue. **(b)** The Hz field profile of the first four modes taken at a β value of $15\,\mu\mathrm{m}^{-1}$.

The dispersion curves of the waveguide modes are partially dependent on thickness d of the waveguide core. For the particular case of the TM$_2$ mode, increasing the core thickness is found to shift the dispersion curve towards higher frequencies and decreases the wavevector where the group velocity goes to zero. This effect has been proposed as a method for stopping and trapping light over a broad range of frequencies in a NRI waveguide with a tapered core [Tsakmakidis *et al.* (2007)]. The thickness can also be varied in order to optimise the structure for stopping light of a particular frequency. Figure 7.2b shows how the velocity of the forwards and backwards going TM$_2$ modes, at a constant frequency of 400 THz, change when the core thickness is varied. It can be seen that the two modes become degenerate at a core width of approximately 140 nm below which neither mode is supported at this frequency. As the core thickness increases the fields of the forwards propagating mode become less confined to the waveguide core until they fully couple into the air cladding (the mode cuts off at a thickness of 194 nm. The backwards mode shows the opposite behaviour becoming increasingly confined to the waveguide core which is reflected in the group velocity asymptotically approaching a value of $0.167c$, equal to the group velocity for a pulse propagating through a homogeneous NRI medium at 400 THz.

The second stopped-light waveguide studied in this work is based on the well known metal-insulator-metal (MIM) design, consisting of a dielectric core of thickness d sandwiched between two metallic layers. As only the permittivity is negative in this structure stopped-light can only be achieved for TM polarised modes. This is a slight disadvantage from the NRI waveguide where the operation was independent of the light polarisation. However as the negative permittivity is inherent to bulk metals, fabrication of the plasmonic waveguide could be more straightforward.

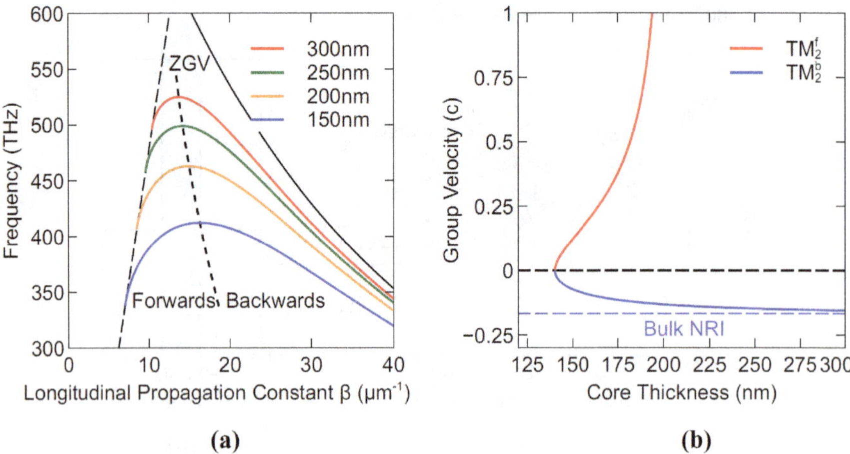

Figure 7.2: (a) Effect of core thickness on the TM_2 modal dispersion, for thickness from 150–300 nm, the line of zero group velocity is marked by a dashed black line showing the degenerate point of the forwards and backwards modes. (b) The change in group velocity at a single frequency (400 THz) caused by varying the core thickness.

It is desirable to achieve stopped light at the telecommunications wavelength of 1.55 μm for compatibility with existing infrastructure. Here indium tin oxide (ITO) has been chosen, over the more traditionally noble metals such as gold in silver, in the present waveguide design as it has a strong plasmonic response in the near infrared resulting from its low plasma frequency. The frequency dependent permittivity of ITO is described using a Drude model with the parameters $\varepsilon_\infty^{(ITO)} = 4$, $\omega_p^{(ITO)} = 3.13 \times 10^{15}$ rad/s and $\Gamma_D^{(ITO)} = 0$ or 1.07×10^{14} rad/s [Noginov et al. (2011)].

The core layer is assumed to be a dielectric with a constant real permittivity of $\varepsilon_D = 11.68$, typical of III-V semiconductors (such as InGaAsP) or silicon, neglecting the small loss of these materials.

The modal field profiles and dispersion curves of the loss free MIM waveguide are plotted in Fig. 7.3 for a core thickness of 290 nm. In general two different behaviours can be observed. First, the two lowest order modes asymptotically approach a frequency $\omega_0 = \omega_p/\sqrt{\varepsilon_\infty + \varepsilon_D}$ at high wavevectors. These are the symmetric and antisymmetric plasmon modes resulting from coupling of SPPs at the two waveguide interfaces. Higher order modes, however, show a different behaviour, where the modal dispersion asymptotically approaches the dielectric light line at high wavevectors. This behaviour is characteristic of waveguide modes that would be supported by just the dielectric core, although here, penetration of the fields into the ITO cladding modifies the modal dispersion at low wavevectors, allowing for extremely small group velocities. Investigations into slow and stopped light in the MIM waveguide shall focus on the specific case of the TM_2 mode as it allows for stopped light and, as shall be shown later, has a lower modal loss than either the TM_0 or TM_1 modes.

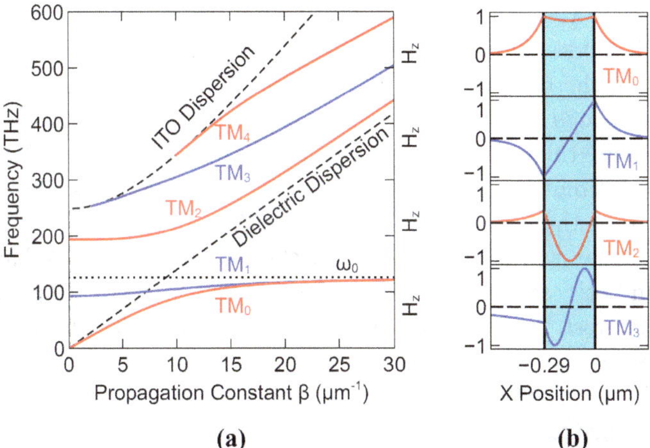

Figure 7.3: (a) Dispersion curves for the first six TM modes of the NRI waveguide, even modes are shown in red while the odd modes are blue. (b) The Hz field profile of the first four modes taken at a β value of 5 μm^{-1}.

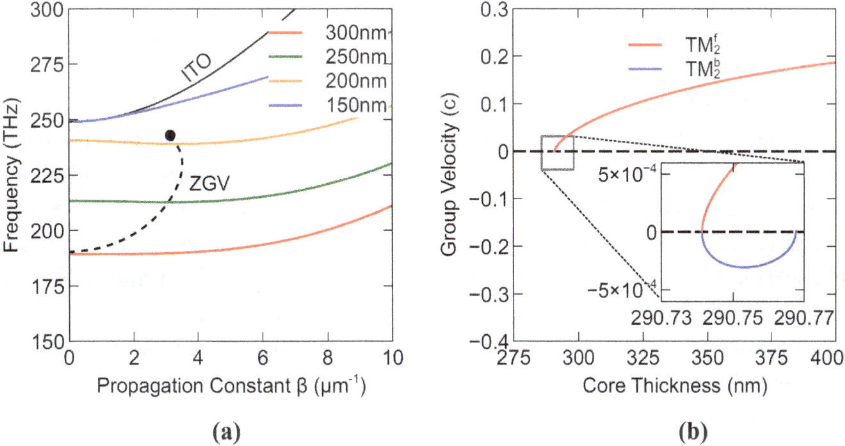

Figure 7.4: (a) Effect of core thickness on the TM_2 modal dispersion, for thickness from 150–300 nm, the line of zero group velocity is marked by a dashed black line showing the degenerate point of the forwards and backwards modes. The stopped-light point cuts off when $d \leq 190$ nm marked by the black cirlce. (b) The change in group velocity at a single frequency ($2\pi c/1.55$ μm) caused by varying the core thickness. Here the backwards mode only exists for a small range of core widths (inset).

Next, the dependence of the group velocity on the core thickness was investigated for the TM_2 mode. The results of these calculations are shown below in Fig. 7.4. It can be seen that increasing the core thickness results in the dispersion curve moving to lower frequencies. This due to the higher field overlap with the dielectric core which shifts the modal dispersion towards the dielectric light line. The stopped-light point of the TM_2 mode only exists over a range of core thickness's from

$d \sim 190\text{--}200$ nm. At the telecommunications wavelength of 1.55 µm, the forwards and backwards modes become degenerate at a core thickness of $d = 290.74$ nm. At this wavelength, the backwards mode only exists for a small range of core widths, becoming cutoff at $d = 290.768$ nm. From these results a value of 290 nm was chosen for the core thickness, as this provided a good match to the FDTD grid used in time domain simulations (as shown later in this chapter) while still achieving stopped light near the desired wavelength of 1.55 µm.

The two planar waveguides that form the basis of this work have now been introduced. In both cases it has been shown that the group velocity, at a particular frequency of interest, can be controlled simply by varying the thickness of the core layer. However, so far the influence of dissipative loss on the stopped light points has been neglected. In the next section the effect of including loss into the material models is investigated.

7.3 Dissipative Loss and Band Splitting

One of the main criticisms regarding stopped light in plasmonic and NRI waveguides is that, when material loss is included, the modal dispersion curve is altered so that the zero group velocity (ZGV) points are removed [Reza *et al.* (2008)]. Unfortunately, loss must be accounted for in the material models as there is a causality imposed relationship between dispersion and absorption. This intractability has led to the belief that stopped light is not achievable without some method for compensating the loss [Genov *et al.* (2007)]. However this issue is only predicted by the complex-k solutions of the dispersion equation, whereas solutions where the frequency is complex retain the ZGV point when loss is included [Yao *et al.* (2009)]. In this section these two solution sets are calculated for the NRI and MIM waveguides in order to investigate the influence of dissipation. Loss is included in the material models for ITO and the NRI metamaterial by setting the damping rates Γ_D to their non-zero values.

7.3.1 *Complex-k*

When the dispersion equation is solved for complex-k, the propagation loss is found from the imaginary component of the propagation constant, $\alpha_k = -2\beta''$, which has units of inverse meters and hence describes a purely spatial loss. Generally, solutions to the dispersion persion equation, found using this method, are reported in the literature as the propagation length, $L_k = 1/\alpha_k$, is of interest for many applications. Using this definition it would be expected that the propagation loss, in a dissipative waveguide, would become infinitely large at the stopped-light point. However, semi-analytic calculations, of modal dispersion in lossy waveguides, have shown that the dispersion curves split and "backbend" around the position on the dispersion diagram where the group velocity would be zero in a lossless waveguide [Reza *et al.* (2008)].

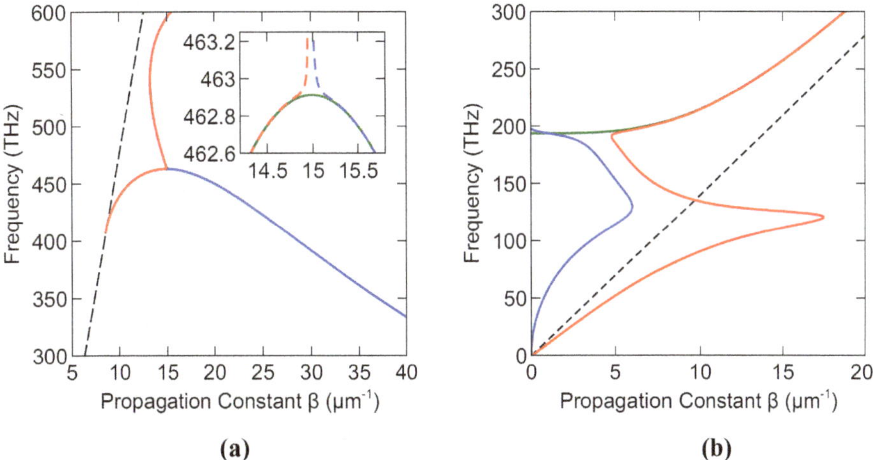

(a)   **(b)**

Figure 7.5: Band splitting and backbending behaviour of the complex-k modes of the lossy NRI (a) and MIM (b) waveguides. The lossy modes are shown in red and blue (forward and backwards propagation) in comparison to the lossless solutions (greeen) curves. The greater loss in the MIM waveguide results in a much stronger splitting of the modes.

Backbending is observed in both the NRI and MIM waveguides considered here (Fig. 7.5), where loss has been included through the Γ_D term of the Drude models used for the NRI and ITO materials,

$$\Gamma_D^{(\mathrm{NRI})} = 0.27 \times 10^{12} \text{ rad/s} \quad \text{and} \quad \Gamma_D^{(\mathrm{ITO})} = 1.07 \times 10^{14} \text{ rad/s}.$$

The larger loss of the ITO material results in a much stronger splitting of the forwards and backwards modes than in the NRI waveguide. It can also be seen that in the MIM waveguide the new forwards TM_2 branch merges with the TM_0 mode at low frequencies which leads to difficulties in properly classifying this dispersion branch. In both waveguides the stopped light points appear to have been removed due to the absence of any region where the gradient of the dispersion curve becomes zero.

When the group velocity $\frac{d\omega}{dk}|_{\mathrm{Re}[k]}$ is calculated for the forwards propagating branch of the TM_2 mode, superluminal and even infinite group velocities are found in the back-bending region, connecting the TM_2 and TM_0 modes, an interesting property considering the original intention of stopping light at these same frequencies. Superluminal group velocities were first discussed in the early 1900s in the context of anomalous dispersive materials [Brillouin (1960)]. As the group velocity is often associated with the velocity of information transmission predictions of superluminal values were of great concern as this would appear to violate the theory of relativity introduced by Einstein. A number of discussions took place before Sommerfeld and Brillouin demonstrated theoretically that the front velocity of a square shaped pulse will always propagate at or below the speed of light in vacuum [Brillouin (1960)]. Furthermore Brillouin argued that in regions of high anomalous dispersion the group velocity can no longer be associated with the velocity of signal transmission due to the initial pulse becoming severely distorted during propagation. Recently these arguments have been questioned due to several experimental works performed using Gaussian pulses that appear to show pulse peak arrival times consistent with superluminal group velocities without significant pulse distortion leading to renewed discussions [Wang *et al.* (2000); Dogariu *et al.* (2001)].

The issues presented by the superluminal group velocity can at least be resolved in the MIM waveguide considered here by noting that in the region of the dispersion curve where superluminal propagation is predicted, the modal decay length $1/2k''$ is shorter than the wavelength in the propagation direction $(2\pi/k')$. Thus the waves will be almost purely evanescent and so cannot combine to form a propagating pulse. As a results the concept of group velocity is no longer a good measure of propagation in this region of the dispersion curve. Interestingly, even if the ITO loss is artificially reduced (by setting a lower damping rate in the Drude equation) the modal decay length always remains shorter than the wavelength in the propagation direction in the backbending region.

Due to this behaviour the only applicable measure of energy transport in this region is the energy velocity which is found by dividing the time averaged energy flux in the propagation direction $\langle S_x \rangle$ by the total energy density of the mode $\langle U \rangle$ spatially integrated over the whole mode profile in the y-direction:

$$v_E = \frac{\int_{-\infty}^{\infty} \langle S_x(y) \rangle \, dy}{\int_{-\infty}^{\infty} \langle U(y) \rangle \, dy}, \tag{7.1}$$

where $\langle \ldots \rangle$ represents time averaging over one cycle in order to remove the effect of fast phase oscillations $\left(\frac{1}{T} \int_t^{t+T} dt \right)$. Care needs to be taken when defining the total energy density in dispersive and dissipative materials such as ITO as energy stored in the polarisation field needs to be added to the electromagnetic energy in order to correctly define the total energy. Following the method shown in [Ruppin (2002)] the time averaged total energy density in the ITO layers, assuming a harmonic time dependence $e^{-i\omega t}$ of the fields, is given by:

$$U^{ITO}(y,\omega) = \frac{\varepsilon_0}{4} \left[\varepsilon_\infty + \frac{\omega_p^2}{\omega^2 + \Gamma_D^2} \right] |\mathbf{E}(y,\omega)|^2 + \frac{\mu_0}{4} |\mathbf{H}(y,\omega)|^2. \tag{7.2}$$

In the range $\omega \gg \Gamma_D$ the term in the square brackets will reduce to $\approx \varepsilon_\infty + \omega_p^2/\omega^2$ given by the more well known expression for the time averaged energy density in dispersive, loss-free media:

$$U(y,\omega)|_{\omega \gg \Gamma} = \frac{\varepsilon_0}{4} \frac{\partial \left(\omega(y,\omega) \right)}{\partial \omega} |\mathbf{E}(y,\omega)|^2 + \frac{\mu_0}{4} |\mathbf{H}(y,\omega)|^2. \tag{7.3}$$

In the dielectric core the energy density is simply given $U^{Core}(y,\omega) = \frac{\varepsilon_0 \varepsilon_D}{4} |\mathbf{E}(y,\omega)|^2 + \frac{\mu_0}{4} |\mathbf{H}(y,\omega)|^2$. Combining the expressions for the energy density in different layers with Eq. (7.1) the energy velocity of the TM$_2$ mode can be calculated and compared to the group velocity. From Fig. 7.6 it can be seen that the

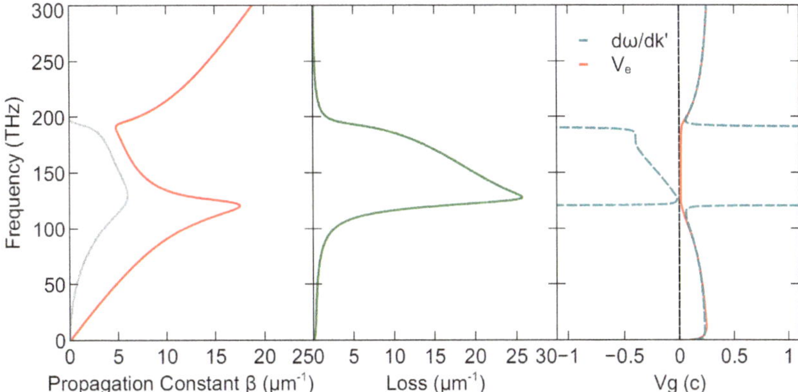

Figure 7.6: (Left) Complex-k dispersion curve of the TM$_2$ mode in the lossy MIM waveguide. **(Center)** the modal propagation loss corresponding to the forwards propagating branch of the TM$_2$ mode. **(Right)** Comparison between the group (blue dashed) and energy (red solid) velocities calculated for the same branch of the TM$_2$ mode.

energy velocity matches the group velocity calculations where the modal loss is relatively small. However, in the backbending region the two velocities differ greatly. In this case the energy velocity becomes close to zero but remains finite and positive at all frequencies, thus avoiding any of the difficulties previously discussed for the group velocity calculation. The energy velocity begins to deviate again from the group velocity in the low frequency region although here, the difference is due to the frequency approaching the scale of the damping rate, thus the influence of Γ_D in Eq. (7.2) becomes non-negligible.

7.3.2 *Complex-ω*

The complex-ω modes are an alternative set of solutions to the dispersion equation where propagation loss is described temporally. In this case a real valued wavenumber enters into the dispersion equation and solutions are sought on the complex-ω plane. The temporal loss is then calculated from the imaginary part of the frequency: $\alpha_\omega = -2\omega''$ and has units of s^{-1}. It can be seen, from Fig. 7.7, that the dispersion curves of these complex-ω solutions are barely changed when loss is included and show none of the backbending behaviour previously observed in the complex-k solutions. As the loss is defined purely in time, it has no dependence on the propagation speed and, subsequently, remains finite at the stopped-light point.

The high loss in the MIM waveguide can be seen to have a small influence on the real part of the dispersion but importantly the point of zero group velocity remains. In this case there is a close match between the group and energy velocities at all frequencies with both going to zero at $\beta \sim 1.4$ μm.

The discrepancy between the complex-ω and complex-k solutions, where one set retain points of zero group velocity while the other exhibits backbending behaviour,

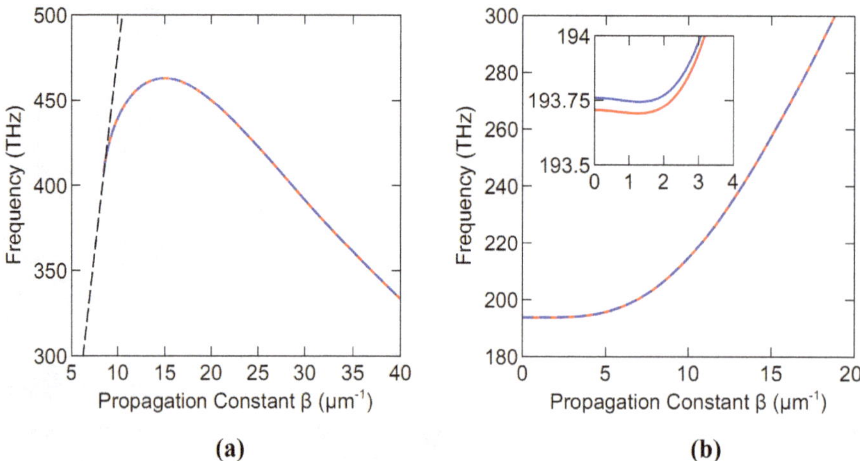

Figure 7.7: Comparison of modal dispersion for lossy (blue) and lossless (red) complex-ω solutions in the **(a)** NRI and **(b)** MIM waveguides under consideration. The higher loss of the MIM waveguide does result in a very small shift in the real part of the dispersion (seen inset) while for the NRI waveguide any variation is below the floating point error of the calculations.

was first discussed in the literature in the 1970s [Alexander *et al.* (1974); Kovener *et al.* (1976)]. At the time several groups had reported experimental measurements for the dispersion curve of an SPP propagating on a metal surface by using attenuated total reflection measurements. Here, minima in the reflectivity were associated with the excitation of SPPs. Issues arose as some groups reported curves exhibiting backbending (corresponding to the complex-k solutions) while others showed curves without this behaviour (complex-ω solutions). In 1974 Alexander *et al.* [Alexander *et al.* (1974)] showed that the difference reported was due to the method used to extract the curves from the reflection data, and that in fact both curves could be derived from one set of measurements. If the reflectivity minima are found by varying the incidence angle at a fixed frequency the extracted dispersion curve would show backbending, while, if the angle was held constant and the frequency varied, the extracted dispersion curve would match the complex-ω solution. Although this explained how the two different curves could be extracted it did not provide physical insight into what they represented, or if indeed a minima in the reflectivity can always be associated with the excitation of a waveguide mode.

In general, the electromagnetic modes of a lossy waveguide will be characterised by a dispersion with both complex frequency and complex wavevector. The complex-ω and complex-k solutions represent two subsets where this generalised complex dispersion curve has been projected onto real wavevector or real frequency spaces, respectively. A physical setup that can excite the complex-ω states of a waveguide, for the study of slow and stopped-light has, to the authors knowledge, not been fully investigated before. In the next section such a scheme is introduced and demonstrated using finite-difference time-domain simulations.

7.4 Incoupling at the Stopped-Light Point

In order to excite the waveguides modes light has to be coupled into the waveguide in a manner that matches the frequency and wavevector to those of the propagating mode. This is usually performed either by out-of-plane or in-plane excitation. Out-of-plane excitation uses an external source incident on the structure that is coupled into the waveguide typically using a prism or grating on the surface [Raether (1988)] in order to momentum-match the component of the incident wavevector in the propagation direction to that of the waveguide mode. This method is required as the bound modes lie below the light line. Alternatively in-plane excitation setups such as endfire coupling [Stegeman *et al.* (1983)] can be used where light is focused onto the edge of a waveguide.

For stopped-light, end-fire coupling is likely to result in the incident wave being reflected at the interface with the waveguide as the dispersion forbids any propagation at the stopped-light point. As such, a scheme based on out-of-plane, prism coupling shall be used here.

In order to couple light into the MIM waveguide, the structure first needs to be modified to reflect a more physical setup, by truncating the upper ITO cladding layer. The thickness of this layer shall be quantified by the parameter t. This modification has two main effects on the modes of the waveguide. The first is the introduction of an SPP mode that is localised to the interface between the upper ITO and the new superstrate (here assumed to be air). From the dispersion curves in Fig. 7.8 it can be seen that the SPP crosses the curve of the TM_2 mode. To avoid

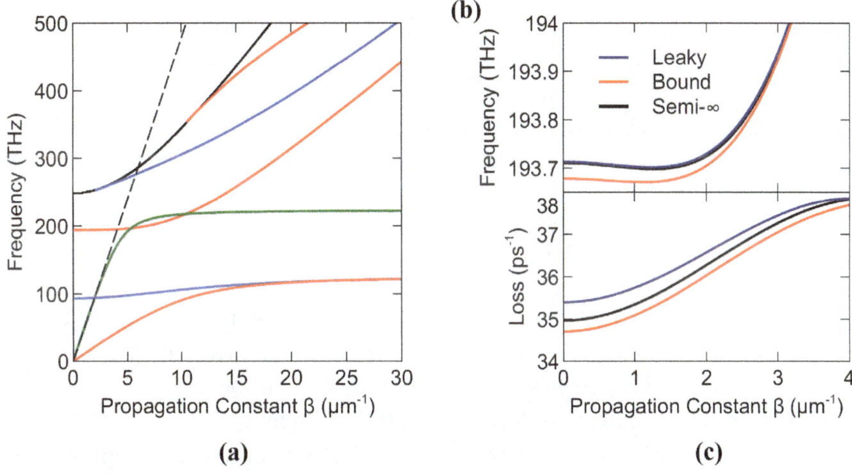

Figure 7.8: (a) Modes of the truncated MIM waveguide highlighting the addition SPP branch (shown in green). (b) the effect on the TM_2 modal dispersion found for leaky and bound solutions in comparison to the previous case of infinite upper ITO thickness (black). (c) Temporal loss predicted by the different solutions.

any coupling between the two modes at this intersection point, the thickness of the ITO layer must be greater than the skin depth of the SPP, here a thickness of 500 nm has been used. The second effect is that the TM_2 mode now lies partially inside the light cone of the air cladding and, crucially, so does the stopped-light point. As a result, the portion of the TM_2 mode inside the light cone can couple to free space radiation modes. This has the benefit that the momentum matching condition required for incoupling light can be achieved simply from air, without needing additional prism, or grating coupling schemes. The downside, is that the reverse is also true, once light is incoupled it can leak back out of the waveguide increasing the modal loss.

When solving the dispersion equation inside the light cone, the bound mode solutions can still be found using the TMM results, however, the leaky mode solution, with a divergent field profile in the air cladding is the more physically appropriate solution in this region (found for $\mathrm{Re}[\gamma_0] < 0$, $\mathrm{Re}[\gamma_{N+1}] > 0$), as shall be shown in the next section where the predicted theoretical results are compared to FDTD simulations.

7.5 Temporal Mode Dynamics

7.5.1 *The FDTD Method*

The finite-difference time-domain (FDTD) method is a numerical technique for simulating electromagnetic interactions in the time domain [Taflove and Hagness (2000)]. Here the time dependent Maxwell's equations are solved by approximating the derivative terms using a central finite difference method. By choosing the positions where the central finite difference of different fields are taken such that they form a special arrangement known as the Yee cell this method can be made second order accurate.

The influence of dispersive material parameters can be accounted for via a time dependent polarisation field that has the Lorentzian form:

$$a_1 \frac{\partial^2 \mathbf{P}}{\partial t^2} + a_2 \frac{\partial \mathbf{P}}{\partial t} + a_3 \mathbf{P} = \varepsilon_0 a_4 \mathbf{E}. \tag{7.4}$$

Several commonly used material models can be expressed in this form by choice of the coefficients, for example the Drude model results when $a_1 = \varepsilon_\infty$, $a_2 = \Gamma$, $a_3 = 0$, and $a_4 = \omega_p^2$. This generalised response model can be included in the FDTD method through the auxiliary differential equation technique.

A concise description of the FDTD method is presented in Chapter 2, along with additional details concerning boundary conditions, plane wave injection, and stability or energy measurement numerical techniques found in [Taflove and Hagness (2000)].

7.5.2 *Excitation Scheme*

The simulation setup, designed to excite the TM_2 with real wavevectors (complex-ω), is shown schematically in Fig. 7.9. Here a monochromatic plane wave (with wavelength $\lambda_0 = 1.55$ μm is launched towards the waveguide from above, with an incident angle around $17°$ in order to match the wavevector in the propagation direction to that corresponding to the point of ZGV. The wave is launched into the air layer towards the structure using the total-field scattered-field injection plane method described in [Taflove and Hagness (2000)]. The injection plane is situated 200 nm above the waveguide in the air cladding. The incident beam has a Gaussian spatial envelope so that the fields decrease nearly to zero at the edges of the injection plane in order to reduce any high frequency contributions that may result from discontinuities in the injected field profile. The source is switched on and smoothly increases to its peak value over a period of 200 fs and is then sustained at its maximum amplitude for a further 800 fs before being switched off smoothly over a decay period of 200 fs. After the incident wave has been switched off and any reflected fields have left the simulation domain, the remaining waveguide excitation is analysed. The simulation domain is surrounded by 11 cells of perfectly matched layer (PML) to approximate an infinite continuation of the materials. For all simulations loss is included in the ITO material model.

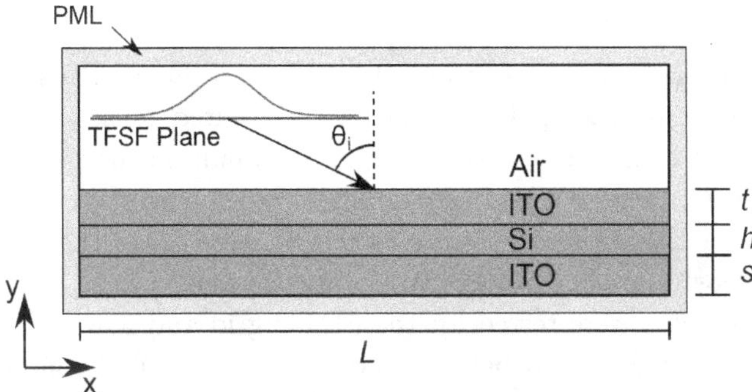

Figure 7.9: FDTD simulation setup for exciting the TM_2 mode in the MIM waveguide. The thickness of the layers used were $t = 500$ nm, $h = 290$ nm, $s = 600$ nm and the air layer was 400 nm. The simulation domain is truncated by PMLs 11 cells thick to approximate open boundary conditions. A TFSF plane is used to inject a plane wave into the simulation domain at an incidence angle θ_i. This setup allows both ω and β to be precisely controlled.

7.5.3 *Extracting the Complex-ω Band*

The excited fields in the FDTD simulations can be used to extract the complex-ω dispersion curve of a waveguide mode by applying a technique called the "Shift Method" [Stuart (1961)]. Here the field amplitude profile is recorded along the length of the waveguide at the centre of the waveguide core at two times, t_1 and $t_2 = t_1 + \Delta t$, where the time interval is short compared to the period of oscillation. This is performed after the exciting plane wave has been switched off and any reflections have left the simulation domain in order to capture the dynamics of the freely decaying waveguide excitation. The spatial Fourier transforms of these two field profiles are then related through the expression:

$$F\left(k, t_2\right) = e^{-i\omega\Delta t} F\left(k, t_1\right), \tag{7.5}$$

where the frequency is assumed to be complex, i.e. $\omega = \omega' + i\omega''$.

The Fourier transform will consist of an amplitude A_j and a phase component θ_j giving:

$$A_2 e^{i\theta_2} = e^{-\left(\omega' + i\omega''\right)\Delta t} A_1 e^{i\theta_1}. \tag{7.6}$$

By taking the logarithm of both sides and rearranging terms the real and imaginary components of ω can be found as:

$$\omega' = \frac{\theta_1 - \theta_2}{\Delta t}, \tag{7.7}$$

$$\omega'' = \frac{\ln(A_1) - \ln(A_2)}{\Delta t}. \tag{7.8}$$

Thus the complex-ω dispersion can be extracted from the FDTD simulations. Furthermore, by reducing the spatial extent of the excited pulse in the propagation direction, a broad range of wavevectors can be simultaneously probed using this technique. However, the FWHM of the pulse can not be reduced arbitrarily, as, below the limit of $1/k_c$, where k_c is the central wavevector of the pulse, the spatial frequencies are lost and the Fourier transform centers at $k = 0$.

Several resolutions were tested for the FDTD grid and it was found that a cell size of 10 nm provided a good balance between accuracy and computational time (see, e.g., [Pickering (2013)]). The complex-ω dispersion curve extracted from the FDTD simulations, matches well to the semi-analytic TMM calculations (Fig. 7.10). There is a slight difference between the two methods but this is likely due to approximations from the FDTD algorithm. Despite the small difference the ZGV point is still present, seen at the minima point of the curve, and occurs at a wavevector of $\beta \approx 1.23~\mu\text{m}^{-1}$, which compares well with the predicted value of $\beta = 1.24~\mu\text{m}^{-1}$, from TMM calculations. This good agreement demonstrates that the out-of-plane incoupling scheme has been successful in exciting modes characterised by real wavevectors and complex frequencies. Next the temporal dynamics and propagation characteristics of wavepackets, excited close to the stopped-light point of the TM_2 mode, are examined.

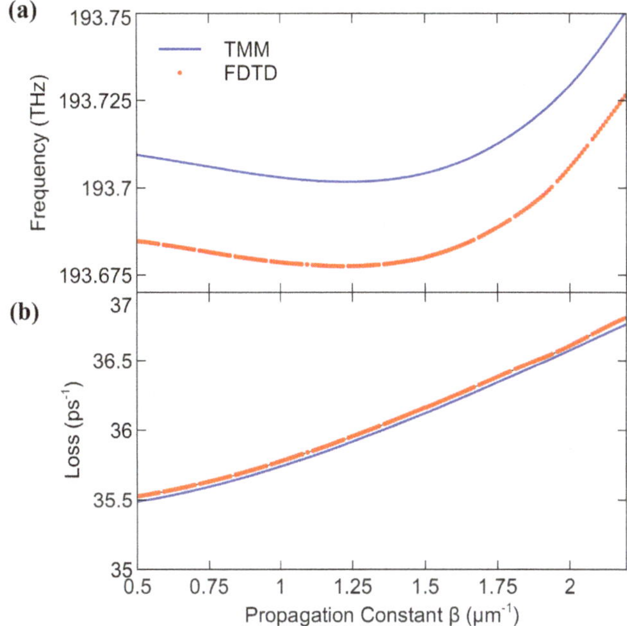

Figure 7.10: Comparison between the complex-ω dispersion curves calculated semi-analytically (blue lines) and extracted from numerical simulations (red dots) for **(a)** the real part of the frequency and **(b)** the temporal loss rate.

7.6 Velocity, Dispersion and Loss in the Time Domain

7.6.1 *Centre of Energy Velocity*

In order to directly test the predicted zero group velocity an accurate method for measuring the propagation of a wavepacket in the time-domain is required. Many methods have been proposed in the literature such as the centrovelocity [Smith (1970)] and correlation velocity techniques. However these methods do not offer adequate precision for measuring extremely low group velocities of the short lived wavepackets in the lossy MIM waveguide. Similarly, well-known methods such as the energy velocity suffer as it would be difficult to distinguish between propagation and dispersion of the pulse using this method.

To overcome these difficulties a spatial equivalent to the centrovelocity, based on the change in the centre of energy (CoE) of the excited wavepacket, shall now be introduced. In analogy to the centre of mass the centre of energy can be defined as:

$$\langle \mathbf{r}_0^m \rangle = \frac{\int \mathbf{r}^m \, U(\mathbf{r}) \, dV}{\int U(\mathbf{r}) \, dV}, \tag{7.9}$$

where m defines the energy moment being extracted. $U(\mathbf{r})$ is the spatially dependent energy density associated with the wavepacket, which, as mentioned previously, has to take account of energy stored in both the electromagnetic fields and the polarisation field in the lossy and dispersive ITO layers. By evaluating Eq. (7.9) over a region that completely envelopes the excited wavepacket the expectation value of the pulse's position in the propagation direction (x) can be recorded over time (with $m = 1$), the velocity is then extracted from the gradient. Higher order moments shall also be used to calculate the full width at half maximum (FWHM= $2\sqrt{2\ln 2}\sqrt{\langle \mathbf{r_0^2} \rangle - \langle \mathbf{r_0} \rangle^2}$) of the wavepacket in order to measure the broadening of the wavepacket over time due to dispersion.

7.6.2   *Extraction of Effective Loss Rates*

Another important parameter to extract from these simulations is the modal loss rate which can be further split into radiative and dissipative contributions in order to provide additional insight into the loss mechanisms involved in the current setup. The loss rate analysis presented here is based on the application of Poynting's theorem which states that the energy flux into or out of a volume is connected to the change of electromagnetic energy in the volume and the work performed by the fields on any charged particles present. In the time-domain simulations the influence of the fields on the ITO layers is represented by the polarisation field $\mathbf{P}^{(\mathrm{ITO})}(\mathbf{r}, t)$ and so Poynting's theorem can be written as:

$$\int \frac{du_{EM}}{dt}\, dV = -\oint \mathbf{S}\, d\mathbf{A} - \int \mathbf{P}^{(\mathrm{ITO})} \cdot \mathbf{E}\, dV, \qquad (7.10)$$

where $u_{EM}(\mathbf{r}, t) = \frac{\varepsilon_0 \varepsilon_r}{2} \mathbf{E}^2(\mathbf{r}, t) + \frac{\mu_0}{2} \mathbf{H}^2(\mathbf{r}, t)$ is the electromagnetic energy density inside the volume V, and $\mathbf{S}(\mathbf{r}, t)$ is the Poynting vector, representing the energy flux through the surface enclosing the chosen volume. As the ITO layers are both dissipative and dispersive the dot product term on the right hand side of Eq. (7.10) does not just represent the work performed by the fields on the charged particles, but also quantifies the change in energy due to dispersion [Ruppin (2002)]. In order to correctly attribute the dispersive contribution to the change in total energy, this term needs to be split into its dispersive and dissipative components. This can be done using the method presented in [Ruppin (2002)], which, for a Drude response, gives:

$$\dot{\mathbf{P}}_D \cdot \mathbf{E} \;=\; \dot{\mathbf{P}}_D \cdot \frac{1}{\varepsilon_0 \omega_p^2} \left(\ddot{\mathbf{P}}_D + \Gamma \dot{\mathbf{P}}_D \right) \qquad (7.11)$$

$$=\; \underbrace{\frac{1}{2\varepsilon_0 \omega_p^2} \frac{d}{dt} \left(\dot{\mathbf{P}}_D^2 \right)}_{=\frac{du_D}{dt}\ dispersive\,term} + \underbrace{\frac{\Gamma}{\varepsilon_0 \omega_p^2} \dot{\mathbf{P}}_D^2}_{dissipative\,term}. \qquad (7.12)$$

With this separation, effective rates can now be defined for the various loss channels that contribute to decay of the waveguide mode by dividing both sides of Eq. (7.10) by the total energy $U = \langle u_{EM} + u_D \rangle$ to give:

$$\frac{dU}{dt} = \Gamma_T U = -\Gamma_R U - \Gamma_D U, \qquad (7.13)$$

where Γ_T is the total energy loss rate, $\Gamma_R = \langle \nabla \cdot \mathbf{S} \rangle / U$ is the radiative loss rate due to flux out of the volume and $\Gamma_D = \langle \dot{\mathbf{P}}_D \cdot \mathbf{E} - \frac{du_D}{dt} \rangle / U$ is the dissipative loss rate. The operator $\langle ... \rangle$ represents both time averaging and integration over the considered volume. This method has been incorporated into the FDTD code via a discretised formulation of Poynting's theorem as, e.g., outlined in [Pickering (2013)].

7.6.3 *Extremely Low Group Velocity and Pulse Broadening in the Lossy MIM Waveguide*

With the centre of energy and loss rate extraction methods, the temporal dynamics, of wavepacket propagation in the waveguide, can be directly studied in the FDTD simulations. Using the same excitation scheme, presented in Fig. 7.9, a wavepacket was excited in the MIM waveguide with a central frequency and wavevector closely matched to the values at the stopped light point in the TM_2 mode found using TMM calculations. Here, the FWHM of the Gaussian input beam was set to $30\lambda = 30\,(1.55\,\mu\text{m})$ to produce a wavepacket with a narrow range of wavevectors in order to accurately probe the dispersion band around the stopped light point.

Several simulations were performed with slightly different angles of incidence corresponding to wavevectors around the predicted ZGV point. In each case the temporal evolution of the centre of energy position and width of the wavepacket was recorded, along with the dissipative and radiative loss rates using the centre of energy and loss rate extraction methods described in the previous section. Here the integration box for both methods was set to completely enclose the wavepacket excited in the waveguide. Furthermore, the upper boundary of the box was aligned with the Air/ITO interface so that the flux recorded through this boundary could be associated with the radiative loss channel.

An example plot of the temporal dynamics of a wavepacket excited close to the stopped light point is shown in Fig. 7.11. Three different regimes of behaviour can be observed from the dynamics. The first shall be referred to as the continuous excitation regime, corresponding to times between 0.7–0.8 ps, here the incident beam is held continuous at its maximum amplitude. As a result the centre of energy and width of the wavepacket are constant in this regime. The source is then switched off, smoothly decreasing in amplitude between 0.8–1 ps, which leads to a sudden change in the measured position and width of the wavepacket. This is due to the plane wave source being incident at an angle to the surface of the waveguide structure. Each wavefront of the incident wave will initially intersect the surface of the waveguide to the left of centre and, as it continues to propagate, the position where the wavefront intersects the waveguide surface will "travel" from left to right. When the source is on continuously this effect is balanced by the arrival of subsequent wavefronts however, during switch-off, the tail end of the

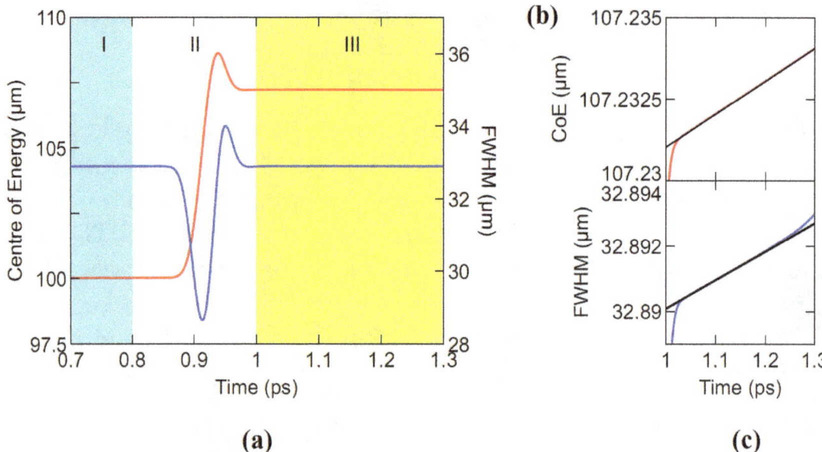

Figure 7.11: An example plot of the temporal evolution of the centre of energy (CoE) (red) and the FWHM (blue) of a wavepacket excited in the MIM waveguide close to the stopped-light point. Three regimes are observed I) where the incident source (black) is in continuous wave operation the centre of energy and FWHM remain constant. II) when the incident source is switched off between 0.8–1 ps the extracted quantities undergo transitory behaviour before settling down into a freely decaying state (III) where linear fits are applied to extract the velocity **(b)** and broadening rate **(c)**.

source wave causes the excited wavepacket to redistribute slightly to the right. Finally for times after ~ 1.05 ps, when the incident plane wave has been switched off and all reflections have left the simulation domain, the remaining wavepacket is seen to propagate with a constant velocity, as evident from the linear dependence of position on time. This shall be referred to as the free decay regime where the dynamics of the wavepacket can be cleanly observed and parameters such as the velocity, dispersion and loss rates can be extracted.

The centre of energy velocity extracted from the FDTD simulations is found to match well to the group velocity predicted from the complex-ω solutions to the dispersion equation (Fig. 7.12), thus directly proving that extremely low group velocities are attainable even in the presence of realistic material loss using the excitation scheme presented in this work. Furthermore, by comparing the FDTD results for both velocity and total loss rate to those predicted by the bound and leaky mode solutions, it can be seen that the leaky modes most accurately describe the propagation characteristics of the excited wavepacket, despite not being part of the complete basis set of modes.

7.6.4 *Loss Induced Spectral Shift*

In a dispersive waveguide it is well known that a wavepacket will broaden during propagation if the group velocity varies with frequency. However, in dissipative waveguides the spectral content of a wavepacket can also change over time as well. This effect is due to the frequency dependent loss rate, as a result some

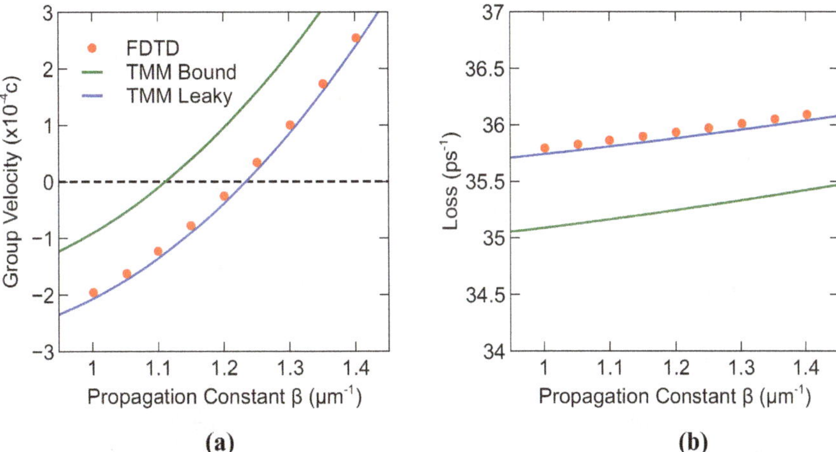

Figure 7.12: Comparing the propagation velocity (a) and total loss rate (b) extracted from FDTD simulations (red) to those predicted by the leaky (blue line) and bound (green line) solutions to the dispersion equation. Demonstrating the close match between the leaky mode solutions and time domain results.

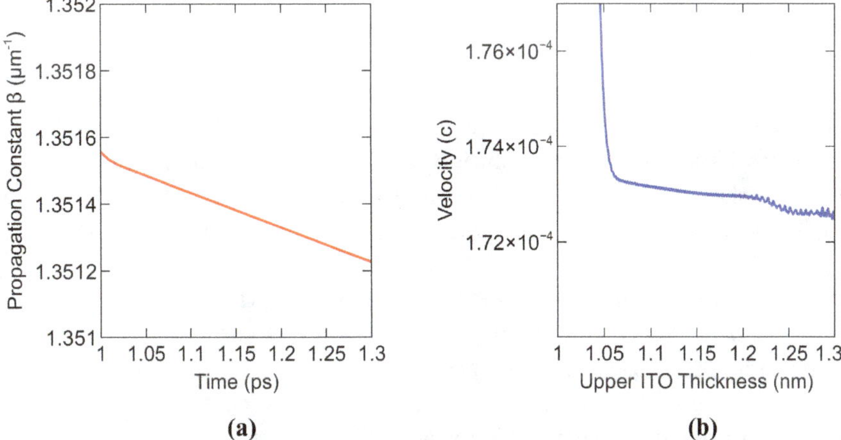

Figure 7.13: (a) Loss induced shift in the wavepacket's central wavevector leading to a small decrease in the centre of energy (CoE) velocity (b).

spectral components of the wavepacket decay at a faster rate than others. In the MIM waveguide considered here it can be seen from Fig. 7.12 that lower β values experience a lower loss rate than higher values. Over time this will eventually lead to a decrease in the central wavevector of the excited wavepacket and hence also changing the group velocity to increasingly negative values. This spectral shift was observed in the FDTD simulations performed in the previous section and an example plot of the temporal evolution of central wavevector and centre of energy velocity is shown in Fig. 7.13.

From this it can be seen that the central wavevector decreases by $\sim 1 \times 10^{-4}$ µm every 100 fs leading to a small decrease in the CoE velocity of approximately $3 \times 10^{-7}c$, which is roughly 0.2% of the mean value of $1.73 \times 10^{-4}c$, measured in the free decay region for times after 1.05 ps. Thus, over the meaningful timescale of the wavepacket lifetime (~ 30 fs), this effect can be neglected in the MIM waveguide studied here.

7.7 Controlling the Radiative Loss

It has been shown that for wavevectors inside the light cone the TM_2 mode can be excited directly from air. This is beneficial as incoupling can be performed without the need for complex prism or grating structures. However the drawback is that the energy of the mode is not completely bound to the waveguide and can outcouple back into free space. The leaky character of the mode results in an increase in the modal attenuation due to this additional radiative loss channel. However, whereas the ohmic losses are an unavoidable consequence of the metallic layers, the radiative loss can be controlled via the thickness of the upper ITO layer. FDTD simulations were performed for a range of thicknesses from 500 nm down to 100 nm for a pulse excited at a wavevector of $k = 1$ µm. For each layer thickness the modal loss rate was recorded from the simulations using the loss rate extraction method. The total loss rate extracted from FDTD simulations matched very well to that predicted by the leaky mode solutions of the dispersion equation. This further confirms that the leaky modes are the correct solutions for describing the properties of the TM_2 mode inside the light cone. The bound and leaky solutions appear to converge as the ITO thickness increases, likely due to the decreased coupling with radiation modes.

From the FDTD simulations, and TMM calculations, the modal loss can be analysed in greater depth by splitting it into the dissipative and radiative components. By doing so it can be seen that the increase in loss at lower thicknesses of the ITO layer, is almost entirely due to the rise in the radiative contribution, whereas the metal loss rate remains nearly constant at ~ 36 ps^{-1}, the rate predicted from the semi-infinite MIM waveguide. The radiative loss however, exponentially increases as the ITO thickness decreases. This behaviour can be understood from the mode profile of the TM_2 mode (Fig. 7.14), as can be seen the field decays exponentially into the upper ITO layer. When this layer is then truncated at a thickness of t the radiated power will be proportional to the intensity of the modal field at the interface with the air cladding. Indeed, fitting the measured radiative loss with a simple exponential gives a decay constant equal to the one describing the decay of the modal fields in the ITO cladding layers, $\gamma = \sqrt{\beta^2 - k_0^2 n_{ITO}^2} \approx 13000$ cm^{-1}.

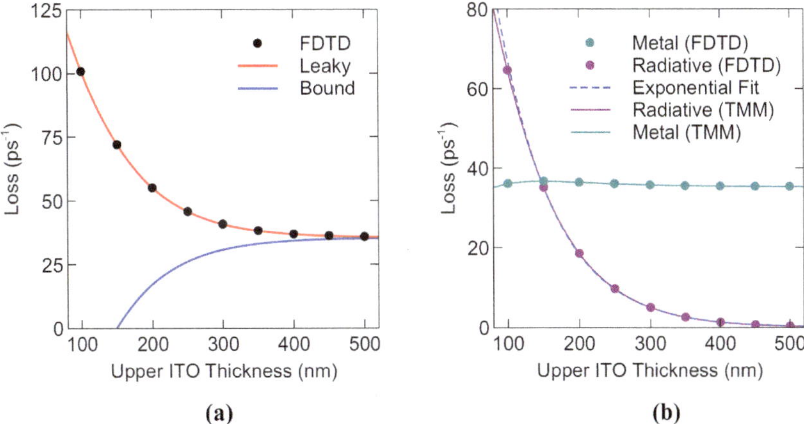

Figure 7.14: **(a)** Dependence of the total modal loss rate on the thickness of the upper ITO layer, by comparing FDTD results (black dots) to the loss rate calculated for bound (blue) and leaky (red) mode solutions it is clear that the leaky solution is the most physically applicable. **(b)** By applying Poynting's theorem it can be seen that the increase in modal loss is almost entirely due to the radiative contribution (magenta) while dissipative loss (cyan) remains constant, excellent agreement is found between analytic leaky mode solutions (lines) to FDTD data (points).

7.8 Conclusion

In this chapter, two waveguide designs have been introduced where opposing flows of energy in layers with positive and negative optical parameters lead to predicted ZGV points. In both cases it was shown how the geometry of the waveguide can be altered to control the group velocity at particular frequencies. Following this loss-free analysis the effect of dissipation on the waveguide modes was investigate solving the dispersion equation for both complex-k and complex-ω solutions. Near the stopped light point the calculated dispersion curves differed strongly depending on which method was used with only complex-ω solutions retaining the stopped light point. Although this difference has noted in the literature generally only the complex-k solutions are reported. Based on these results it had been argued that stopped light is not possible in the presence of loss [Reza *et al.* (2008)]. Here FDTD simulations based on the MIM waveguide were used clarify the argument, finding that complex-ω solutions allow for extremely low group velocities even when dissipation is taken into account. An out-of-plane excitation scheme was used to excite wavepackets of the TM_2 mode and using a new method the velocity was measured directly with the results closely matching those predicted by the complex-ω solutions. The change in central wavevector of the wavepacket due to the frequency dependent loss rate was quantified and found to have a negligible effect on the group velocity over the lifetime of the pulse although any future designs will have to consider this effect. Finally the modal loss rate was considered finding that both radiative and dissipative processes contributed to the overall loss due to the leaky nature of the TM_2 mode. By increasing the thickness of the upper ITO layer it was found that the radiative loss could be decreased exponentially.

Chapter 8

Impact of Surface Roughness on Stopped-Light

8.1 Introduction

In the previous chapter, stopped-light was investigated in NRI and MIM waveguides assuming idealised structures formed of homogeneous layers with perfectly smooth interfaces. However in practice geometric imperfections such as surface roughness can be introduced during fabrication. In light of this, it is important to determine what impact surface roughness will have on device performance if a desired application is to be realised.

The effect of geometric imperfections on stopped-light has been studied previously for photonic-crystal waveguides (PCW) [Engelen *et al.* (2008)]. These structures consist of a periodic array of holes in a dielectric medium which can be arranged to produce specific dispersion characteristics, including low group velocities [Inoue (2002); Notomi *et al.* (2001)]. As the propagation properties are determined by the geometry, slow light in PCWs can be achieved without using metals thus avoiding the associated losses. Unfortunately, the strong dependence on structural dispersion also means that geometric disorder and surface roughness have a large impact on modal propagation. Experimental studies have shown that the presence of geometric imperfections in PCWs currently limits the minimum achievable group velocity to only a few hundredths of the speed of light [Baba (2008); Engelen *et al.* (2008)].

Although similar studies on the impact of roughness on stopped-light in MIM and NRI waveguides are lacking, insight can be gained from the numerous investigations conducted on the propagation of SPPs at single rough interfaces [Raether (1988)] and in multi-layer structures [Leong *et al.* (2011); Min and Veronis (2010)]. For example it is well known that SPPs scattered on rough surfaces can couple to free space radiation, emitting light and thus increasing propagation loss. Other effects such as a shift in the SPP dispersion curve to higher wavevectors [Kolomenski *et al.* (2009)] and the formation of localised surface plasmons [Bozhevolnyi and Vohnsen (1995)] have been observed on surfaces with large amplitude roughness. Strong field enhancements on rough metal surfaces are useful for techniques such as surface-enhanced Raman spectroscopy (SERS) and second-harmonic generation (SHG) [O'Donnell *et al.* (1997)]. However, in most applications, including waveguiding, optical cloaking and perfect lenses [Lee *et al.* (2005)], scattering loss is detrimental and so there has been a growing interest in developing methods of fabricating smoother metal layers [Nagpal *et al.* (2009)]. By using new techniques silver films with subnanometer scale roughness have been demonstrated [Logeeswaran *et al.* (2007)], although further work will be required to improve the reliability of these techniques.

Over the past several decades a number of different models have been developed for understanding and predicting the effects of surface roughness. For example first order approximations of the dispersion equation have proven successful at predicting the observed pattern of light emission from surfaces with low amplitude roughness. Higher order approximations have been applied to explain shifts in the SPP dispersion curve although it can be difficult to properly assess the regimes of validity for these methods. Advances in computer modeling have made it possible to apply new approaches, such as finite element (FEM) and FDTD simulations. As these methods directly solve Maxwell's equations they do not have the same issues as analytic approximations in terms of their regimes of validity.

This chapter investigates the impact of surface roughness on stopped-light in a MIM waveguide using FDTD simulations. In Section 8.2 an overview of previous research into the effect of surface roughness on SPPs relevant to this investigation is presented. Next, a method for generating statistically random rough interfaces and incorporating them into FDTD simulations is introduced in Section 8.3. The impact of roughness on wavepacket propagation in the MIM waveguide near the stopped-light point is then examined in Section 8.4. The effect on dissipative and radiative loss rates is analysed in Section 8.5. Finally the main results are summarised in Section 8.6.

8.2   Scattering Theory

Insight into the effect of surface roughness on the MIM waveguide modes can be gained from the technique of grating coupling, used to excite SPPs in plasmonic waveguides (shown schematically in Fig. 8.1). Due to their bound nature, SPPs cannot be coupled to by EM waves incident from free space as the SPP dispersion lies below the light line. In order to excite SPPs, additional momentum $\Delta k_x = k_x^{(SPP)} - k_x^{(p)}$ has to be provided to match the wavevector $(k_x^{(p)})$ of the incident photons to the SPP wavevector $(k_x^{(SPP)})$. This additional momentum can be provided by incorporating a periodic grating on the metal surface. When incident photons are scattered by the grating their wavevector is increased or decreased by integer multiples of the grating wavevector $(k_g = 2\pi/a)$, where a is the grating constant. Thus the momentum matching required to excite bound SPPs can be achieved by

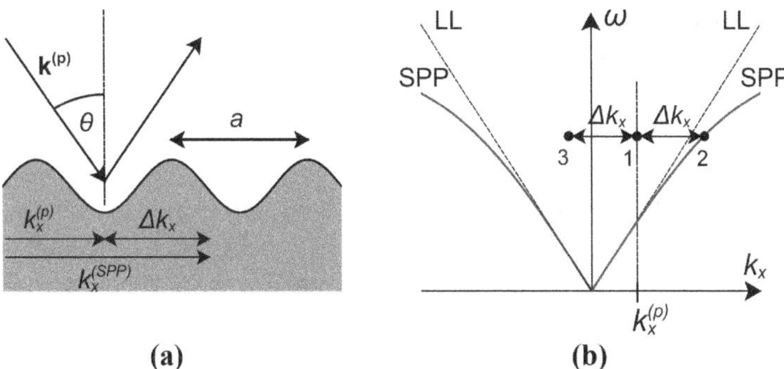

Figure 8.1: **(a)** Schematic of a basic sinusoidal grating with period a that can be used to provide the additional momentum Δk_x required to match the incident wavevector $k_{x(inc)}$ to the SPP wavevector k_{SPP}. **(b)** Dispersion diagram representation of the coupling process.

satisfying the following condition:

$$k_x = k_x^{(p)} + \Delta k_x = k_x^{(SPP)}, \tag{8.1}$$

$$k_x = \frac{\omega}{c}\sin\theta + m\frac{2\pi}{a} = k_x^{(SPP)}, \tag{8.2}$$

where θ is the incidence angle of a photon with frequency ω hitting a grating with grating constant a, and m is an integer representing the diffraction order.

For the correct incidence angle and grating constant SPPs can then be excited from free space. It is important to note that this process is reversible meaning SPPs can couple back to free space radiation thus increasing the modal loss rate. As a rough surface can be described in terms of a superposition of many sinusoidal gratings it follows from Eq. (8.1) that a rough surface will also transform the wavevector of incident light as well as that of waveguide modes. Indeed it is well known that free space radiation can couple to SPPs on rough surfaces. However, whereas the simple sinusoidal grating only provides additional momentum in discrete amounts $\Delta k_x = n\frac{2\pi}{a}$, a statistically rough surface provides a continuum of Δk_x values.

Since the roughness profile of a sample is generally not known and cannot be described by an analytic function, roughness profiles are typically characterised by their statistical autocorrelation function. For a roughness profile $y_r(x) = y_0 + \Delta y(x)$ describing a height perturbation $\Delta y(x)$ along the surface with random amplitudes normally distributed around the mean interface position, y_0, such that $\int \Delta y(x)\,dx = 0$, the autocorrelation function is given as:

$$G(x) = \frac{1}{L}\int_L \Delta y(x')\,\Delta y(x' - x)\,dx', \tag{8.3}$$

where L is the length of the rough surface segment. At $x = 0$ it is $G(0) = \overline{\Delta y^2}$ where $\left(\overline{\Delta y^2}\right)^{1/2} = \delta$ is the root mean square (RMS) height of the roughness which quantifies the magnitude of the roughness. The Wiener–Khinchin theorem states that the Fourier transform of the autocorrelation function will give the corresponding power spectral density (PSD) function of the rough surface:

$$\frac{1}{2\pi} \int_L G(x) \exp(-i\Delta k_x x)\, dx = |S(\Delta k_x)|^2. \tag{8.4}$$

The power spectral density function $|S(\Delta k_x)|^2$ is an important quantity in the study of roughness as it describes the spectrum of Δk_x momenta that can transferred from the surface to the waveguide modes or incident waves. When determining the PSD of unknown roughness profiles it is often assumed that the autocorrelation approximately follows a Gaussian distribution:

$$G(x) = \delta^2 \exp\left(-\frac{x^2}{\sigma^2}\right), \tag{8.5}$$

where δ is the RMS and σ is the autocorrelation length. In this case the PSD is found to be:

$$|S(\Delta k_x)|^2 = \frac{\sigma \delta^2}{2\sqrt{\pi}} \exp\left(-\frac{\sigma^2(\Delta k_x)^2}{4}\right). \tag{8.6}$$

Experimental measurements using atomic force microscopy have shown that the Gaussian function is generally a good approximation, although in some cases, such as polished surfaces, an exponential autocorrelation function is more appropriate. In cases where it is not possible to directly measure the surface topology, the PSD function can be found from the angular dependent emission intensity resulting from SPP scattering. By fitting the analytic PSD expression (Eq. (8.6)) surface roughness characteristics such as the correlation length and RMS can then be extracted. Knowledge of the PSD should prove useful for studying the effect of roughness on the MIM waveguide modes by determining the range of momenta that can be transferred to and from the interfaces. Additionally the PSD can be used to compare the roughness parameters used in simulations to realistic values determined from experiments.

8.3   Implementing Roughness in FDTD

In order to investigate the effect of surface roughness on wavepacket propagation in the MIM waveguide, a method for generating statistically random rough interfaces and incorporating them into FDTD simulations was required. Following from the previous section it is assumed that the roughness profile can be described as a random height perturbation with amplitudes following a normal distribution around the average interface position between adjacent layers in the waveguide. This is represented by the roughness function $\Delta y(x)$ with a zero mean value and a magnitude characterized by the root mean square height δ. Each individual realisation of the roughness function $\Delta y(x)$ is created by choosing a random displacement of the interface, in the vertical direction, at a single point every 10 nm along the x-axis,

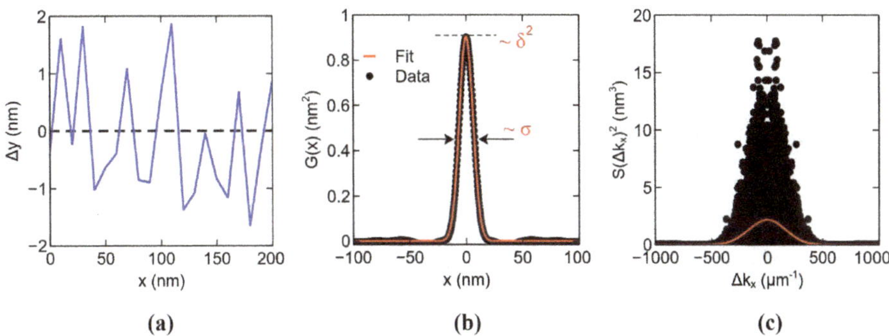

Figure 8.2: **(a)** 200 nm section of a typical roughness profile generated with an RMS height of 1 nm, **(b)** the corresponding autocorrelation function of the full 200 µm long roughness profile calculated numerically (black dots) and fit with a Gaussian function (red line) used to extract the RMS height and correlation length. **(c)** The numerically calculated (black dots) PSD compared to the analytically predicted form (red line).

uniformly distributed in the range $\pm h$ where $h \approx 2\delta$ approximately twice the desired RMS height. Linear interpolation is then performed to more accurately resolve $\Delta y(x)$. A short 200 nm section of an example roughness profile $\Delta y(x)$ generated for $h = 2$ nm is shown in Fig. 8.2a, in practice $\Delta y(x)$ extends along the whole length of the computational domain which for these simulations is 200 µm long.

As this is a computational implementation, where the roughness profile has been explicitly defined, the autocorrelation function can be directly computed by:

$$G(x) = \mathrm{FT}^{-1}\left[\mathrm{FT}^{-1}[\Delta y(x)]\,\mathrm{FT}[\Delta y(x)]\right], \qquad (8.7)$$

where FT anf FT^{-1} represent the discrete Fourier and inverse Fourier transforms respectively. The power spectral density function can then be found by taking the Fourier transform of $G(x)$. An example plot of both the autocorrelation function and the PSD for a typical roughness profile generated with $h = 2$ nm is shown in Figs. 8.2b,c. By fitting a Gaussian function to the calculated autocorrelation distribution, the RMS height and correlation lengths can be extracted. For a sample size of 100 different roughness profiles, generated with $h = 2$ nm, the average RMS height was found to be $\delta = (0.946 \pm 0.005)$ nm with an average correlation length of $\sigma = (8.5 \pm 0.1)$ nm which compares well to experimental studies of roughness on ITO thin films [Raoufi *et al.* (2007); Kim and McCall (2003)]. The RMS value can be increased via the peak height h while the correlation length is dependent on the distance between each randomly chosen height perturbation along the x-axis (here always set at 10 nm).

The PSD of the roughness profile was then calculated by taking the discrete Fourier transform of the autocorrelation function and compared to the analytic expression, Eq. (8.6) using the RMS height and correlation length extracted from the autocorrelation function. In terms of the Δk_x distribution good agreement is observed between the two sets of results, although it was found that the analytic expression underestimates the amplitude by a factor of approximately 4–6 times.

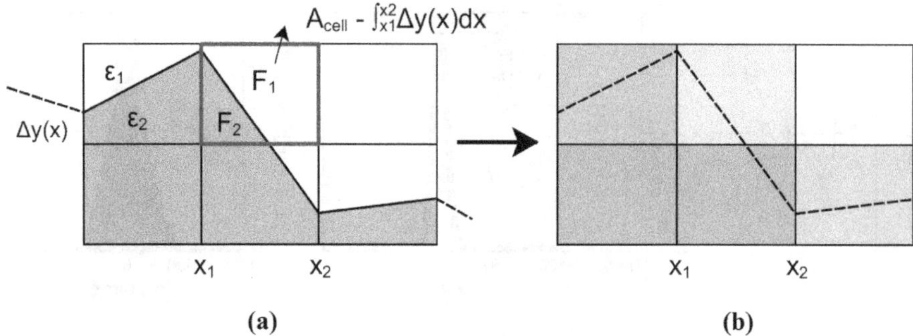

Figure 8.3: **(a)** Process of applying surface roughness to the discretised FDTD grid using integration of the roughness function to determine material filling factors $F_{1,2}$. **(b)** Permittivity of grid cells near the rough interfaces are then calculated via the weighted sum of the two material permittivites $\varepsilon_{1,2}$.

This is due to the contribution of the random non-zero side tails of the numerically determined autocorrelation function. From the PSD it can be seen that a very wide range of Δk_x momenta can be transferred from the rough surface to both waveguide modes and free space radiation. This indicates that rough interfaces could potentially lead to coupling between different waveguide modes as well as increasing the radiative loss, which will be studied in a later section.

Next the statistically rough interfaces have to be incorporated into the FDTD simulations. Unfortunately due to computational limitations the surface roughness profile can not be fully resolved in the simulations as there is a large discrepancy in scales between the wavepacket envelope (≈ 100 µm) and roughness features ≈ 1–10 nm. To overcome this problem, the FDTD grid cells around the rough interfaces are assigned an effective permittivity based on the volume effective permittivity (VEP) method. Here, the roughness profile $y_r(x) = y_0 + \Delta y(x)$ is numerically integrated inside each cell it passes through, to find the filling factors F_1 and F_2 of the materials above and below the interface (see Fig. 8.3). The effective permittivity of these grid cells are then determined by the weighted sum $\varepsilon_{eff} = \varepsilon_1 F_1 + (1 - F_1)\varepsilon_2$. Using this method the rough interfaces are approximated as thin layers exhibiting a randomly varying permittivity along the length of the waveguide similar to the method used in a number of published works. With roughness applied to all interfaces in the MIM waveguide simulations could then be performed to determine the effect on the propagation and loss characteristics.

8.4   Effect on Velocity

Propagation of wavepackets in the MIM waveguide will be influenced by four main factors:

(a) The underlying dispersion of the smooth waveguide structure.

(b) The angle of incidence, which determines where on the dispersion curve the wavepacket will lie.

(c) The RMS height of the surface roughness.

(d) The local topology of surface roughness in the region of the wavepacket.

In the case of smooth interfaces the influence of waveguide dispersion can be revealed via the angular dependence of propagation; for example wavepackets excited with angle of incidence $\theta_{inc} = 21.6°$ will propagate with a positive group velocity while at $13.6°$ the group velocity is negative. As the group velocity of the complex-ω mode varies continuously, as shown in the previous chapter, then there is an intermediate angle where the group velocity goes to zero and the wavepacket is stopped, see Fig. 8.5a.

In rough waveguides, to separate the four processes influencing propagation, identified above, wavepackets were excited in the MIM waveguide at two incident angles for a variety of RMS heights and repeated 20 times for different roughness profiles. The change in position of the wavepacket's centre of energy over time was recorded in each case (see Fig. 8.4). As in the smooth case, comparing the propagation at different angles of incidence will reveal any influence of the underlying modal dispersion. Repeating the simulations 20 times for each RMS height should provide a reasonable sample size in order to separate the systematic impact of increased rms height from the random nature of the particular local roughness topology.

In these simulations an incident plane wave with a Gaussian envelope was used to excite the TM_2 mode with the out-of-plane excitation scheme described in the

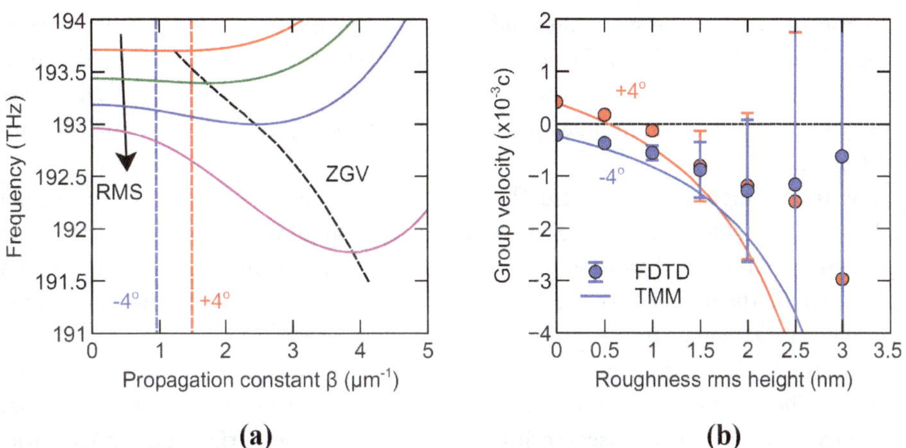

Figure 8.4: **(a)** Dispersion curves of the TM₂ mode in MIM waveguides with smooth (red) and rough interfaces where RMS = 1 nm (green), 2 nm (blue) and 3 nm (pink). Roughness is modelled as a slight variation of permittivity near the interfaces proportional to RMS height. The changing position of the ZGV point is marked by the black dashed line while red and blue dashed lines highlight wavevectors corresponding to the two angles of incidence used in simulations. **(b)** The average group velocity (points) extracted from simulations compared to group velocity calculated from dispersion curves **(a)** for angles of incidence $\theta_{ZGV} \pm 4°$ (red, blue respectively).

previous chapter. The incident frequency was set to correspond to $\lambda_0 = 1.55\,\mu\text{m}$ and incidence angles of $\theta_i = (17.6 \pm 4)^\circ$ were used with the beam FWHM set equal to $30\lambda_0$. The incident wave was switched on smoothly over a period of 200 fs, then sustained at its peak amplitude for 800 fs before being switched off over another 200 fs. Propagation of the wavepacket was then measured using the centre of energy method discussed in the previous chapter. Each trace of the centre of energy was normalised to zero at $t = 1.1$ ps so that the change in position could be easily compared between simulations. Roughness RMS heights between 0.5–3 nm were investigated increasing in 0.5 nm increments. The combined results of 20 simulations at each RMS height and incidence angle are shown in Figs. 8.5b–g.

Remarkably, in simulations with low levels of surface roughness (0.5–1 nm RMS) it was found that the wavepackets continue to propagate with a constant velocity, shown by the linear evolution of the centre of energy position with time. Additionally, the propagation, and hence group velocity, still shows a clear dependence on the angle of incidence. This indicates that the underlying waveguide dispersion still has a strong influence on wavepacket propagation and so, assuming the group velocity varies continuously, there should still be an angle of incidence which corresponds to a zero group velocity. To test this, the angle of incidence was optimised for MIM waveguides with RMS heights of 0.5 and 1 nm. In both cases it was found that near zero propagation was indeed possible at shifted incidence angles of $\theta = 19.58^\circ$ and $\theta = 22.67^\circ$. In comparison, wavepackets are stopped in the smooth structure at an angle of $\theta = 17.6^\circ$. Although it is still possible to achieve near zero group velocity the surface roughness does have an increasing influence on propagation. This appears as both a systematic effect, whereby the group velocity becomes increasingly negative as the RMS height increases (shown by the negative shift in the average end position of the wavepacket) and a random effect observed in the spread of propagation paths for simulations with different roughness profiles but the same RMS height.

For RMS heights above 1 nm the propagation characteristics change dramatically. There is no longer a clear angular dependency, showing that the local roughness profile now has a stronger influence on propagation than the underlying waveguide dispersion. As the RMS height increases up to 3 nm the propagation of wavepackets begins to resemble a diffusion-like process, being equally likely to travel in the positive or negative x-directions and in many cases with a non-constant velocity. The random motion of the wavepacket in the presence of large roughness is due to an increased backscattering of energy from the surface inhomogeneities. This is clearly evident in the x-component of the electromagnetic Poynting vector shown in Fig. 8.6. In the smooth waveguide, energy in the dielectric core only travels in the positive x-direction whereas at 3 nm RMS pockets of negative energy flow in the core are observed. This enhanced backscattering leads to a breakup of the pulse envelope and so the concept of group velocity is no longer applicable in this situation. Field hotspots along the rough interfaces can be seen in plots of

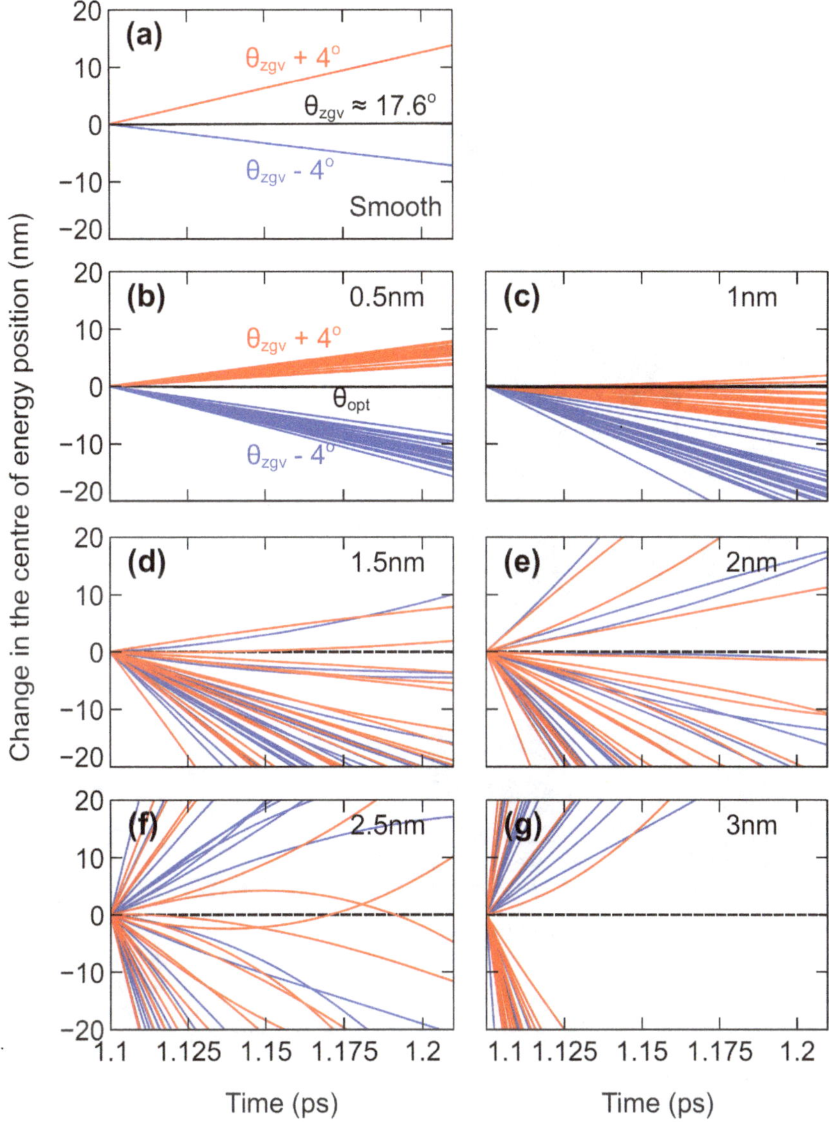

Figure 8.5: Change in the centre of energy position of wavepackets excited in MIM waveguides with smooth **(a)**, and rough interfaces (0.5–3 nm RMS height) **(b–g)**. Red lines indicate wavepackets incoupled at an angle of incidence $\theta = \theta_{ZGV} + 4°$, while blue lines correspond to an incidence angle $\theta = \theta_{ZGV} - 4°$. For cases **(a–c)** an optimal angle has been found corresponding to near zero propagation shown by the solid black lines. The dashed black lines in **(d–g)** mark the zero position.

the electric field amplitude. At these positions the electric field shows a strong enhancement in comparison to the smooth waveguide reaching up to 3.5 times higher in structures with 3 nm RMS. As mentioned previously, enhancement of the electric field on rough surfaces is well known and can be used for different applications

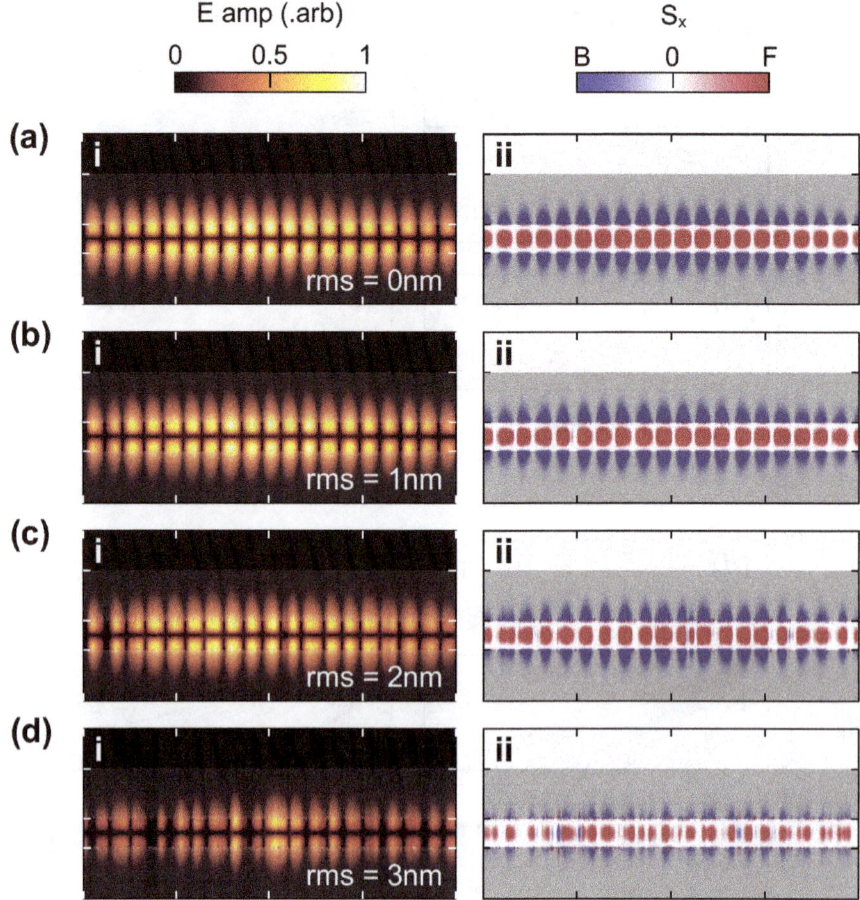

Figure 8.6: Electric field amplitude (**a.i–d.i**) and the x-component of the Poynting vector (**a.ii–d.ii**) in the MIM waveguide with smooth interfaces (**a**), and rough interfaces with RMS = 1 nm (**b**), 2 nm (**c**), and 3 nm (**d**). Higher electric field amplitudes are shown in lighter colors. For the Poynting vector energy flow in the positive x-direction is shown in red while blue denotes energy flow in the negative x-direction. Grey boxes highlight the positions of ITO layers.

such as SERS and second harmonic generation. However, for stopped-light, the change in behaviour from waveguide mode to localised plasmons is detrimental, as the propagation can no longer be controlled by the waveguide dispersion.

To better understand the systematic negative shift in the wavepacket finishing position observed in the MIM waveguide with low roughness (0.5–1 nm), TMM calculations are performed. Discretising the roughness profile onto the FDTD grid only affects grid cells directly above and below an interface. The permittivity of these grid cells are then given by the weighted average of the permittivity of the two materials either side of the interface. As the wavelength inside the waveguide is much longer than the roughness correlation length $\lambda_{eff} = \lambda_0/n_{eff} \approx 5090$ nm $\approx 600\sigma$, the two thin strips (of grid cells) either side of an interface can be treated as an effective medium in the propagation direction. The permittivity of these effective layers is assumed to be equal to the average effective permittivity of the FDTD grid cells next to the waveguide interfaces, averaged along the whole length of the waveguide. The dispersion and modal loss of the MIM waveguide, incorporating these effective layers (10 nm in height) near the interfaces, was then calculated using the TMM. It was found that, as the RMS height increased, the zero group velocity point shifted to a higher wavevector and lower frequencies. This means that the group velocity at wavevectors corresponding to the two incident angles used in the FDTD simulations become increasingly negative, qualitatively matching the behaviour observed in FDTD simulations up to RMS heights of around 1.5 nm. Above this RMS height the group velocity extracted from FDTD simulations is no longer meaningful in describing propagation due to breakup of the pulse. This comparison between FDTD simulations and TMM calculations suggest that the systematic effect of roughness at low rms heights is primarily due to the average change in permittivity around the rough interface while scattering effects become increasingly dominant as the RMS height increases.

8.5 Modal Loss

The effect of surface roughness on modal loss was investigated in FDTD simulations using the loss rate extraction method introduced in the previous chapter. In order to accurately capture the radiative loss rate, the upper edge of the integration contour was set to be 20 nm above the waveguide surface. Loss rate measurements were taken from the same simulations used in the previous section. From Fig. 8.7 it can be seen that the total loss rate increases approximately exponentially with the roughness RMS height. Interestingly when the total loss rate is separated into radiative and dissipative contributions it was found that the increased loss came solely from a rise in dissipative losses while the radiative loss actually fell, in contrast to studies of SPPs on single interfaces where radiative loss is generally increased by roughness. These results are in good agreement with a previous study conducted on the effect of roughness on modal loss in a similar metal-insulator-metal waveguide. In [Min and Veronis (2010)] the authors also found that losses in a rough MIM waveguide increase almost exponentially with the rms height which was attributed to reflections from the rough surface and enhanced dissipation in the metal layers. For correlation lengths below $\sim$ 20 nm (here 8 nm is used) it was demonstrated that the increased loss rate was primarily due to a rise in dissipative absorption, while at higher correlation lengths reflection from the surfaces came to dominate. Scattering losses peaked at correlation lengths equal to $\sigma = \lambda_{eff}/2\sqrt{2}$ (where λ_{eff} is the modal wavelength in propagation direction), associated with Bragg reflection from the interfaces. For the TM_2 mode at the stopped light point the effective wavelength is $\lambda_{eff} \approx 4.5\,\mu m$ and so Bragg reflection would not be expected to occur until the roughness correlation length approaches $\sigma \approx 1.6\,\mu m$.

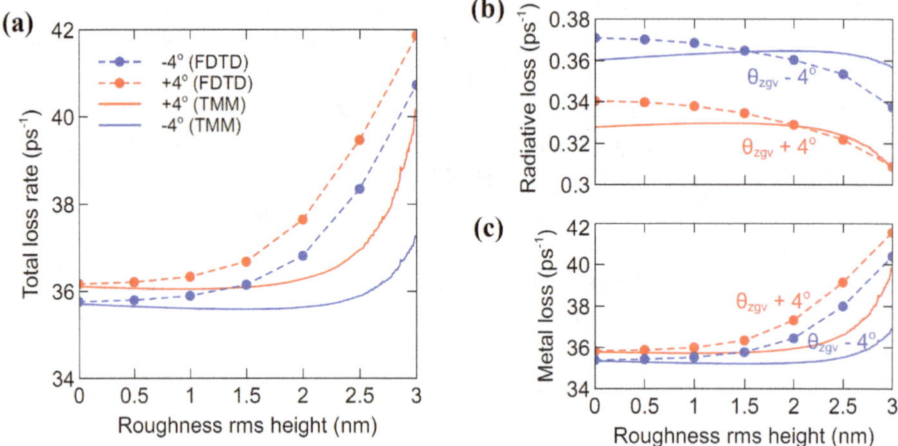

Figure 8.7: The change in total **(a)**, radiative **(b)**, and dissipative **(c)** loss rates with increasing RMS height extracted from FDTD simulations (points) compared to rates calculated from the TMM solver (lines) for two incidence angles $\theta_{ZGV} \pm 4°$ (red, blue respectively).

To gain greater insight into the effect of roughness on absorption, loss rates were also calculated using the TMM, assuming the inclusion of thin effective medium layers as discussed in the last section. These loss rates are compared to the loss rate extracted from FDTD simulation in Fig. 8.7. In this case the TMM results actually predict an initial drop in the total loss rate before increasing sharply for RMS heights above 2 nm primarily due to the change in dissipative losses. The small contribution from radiative loss is observed to rise slightly before decreasing again above 2 nm RMS height. Differences between the loss rates calculated from the TMM and those extracted from FDTD simulation are likely due to scattering effects.

Previous works investigating SPP propagation on rough metal surfaces found that increased radiative loss was caused by momenta being transferred from the metal surface to the SPP mode reducing the wavevector inside the light cone and transforming the SPP into freely propagating photons. Due to the flatness of the TM$_2$ modal dispersion studied here, scattering to lower wavevectors will only result in transformation into the same TM$_2$ mode but at a different wavevector. As the dissipative and radiative loss rates of the TM$_2$ mode are relatively uniform over the wavevector range 0 to $5\,\mu m^{-1}$, where there is a frequency overlap with the incident frequency, transformation to the TM$_2$ mode has only a small impact on the loss rate. Larger transfers of momentum could potentially transform the TM$_2$ mode into a SPP on the waveguide surface due to the frequency overlap of the dispersion curves (see Fig. 8.8). This would require a momentum transfer equivalent to $\Delta k_x \approx 3.9\,\mu m^{-1}$ which is well within the roughness PSD half width (HWHM $\approx 200\,\mu m^{-1}$) calculated in Section 8.3. TMM calculations show that dissipative losses of the SPP are approximately 1.5 times higher at the incident

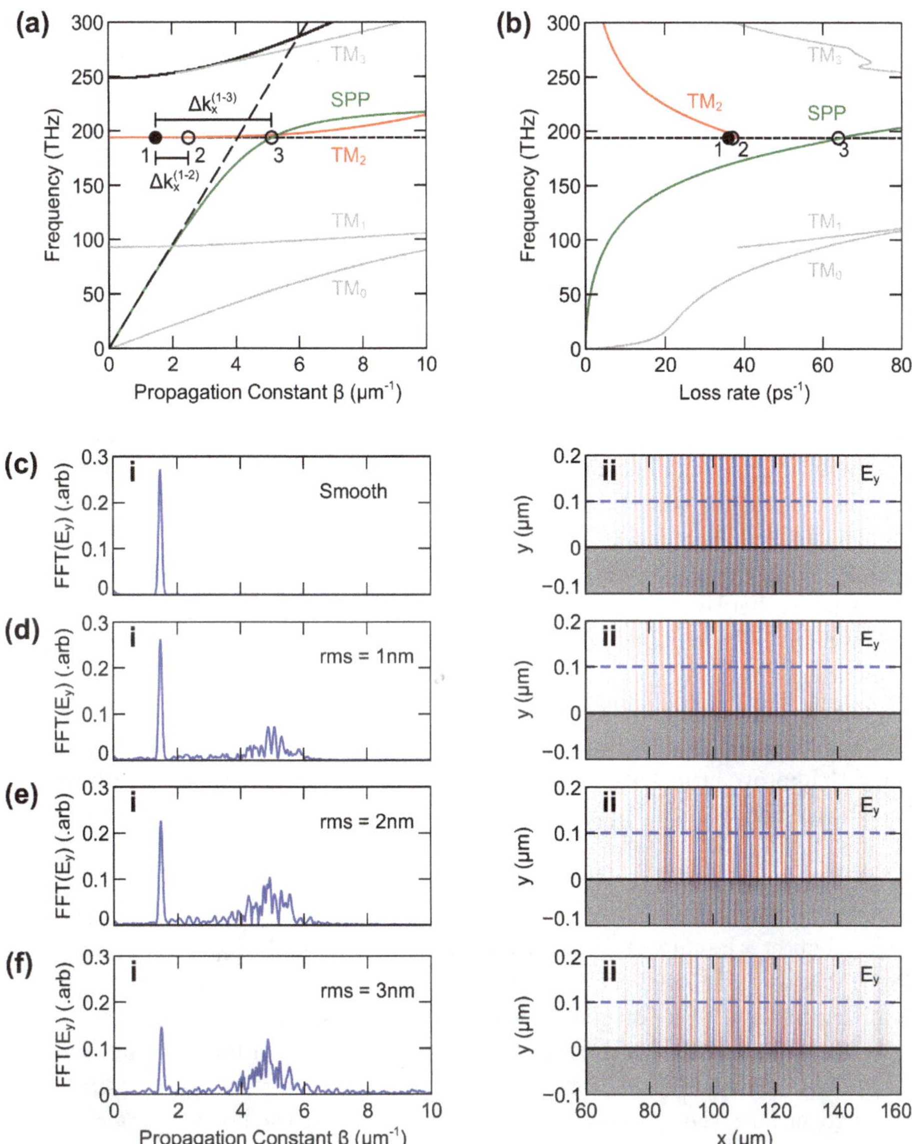

Figure 8.8: **(a)** Dispersion diagram of TM modes in the MIM waveguide highlighting the potential transformation of momentum from the incident wavevector (point 1) to higher wavevectors (point 2) or to the SPP mode (point 3). **(b)** The corresponding loss rates of waveguide modes. **(c.i–f.i)** k_x-spectra of the E_y field component **(c.ii–f.ii)** 100 nm above the waveguide surface (blue dashed line) for smooth **(c)**, and rough interfaces with RMS = 1 nm **(d)**, 2 nm **(e)**, and 3 nm **(f)**.

frequency ($\lambda_0 = 1.55\,\mu\mathrm{m}$) than for the TM_2 mode, while the radiative losses are zero as the SPP is a bound mode. Thus coupling to the SPP could explain the difference in behaviour of the dissipative and radiative losses observed in FDTD simulations compared to the TMM calculations Fig. 8.7.

In order to test whether the rough interfaces were causing a transformation between the TM_2 and SPP modes the electric field k-spectrum near the waveguide surface was extracted. This was done by taking the Fourier transform of the y-component of the electric field 100 nm above the upper ITO surface along the length of the waveguide (see Fig. 8.8 (c.ii–f.ii)). The extracted k-spectrum was found to peak at a value of $k \approx 1.49\,\mu m^{-1}$ in the absence of roughness, which corresponds simply to the incidence angle used in these simulations $k = k_0 \sin(21.56°) = 1.49\,\mu m^{-1}$. When roughness is introduced the k-spectra still peak strongly at $k \approx 5\,\mu m^{-1}$ which grows in amplitude as the roughness RMS height increases. The central wavevector of this second peak closely matches the wavevector of the SPP mode at the incidence frequency $k_{SPP}(\lambda_0) \approx 5.15\,\mu m^{-1}$, giving strong evidence that the SPP mode has been excited, which can only be a result of momentum transfer from the surface (due to the bound nature of the SPP). Although the SPP peak becomes comparable to the TM_2 peak for large rms heights it is still much lower than the TM_2 fields at the centre of the waveguide which are roughly 12 times higher than at the surface. Hence, only a small amount of energy is being lost by the TM_2 mode to the SPP although this could be enough to explain the difference between the loss rate extracted from FDTD simulations and those predicted by TMM calculations. Coupling to SPPs could potentially be mitigated by including a thin layer of high dielectric above the upper ITO layer which would shift the plasmon resonance to lower frequencies without significantly affecting the TM_2 modal dispersion. This could reduce the loss rate, although it would not be expected to improve propagation characteristics of the wavepacket, as this is primarily affected by energy localising on the rough interfaces of the waveguide core.

8.6 Conclusion

In this chapter, the impact of surface roughness on modal propagation and loss in the MIM waveguide has been investigated. Initially, background theory regarding the propagation of SPPs on rough surfaces was discussed. It was shown that rough surfaces can be characterised by a Gaussian autocorrelation function and that momentum can be transferred between waveguide modes and free-space photons via the rough surface, the range of momenta being given by the power spectral density.

Following this introduction, a method for generating suitable roughness profiles and applying them to the discrete FDTD grid was presented. Surface characteristics such as rms height and correlation length were extracted by calculating the autocorrelation function. These results compared well with experimentally reported measurements of ITO surfaces. Finally, the power spectral density was calculated, indicating that rough interfaces could transfer a very broad spectrum of momenta to the waveguide modes.

Next, wavepackets, corresponding to the TM_2 mode near the stopped light point, were injected into the rough MIM waveguide. It was shown that surface roughness had both a systematic effect on propagation, resulting from the average change in permittivity near the interfaces, and random effects, due to scattering from surface inhomogeneities. Remarkably, it was found that extremely low group velocities could still be achieved in the MIM waveguide when the RMS height of the rough surfaces was below 1.5 nm. The resilience of stopped light to surface roughness can be attributed to the fact that, in the MIM waveguide, stopped light arises from counter propagating energy flows in the metal and dielectric layers. As this mechanism is primarily related to the bulk material dispersion of the metallic layers rather than, for example, coherent backscattering in photonic crystal waveguides, the opposing energy flows are still observed in the presence of surface roughness and thus light can still be decelerated to extremely low group velocities.

For very rough interfaces, corresponding to rms heights of 1.5–3 nm, the modal fields began to show strong localisation to surface inhomogeneities. This resulted in a breakup of the pulse envelope and so the concept of group velocity could no longer be strictly applied in the FDTD simulations. Propagation became increasingly random with no apparent dependence on incident angle showing that the influence of roughness was greater than the underlying waveguide dispersion.

The rough interfaces also affected modal attenuation, leading to an increase of dissipative loss, while radiative losses decreased. This could not be fully explained by the change in average permittivity near interfaces which only lead to an increase in the loss rate at very high RMS heights. It was found that due to the transfer of momentum from the rough waveguide surface some of the TM_2 modal energy was being converted into SPPs confined to the Air/ITO interface. This lead to an increase in the overall dissipative loss increased due to the higher loss rate of the SPP, while radiative losses decreased, reflecting the bound nature of the SPP mode. These results are in good agreement with a previous study investigating modal loss in a similar MIM waveguide design. In [Min and Veronis (2010)] the authors reported an increase in dissipative loss as the RMS height of the surface roughness was raised, which was primarily due to increased dissipation in the metal when the roughness correlation length was below 20 nm as is the case for results presented in this chapter.

This chapter concludes the portion of this thesis devoted to studying stopped light in passive NRI and MIM waveguides. Through semi-analytic TMM calculations and numerical FDTD simulations it has been shown that zero group velocity can still be achieved even in the presence of dissipative loss and low levels of surface roughness. Thus showing that light can indeed be stopped in realistic structures, potentially providing a route to realising some of the proposed applications of stopped light. In the next half of this work the interaction of stopped-light with gain media is investigated in active NRI and MIM waveguides leading to the new concept of stopped-light lasing.

Chapter 9

Maxwell–Bloch Theory and Gain

In this chapter we briefly outline some method to compensate losses in metamaterials. It is based on introducing gain element. Such system has been modelled by employing Maxwell–Bloch approach.

Computational studies of the gain-enhanged metamaterials have a long history, see review by Wuestner and Hess [Wuestner and Hess (2014)]. The simplest model was based on a two-level system coupled to a plasmonic resonance [Wegener *et al.* (2008)]. In this approach spatial dependence of the plasmon-gain interaction was neglected. However, based on finite element frequency-domain calculations it was shown later that spatial dependence of the coupling affects loss compensation properties [Sivan *et al.* (2009)]. Similar simulations on loss compensation based on spatially resolved time-domain were reported by Fang *et al.* [Fang *et al.* (2009)]. Therefore more realistic modeling was needed and this approach is summarised here. First, we will illustrate main concept on an example of two-level (TLS) system.

9.1 Preliminaries. Two-Level System

Maxwell equations were introduced in Chapter 1. Combining Maxwell's equations one obtains

$$\nabla^2 - \frac{1}{c^2}\frac{\partial^2}{\partial t^2}\mathbf{E} = \mu_0 \frac{\partial^2}{\partial t^2}\mathbf{P}. \tag{9.1}$$

The above is the wave equation driven by the polarisation of the medium. Polarisation $\mathbf{P}$ is introduced as

$$\mathbf{D} = \varepsilon_0 \mathbf{E} + \mathbf{P}. \tag{9.2}$$

For linear medium one obtains in the frequency domain

$$\mathbf{P}(\omega) = \varepsilon_0 \chi(\omega)\mathbf{E}(\omega). \tag{9.3}$$

One finally obtains

$$\left(\nabla^2 + \frac{\omega^2}{v^2}\right) \mathbf{E}\left(\omega\right) = 0 \tag{9.4}$$

where $1 + \chi\left(\omega\right) = n^2$, where n is the refractive index and $v = c/n$ is the velocity of light in a medium.

9.1.1 *Bloch Equations*

We will illustrate the origin of Bloch equation using two-level system (TLS) interacting with the electric or magnetic field as an example. Bloch equations are used to describe atoms which show many discrete energy states. There are possible transitions between those states. In many applications, say related to laser transitions it is enough to consider only two energy levels (such an approach is known as TLS). More complicated models also exist.

Lets introduce two energy eigenstates $|1\rangle$ and $|2\rangle$ corresponding to two energy levels E_1 and E_2, see Fig. 9.1. They are known as a ground state and an excited state. We disregard all the remaining energy levels. The following relations exist (in the following we use Dirac notation)

$$H_0 |1\rangle = E_1 |1\rangle \tag{9.5}$$

and

$$H_0 |2\rangle = E_2 |2\rangle \tag{9.6}$$

where H_0 is the Hamiltonian. In the selected basis it can be expressed as

$$H_0 = E_1 |1\rangle \langle 1| + E_2 |2\rangle \langle 2| . \tag{9.7}$$

Link with 'ordinary' wave functions in position representation is established as

$$\psi_1(x) = E_1 \langle x | 1\rangle \quad \text{and} \quad \psi_2(x) = E_2 \langle x | 2\rangle \tag{9.8}$$

One thus works in two-dimensional (2D) Hilbert space. In this space eigenstates $|1\rangle$ and $|2\rangle$ are selected as

$$|1\rangle = \begin{pmatrix} 1 \\ 0 \end{pmatrix}, |2\rangle = \begin{pmatrix} 0 \\ 1 \end{pmatrix} . \tag{9.9}$$

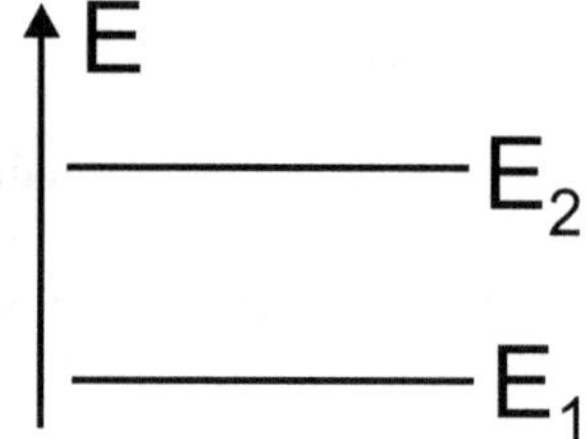

Figure 9.1: Two-level system.

In this space there are 4 linearly independent operators which are

$$1 = |1\rangle\langle 1| + |2\rangle\langle 2|$$

$$\sigma_x = |1\rangle\langle 2| + |2\rangle\langle 1|$$

$$\sigma_y = i\left(|2\rangle\langle 1| - |1\rangle\langle 2|\right)$$

$$\sigma_z = |1\rangle\langle 1| - |2\rangle\langle 2|.$$

The above objects are known as Pauli matrices and in a matrix form are

$$\sigma_x = \begin{pmatrix} 0 & 1 \\ 1 & 0 \end{pmatrix}, \sigma_y = \begin{pmatrix} 0 & -i \\ i & 0 \end{pmatrix}, \sigma_z = \begin{pmatrix} 1 & 0 \\ 0 & -1 \end{pmatrix}. \tag{9.10}$$

9.1.2 *Interaction of TLS with Electromagnetic Field*

The interaction of TLS with electromagnetic field $\mathbf{E}$ is described by the Hamiltonian [Scully and Zubairy (1997)]

$$H_{int} = -\mathbf{E} \cdot \boldsymbol{\mu}_e \tag{9.11}$$

where $\boldsymbol{\mu}_e$ is the atomic electric dipole moment. Dipole moment is

$$\boldsymbol{\mu}_e = e\mathbf{r} = \mathbf{d}. \tag{9.12}$$

Assume that electric field is polarised along x-axis. Within dipole approximation the interaction Hamiltonian becomes

$$H_1 = -exE(t) = -exE_0 \cos\omega t. \tag{9.13}$$

In our basis it is evaluated as follows

$$\begin{aligned}
H_1 &= -exE(t) = -eE(t)\left(|2\rangle\langle 2| + |1\rangle\langle 1|\right)x\left(|2\rangle\langle 2| + |1\rangle\langle 1|\right) \\
&= -E(t)\left\{p_{21}|2\rangle\langle 1| + p_{12}|1\rangle\langle 2|\right\}.
\end{aligned} \tag{9.14}$$

We have defined matrix element of the transition as

$$p_{21} = e\langle 2|x|1\rangle = p_{12}^*. \tag{9.15}$$

Observe that there are no transitions between the same levels, since

$$e\langle i|x|i\rangle = e\int x\,|\psi_i(x)|^2\,dx = 0. \tag{9.16}$$

Matrix elements vanish because x is an odd function while $|\psi_i(x)|^2$ is even. From the above Hamiltonian one obtains two equations of motion

$$\dot{C}_2(t) = -i\omega_2 C_2(t) + i\Omega_R e^{-i\phi}\cos\omega t C_1(t) \tag{9.17}$$

$$\dot{C}_1(t) = -i\omega_1 C_1(t) + i\Omega_R e^{i\phi}\cos\omega t C_2(t) \tag{9.18}$$

where Ω_R is the Rabi frequency.

9.1.3 *Rotating Wave Approximation (RWA)*

Introduce new (slowly varying) amplitudes $\overline{C}_2(t)$ and $\overline{C}_g(t)$

$$\overline{C}_e(t) = C_2(t)e^{i\omega_2 t} \tag{9.19}$$

and

$$\overline{C}_1(t) = C_1(t)e^{i\omega_1 t} \tag{9.20}$$

where

$$E_i = \hbar\omega_i, i = 1, 2. \tag{9.21}$$

Taking time derivatives, substituting into (9.17) and multiplying by $e^{i\omega_2 t}$ gives

$$\dot{\overline{C}}_2(t) = i\Omega_R e^{-i\phi} \cos{(\omega t)}\,\overline{C}_2(t)e^{i(\omega_2 - \omega_1)t}. \tag{9.22}$$

In a similar way one obtains

$$\dot{\overline{C}}_1(t) = i\Omega_R e^{i\phi} \cos{(\omega t)}\,\overline{C}_2(t)e^{-i(\omega_2 - \omega_1)t}. \tag{9.23}$$

The time dependent exponential terms can be rewritten using

$$\cos{(\omega t)} = \frac{1}{2}\left(e^{i\omega t} + e^{-i\omega t}\right)$$

as

$$\cos{(\omega t)}\,e^{i(\omega_2 - \omega_1)t} = \cos{(\omega t)}\,e^{i\omega_0 t} = \frac{1}{2}\left[e^{i(\omega_0 + \omega)t} + e^{i(\omega_0 - \omega)t}\right]$$

where we have defined

$$\omega_0 = \omega_2 - \omega_1. \tag{9.24}$$

For $|\omega - \omega_0| << \omega$ we can neglect fast oscillating terms $(\omega + \omega_0)\,t$. The evolution of the system will be determined by the slow oscillating term. This approximation is known as the *rotating wave approximation (RWA)*. Within the RWA the resulting equations are

$$\dot{\overline{C}}_2(t) = i\frac{1}{2}\Omega_R e^{-i\phi}\overline{C}_1(t)e^{i(\omega_0 - \omega)t} \tag{9.25}$$

and

$$\dot{\overline{C}}_1(t) = i\frac{1}{2}\Omega_R e^{i\phi}\overline{C}_2(t)e^{-i(\omega_0 - \omega)t}. \tag{9.26}$$

The difference $\omega_0 - \omega = \delta$ is known as detuning. For zero detuning, i.e. $\omega = \omega_0$ the system performs the so-called Rabi oscillations between the ground and excited state of the driven TLS.

Lets assume that at $t = 0$ the system is in the ground state with $|C_1|^2 = 1$ and $|C_2|^2 = 0$. Solving the above equations one finds

$$|C_1|^2 = \cos^2\left(\frac{1}{2}\Omega_R t\right) \tag{9.27}$$

$$|C_2|^2 = \sin^2\left(\frac{1}{2}\Omega_R t\right). \tag{9.28}$$

9.1.4 *The Density Operator*

To describe dissipative processes it is useful to switch to a statistical description using the density operator instead of wave function. The density operator of a pure state is defined as

$$\rho = |\psi\rangle \langle\psi| . \tag{9.29}$$

In matrix form it is

$$\rho = \begin{pmatrix} \rho_{22} & \rho_{21} \\ \rho_{12} & \rho_{11} \end{pmatrix} . \tag{9.30}$$

It obeys the following equation of motion [Scully and Zubairy (1997)]

$$\frac{\partial \rho}{\partial t} = -\frac{i}{\hbar} [H, \rho] - \frac{1}{2} \{\Gamma, \rho\} \tag{9.31}$$

where $\{\Gamma, \rho\} = \Gamma\rho + \rho\Gamma$. Here matrix Γ is responsible for damping. Its matrix elements are

$$\langle n| \Gamma |m\rangle = \gamma_n \delta_{nm}. \tag{9.32}$$

Here, $n, m = 2, 1$. From above, one obtains

$$\frac{d\rho_{22}}{dt} = -\gamma_2 \rho_{22} + \frac{i}{\hbar} \left(p_{21} E \rho_{12} - c.c. \right) \tag{9.33}$$

$$\frac{d\rho_{11}}{dt} = -\gamma_1 \rho_{11} - \frac{i}{\hbar} \left(p_{21} E \rho_{12} - c.c. \right) \tag{9.34}$$

$$\frac{d\rho_{21}}{dt} = -\left(i\omega + \gamma_{21} \right) \rho_{21} - \frac{i}{\hbar} p_{21} E \left(\rho_{22} - \rho_{11} \right) \tag{9.35}$$

where $\gamma_{21} = (\gamma_2 + \gamma_1)/2$.

9.2 Optical Bloch Equations for a Two-Level System

With electric field and using dipole approximation the real-valued probabilities of occupation ρ_{11} and $\rho_{22} = 1 - \rho_{11}$, are given by [Boyd (2008)]

$$\frac{\partial \rho_{12}}{\partial t} = \frac{\partial \rho_{21}^*}{\partial t} = -\left(i\omega_0 + \Gamma \right) \rho_{12} - i\frac{\mu \cdot \mathbf{E}}{\hbar} \left(\rho_{22} - \rho_{11} \right) \tag{9.36}$$

$$\frac{\partial \rho_{22}}{\partial t} = -\frac{\partial \rho_{11}^*}{\partial t} = -\gamma\rho_{22} + i\frac{\mu \cdot \mathbf{E}}{\hbar} \left(\rho_{12}^* - \rho_{12} \right) = -\gamma\rho_{22} + i\frac{\mu \cdot \mathbf{E}}{\hbar} + Im \left(\rho_{12} \right) \tag{9.37}$$

where $\omega_0 = (E_2 - E_1)/\hbar$ is the transition resonance frequency between the two energy levels E_2 and E_1 and μ is the dipole matrix element of the electronic transition. Also, $\Gamma = 1/T_2$ is the dephasing rate is responsible for the decoherence of the polarisation ρ_{12} and time T_2 describes decay of the dipole moment of an ensemble of undriven, excited atom. The relaxation rate $\gamma = 1/T_1$ accounts for nonradiative and radiative (spontaneous) decay from the upper to the lower level.

9.3 Loss Compensation in Metamaterials

We concentrate on the double-fishnet metamaterial, see Fig. 9.2. Loss compensation is such system was demonstrated experimentally by [Xiao *et al.* (2010b)] using gain from optically pumped fluorescent dye molecules.

Here, we summarise the analysis of using the self-consistent Maxwell–Bloch approach to compensate losses in metamaterials [Wuestner *et al.* (2010)], [Wuestner *et al.* (2011)]. Obtained results are in good agreement with experiments reported by Xiao *et al.* [Xiao *et al.* (2010b)].

9.4 Four-Level System as Model of Gain Medium

The approach based on TLS can be generalised to a four-level system and illustrated of how it is used for loss compensation. We summarise, following Wuestner *et al.* [Wuestner *et al.* (2011)] an approach used to study dynamical processes in metamaterials. In this model surface plasmons (SP) are considered as combination of electromagnetic waves interacting with plasma formed by free electrons. SP in turn interact with gain medium, here rhodamine 800 molecules.

Wuestner *et al.* [Wuestner *et al.* (2010)] applied MB approach to analyze the possibility of using gain to compensate losses in a negative refractive index fishnet metamaterial. The analysis was done on the basis of a full-vectorial three-dimensional (3D) Maxwell–Bloch approach.

They considered two configurations, passive and active. In the passive configuration two silver fishnet films were embedded inside a dielectric host with a value of refractive index $n_h = 1.62$. The permittivity of silver was modeled by Drude approach corrected by two Lorentzian resonances to match experimental data at visible wavelengths. In the active configuration it is shown that incorporation of a gain medium in a structure of a double-fishnet nonbianisotropic metamaterial can fully compensate losses in the regime where the real part of the refractive index is negative.

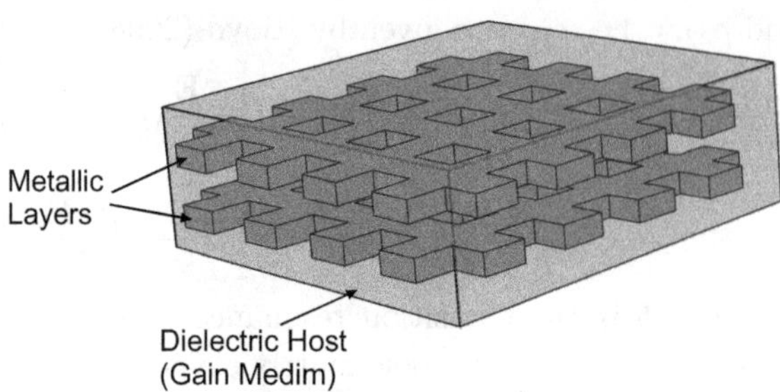

Figure 9.2: Two silver fishnet films embedded in a dielectric host consisting of gain material (dye molecules).

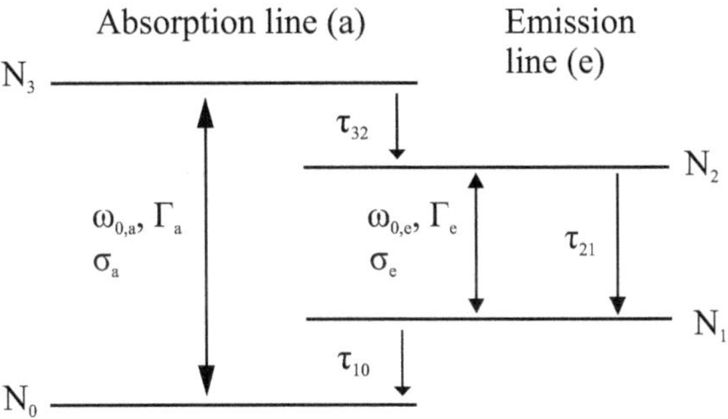

Figure 9.3: Schematic of the four-level system and its parameters.

Gain medium is modelled as a four-level system see Fig. 9.3. It can coherently absorb energy at the pump wavelength and emit into the signal field at a longer wavelength. To model it in a self-consistent way at least two optical transitions are required. Referring to Fig. 9.3 [Pusch *et al.* (2012)], for pump transitions are $(0 \longleftrightarrow 3)$ and for signal $(1 \longleftrightarrow 2)$. These transitions are coupled (in a phenomenological way) to a four-level system by adding non-radiative carrier relaxation processes $(3 \to 2)$ and $(1 \to 0)$. Optical transitions between the subsystems $(0,3)$ and $(1,2)$ are forbiden [Wuestner *et al.* (2011)].

Dynamical equations which describe evolution of densities of carriers within this four-level system are [Wuestner *et al.* (2011)]

$$\frac{\partial N_3}{\partial t} = \frac{1}{\hbar \omega_{r,a}} \left(\frac{\partial \mathbf{P}_a}{\partial t} + \Gamma_a \mathbf{P}_a \right) \cdot \mathbf{E}_{loc} - \frac{N_3}{\tau_{32}} \tag{9.38}$$

$$\frac{\partial N_2}{\partial t} = \frac{N_3}{\tau_{32}} + \frac{1}{\hbar \omega_{r,e}} \left(\frac{\partial \mathbf{P}_e}{\partial t} + \Gamma_e \mathbf{P}_e \right) \cdot \mathbf{E}_{loc} - \frac{N_2}{\tau_{21}} \tag{9.39}$$

$$\frac{\partial N_1}{\partial t} = \frac{N_2}{\tau_{21}} - \frac{1}{\hbar \omega_{r,e}} \left(\frac{\partial \mathbf{P}_e}{\partial t} + \Gamma_e \mathbf{P}_e \right) \cdot \mathbf{E}_{loc} - \frac{N_1}{\tau_{10}} \tag{9.40}$$

$$\frac{\partial N_0}{\partial t} = \frac{N_1}{\tau_{10}} - \frac{1}{\hbar \omega_{r,a}} \left(\frac{\partial \mathbf{P}_a}{\partial t} + \Gamma_a \mathbf{P}_a \right) \cdot \mathbf{E}_{loc}. \tag{9.41}$$

The evolution of polarisation densities $\mathbf{P}_a = \mathbf{P}_a(\mathbf{r}, t)$ of the transition $0 \leftrightarrow 3$ and $\mathbf{P}_e = \mathbf{P}_e(\mathbf{r}, t)$ of the transition $1 \leftrightarrow 2$ under the local electric field $\mathbf{E}(\mathbf{r}, t)$ is described by the differential equations $(i = a, e)$

$$\frac{\partial^2 \mathbf{P}_i}{\partial t^2} + 2\Gamma_i \frac{\partial \mathbf{P}_i}{\partial t} + \omega_{0,i}^2 \mathbf{P}_i = -\sigma_i \Delta N_i \mathbf{E}_i. \tag{9.42}$$

The resonance frequencies are defined as $\omega_{0,i} = \left(\omega_{r,i}^2 + \Gamma_i^2 \right)^{1/2}$, $\Delta N_a(\mathbf{r}, t) = N_3(\mathbf{r}, t) - N_0(\mathbf{r}, t)$ is the inversion of the pump transition and $\Delta N_e(\mathbf{r}, t) = N_2(\mathbf{r}, t) - N_1(\mathbf{r}, t)$ is the inversion of the probe transition, σ_i is a phenomenological coupling constant.

Details of the numerical implementation are provided in [Wuestner *et al.* (2011)].

9.4.1 *Results*

In Fig. 9.4 we reproduce some of the results. Refractive index was determined using method described by Smith *et al.* [Smith *et al.* (2008)] which is based on the analogy between a metamaterial slab and a dielectric Fabry–Perot. Effective refractive index for slab of thickness d was determine using the following expression

$$n_{eff} = \pm \frac{\arccos\left[\left(1 - r^2 + t'^2\right)/\left(2t'\right)\right] + 2\pi m}{kd} \tag{9.43}$$

where r is the reflection coefficient and the modified transmission coefficient $t' = \exp\left(ikd\right)t$. The value of m originates from the phase uncertainty of 2π.

In Fig. 9.4 the real and imaginary parts of the effective refractive index of the double fishnet structure is shown for different pump amplitudes. Increasing pump intensity (and gain) results in a decrease of the imaginary part of the refractive index in the region around the maximum emission cross section of the dye, i.e. around 710 nm. Figure-of-merit (FOM) diverges at two wavelengths bounding the loss compensation region.

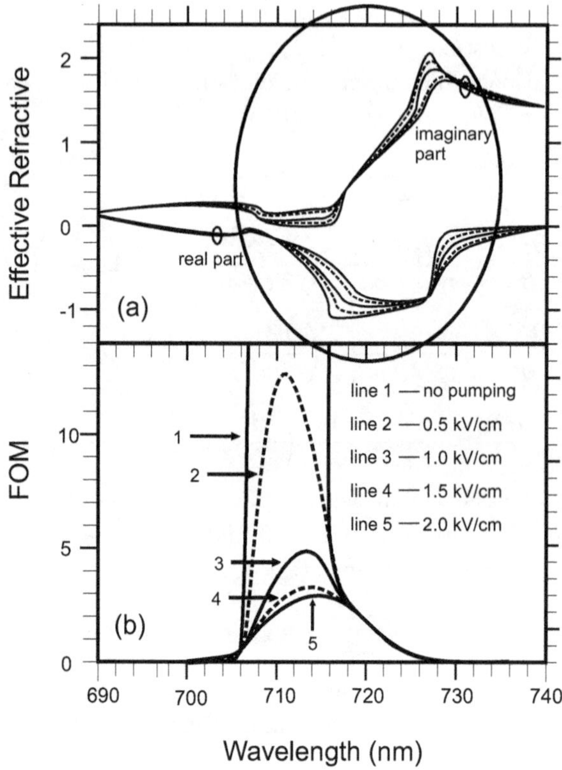

Figure 9.4: Real and imaginary part of the retrieved effective refractive indices of the double-fishnet structure for different pump amplitudes.

Chapter 10

Summary of Surface Plasmons and Active Plasmonics

In recent years we have witnessed the emergence of the field of plasmonics. It deals with the interactions of free electron gas and light and manifests, say in the brilliant colors of stained glass. The effect was known centuries ago [Maier and Atwater (2005)]. An example is the Lycurgus Cup (glass; British Museum; 4th century A.D.). When illuminated from outside, it appears green. However, when illuminated from within the cup, it glows red. The red color is due to tiny gold particles embedded in the glass, which have an absorption peak at around 520 nm.

At such length scales, collective oscillations of the electron density known as surface plasmons can arise. Those excitations found a variety of applications in the natural sciences, engineering, and even medicine. Examples include surface-enhanced molecular spectroscopy, plasmon-driven photochemistry, sensing to photothermal cancer therapy, solar-energy conversion, plasmonic lasers and waveguides.

In short, surface plasmons (SP) are oscillations of conduction band electrons close to the interface between conductor (metal or semiconductor) and dielectric. Both materials forming the interface have different dielectric permittivities. SP are coupled to electromagnetic waves (EM) (like light) propagating near the interface. They can be classified as bosons (polaritons) showing experimental consequences similar to photons (more on the experimental evidence later), showing e.g. stimulated emission.

As plasmonic is a multidisciplinary field, lets briefly explain what are SP from different perspectives; from the point of view of solid state physics plasmon is a collective oscillation of electrons, from the point of view of optics, SPs are modes of an interface, and from the view of electrodynamics SPs are some type of surface waves.

SP exist in two forms: as surface plasmon polaritons (SPP) and as localized surface plasmons (LSP). SPP *propagate* (in space) near metallic (or semiconductor) surfaces. Precisely, SPP are quantized, transverse magnetic (TM) waves.They can be excited by light. Their properties are determined by the complex electrical permitivity of the metal. The effects are numerically analysed by various methods, including FDTD.

LSP can also be excited by light. They exist on the metalic particles which are much smaller than the excitation wavelength and can be considered as local excitations.

In recent years SPP were considered as the replacement of electrical interconnects in electronic integrated circuits [Thraskias *et al.* (2018)]. The option of using optical interconnects is not viable as the optoelectronic devices are generally large in size, hence restricting electronic–photonic integration. Plasmonics, which allows manipulation of light at the subwavelength scale would be the key technology to provide the integration platform [Ooi (2013)].

We start this chapter with providing mathematical background based on Maxwell equations. Next, we analyse SPP propagating along a single interface and double interface and stripe waveguides. We finish with the discussion of losses and some methods of compensating them.

10.1 Mathematical Formulation

In this section we summarise electromagnetic properties based on Maxwell's equations and also two important classical models developed by Drude and Lorentz which play an important role in the determining optical properties of materials. Drude model describes metals and semiconductors whereas Lorentz model describes mostly insulators.

10.1.1 *Dielectric Constant*

We start with Maxwell's equations in differential form (SI system) (they were introduced in Chapter 1 and here we summarise them for convenience)

$$\nabla \times \mathbf{E} = -\frac{\partial \mathbf{B}}{\partial t} \tag{10.1}$$

$$\nabla \times \mathbf{H} = \mathbf{J} + \frac{\partial \mathbf{D}}{\partial t} \tag{10.2}$$

$$\nabla \cdot \mathbf{D} = \rho \tag{10.3}$$

$$\nabla \cdot \mathbf{B} = 0 \tag{10.4}$$

where $\mathbf{E}$ is the electric field intensity, $\mathbf{B}$ is the magnetic flux density, $\mathbf{H}$ is the magnetic field intensity, $\mathbf{D}$ is electric flux density, $\mathbf{J}$ is the electric current density and ρ is volume charge density. The above relations are supplemented with constitutive relations

$$\mathbf{D} = \varepsilon_0 \varepsilon_r \mathbf{E} \tag{10.5}$$

$$\mathbf{B} = \mu_0 \mu \mathbf{H} \tag{10.6}$$

and Ohm's law

$$\mathbf{J} = \sigma \mathbf{E}. \tag{10.7}$$

Here ε_0 and μ_0 are fundamental constants, ε_r is the relative dielectric constant and σ conductivity. We also assume nonmagnetic materials, i.e. $\mu = 1$.

We concentrate on metals where there are no free charges, i.e. $\rho = 0$. In the following we assume harmonic time dependence of all quantities to be of the form $e^{-i\omega t}$. Operate with $\nabla \times$ on Eq. (10.1), use Eq. (10.2), constitutive relations and Ohm's law to obtain

$$\nabla \times \nabla \times \mathbf{E} = \left(i\omega\mu_0\sigma + \omega^2\mu_0\varepsilon_0\varepsilon_r \right) \mathbf{E}. \tag{10.8}$$

General mathematical relation allows to express term on the RHS as

$$\nabla \times \nabla \times \mathbf{E} = \nabla \left(\nabla \cdot \mathbf{E} \right) - \nabla^2 \mathbf{E}.$$

Applying the above relation to (10.8), using Eq. (10.3) one obtains

$$\nabla^2 \mathbf{E} = \frac{\omega^2}{c^2} \left(\varepsilon_r + i\frac{\sigma}{\omega\varepsilon_0} \right) \mathbf{E} \tag{10.9}$$

where

$$c^2 = \frac{1}{\mu_0\varepsilon_0} \tag{10.10}$$

is the velocity of light in a vacuum. This has to be compared with the wave equation [Cheng (1993)]

$$\nabla^2 \mathbf{E} + \frac{\omega^2}{c^2}\varepsilon(\omega)\mathbf{E} = 0. \tag{10.11}$$

Comparison allows for an identification of a frequency dependent dielectric constant to be

$$\varepsilon(\omega) = \varepsilon_r + i\frac{\sigma}{\omega\varepsilon_0}. \tag{10.12}$$

In the above ε_r is also known as ε_∞ which is high frequency permittivity. To obtain final expression for $\varepsilon(\omega)$ one needs to determine conductivity. This will be done in a specific models.

10.1.2 *Classical Models for Plasmons*

There are several models of increased complexity which we will summarised now. Here we concentrate on classical approaches. Quantum models will be discussed later.

The link (in frequency domain) between displacement vector $\mathbf{D}$ and electric field $\mathbf{E}$ is

$$\mathbf{D}\left(\mathbf{r},\omega\right) = \varepsilon_0 \int \varepsilon\left(\mathbf{r},\mathbf{r}',\omega\right) \mathbf{E}\left(\mathbf{r}',\omega\right) d\mathbf{r}'. \tag{10.13}$$

In writing the above we implied that the dependence is nonlocal, i.e. $\mathbf{D}$ is a function of the electric field at all positions. The relative dielectric constant $\varepsilon\left(\mathbf{r}, \mathbf{r}', \omega\right)$ is inhomogeneous. Its form is complicated and often unknown. To simplify this problem one often introduces local-response approximation where nonlocal effects are neglected and writes

$$\varepsilon\left(\mathbf{r}, \mathbf{r}', \omega\right) = \varepsilon_{local}\left(\omega\right) \delta\left(\mathbf{r} - \mathbf{r}'\right). \tag{10.14}$$

In this approximation the constitutive relation becomes

$$\mathbf{D}\left(\mathbf{r}, \omega\right) = \varepsilon_0 \varepsilon_{local}\left(\omega\right) \mathbf{E}\left(\mathbf{r}, \omega\right). \tag{10.15}$$

In the following we will outline determination of relative dielectric constant in simple classical models.

10.1.2.1 *Drude model*

In this approach metal is considered as a free electron gas system. Not included are interactions between electrons and non-local effects. Dynamics of electrons is described by Newton's second law with collisions [Ashroft and Mermin (1976)]

$$\frac{d\mathbf{p}(t)}{dt} = -e\mathbf{E}(t) - \frac{1}{\tau}\mathbf{p}(t) \tag{10.16}$$

where $\mathbf{p}$ is the momentum of the electron and τ is the relaxation time (average time between collisions). Assuming harmonic variations of $\mathbf{p}$ and $\mathbf{E}$, i.e. of the form

$$\mathbf{E}\left(t\right) = Re\left\{\mathbf{E}\left(\omega\right) e^{-i\omega t}\right\}. \tag{10.17}$$

Reaction of the system is at the same frequency and therefore we seek the solution in the form

$$\mathbf{p}\left(t\right) = Re\left\{\mathbf{p}\left(\omega\right) e^{-i\omega t}\right\}. \tag{10.18}$$

Substitution into equation of motion gives

$$\mathbf{p}\left(\omega\right) = \frac{e}{i\omega - \gamma}\mathbf{E}\left(\omega\right) \tag{10.19}$$

where $\gamma = 1/\tau$.

One need to evaluate electrical current which is

$$\mathbf{j}\left(t\right) = -en\mathbf{v} = -en\frac{\mathbf{p}\left(t\right)}{m}. \tag{10.20}$$

Its frequency dependence is

$$\mathbf{j}\left(t\right) = Re\left\{\mathbf{j}\left(\omega\right) e^{-i\omega t}\right\}. \tag{10.21}$$

Combining the above gives the following expression for current

$$\mathbf{j}\left(\omega\right) = \frac{ne^2}{m}\frac{\mathbf{E}\left(\omega\right)}{1/\tau - i\omega}$$
$$= \sigma(\omega)\mathbf{E}\left(\omega\right). \tag{10.22}$$

Frequency-dependent conductivity is

$$\sigma(\omega) = \frac{\sigma_0}{1 - i\omega\tau} \tag{10.23}$$

where $\sigma_0 = \frac{ne^2\tau}{m}$ is dc conductivity.

10.1.2.2 *AC dielectric constant*

Using previously established relation for dielectric constant (10.12) and expression for conductivity one has

$$\varepsilon(\omega) = \varepsilon_r + i\frac{1}{\omega\varepsilon_0}\frac{\sigma_0}{1 - i\omega\tau}$$
$$= \varepsilon_r - \frac{\omega_p^2}{\omega^2 + i\gamma\omega}. \tag{10.24}$$

In the limit $\omega\tau \gg 1$ it reduces to

$$\varepsilon(\omega) = \varepsilon_r - \frac{\omega_p^2}{\omega^2} \tag{10.25}$$

where ω_p is the plasma frequency

$$\omega_p^2 = \frac{ne^2}{m\varepsilon_0}. \tag{10.26}$$

In Fig. 10.1 we show dielectric constant for gold, adopted from [Derkachova *et al.* (2016)]. Experimental date from Johnson and Christy [Johnson and Christy (1972)]. Parameters used $\varepsilon_0 = 9.84$, $\omega_p = 9.010\,\mathrm{eV}$, $\gamma = 0.072\,\mathrm{eV}$.

10.1.2.3 *Lorentz model*

It is based on multi-oscillation model. It represents electrons bound to the nuclei. The Lorentz model [Li (2017)] describes dielectric response of an insulator in which electrons are bound to positive ions.

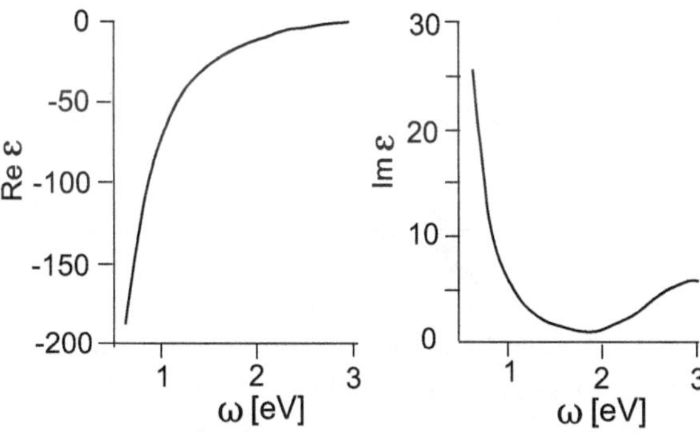

Figure 10.1: Real and imaginary parts of the dielectric function of gold. Experimental data from Johnson and Christy. Adapted from Derkachova [Derkachova *et al.* (2016)].

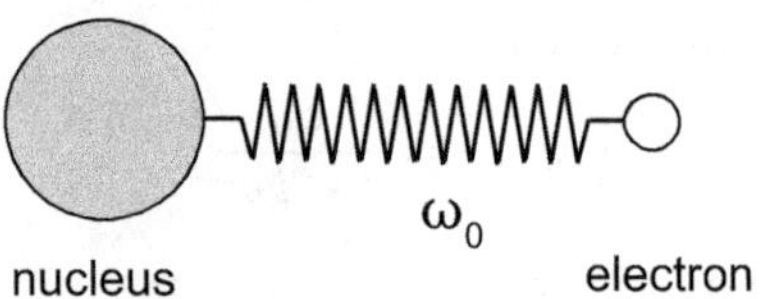

Figure 10.2: Electron coupled to a nucleus.

Consider first a *single* damped oscillator, see Fig. 10.2 [Locharoenrat (2015)], [Marasinghe (2018)]. Equation of motion of an electron in an oscillating electric field **E** with frequency ω is

$$m\frac{d^2x}{dt^2} = F_{driving} + F_{damping} + F_{elastic}. \tag{10.27}$$

One assumes velocity-dependent damping as $F_{damping} = m\Gamma_0\frac{dx}{dt}$, elastic force as $F_{elastic} = -m\omega_0^2 x$ and external driving field as $F_{driving} = qE_0 e^{-i\omega t}$. Equation of motion becomes

$$m\left(\frac{d^2x}{dt^2} + \Gamma_0\frac{dx}{dt} + \omega_0^2 x\right) = qE_0 e^{-i\omega t}. \tag{10.28}$$

In the above m is electron's mass, x is the position of an electron relative to the atom, Γ_0 is the damping coefficient, q is the absolute value of the electron charge, ω_0 is the resonance angular frequency of an electron and E_0 is the amplitude of the electric field. Assume harmonic solution of the form

$$x = x_0 e^{-i\omega t} \tag{10.29}$$

substitute into equation of motion and have

$$m\left(-\omega^2 - i\omega\Gamma_0 + \omega_0^2\right)x_0 e^{-i\omega t} = qE_0 e^{-i\omega t}. \tag{10.30}$$

Solution is

$$x_0 = \frac{qE_0}{m\left(\omega^2 - i\omega\Gamma_0 - \omega_0^2\right)}. \tag{10.31}$$

We are interested in the optical properties. For that, we need to evaluate polarisation **P** which is defined as

$$\mathbf{P} = Nq\mathbf{x} = (\varepsilon - \varepsilon_0)\,\mathbf{E}$$

where N is the number of oscillators (or electrons) in a unit volume. Combining the above results gives

$$Nq\frac{qE_0}{m\left(\omega_0^2 - \omega^2 - i\omega\Gamma_0\right)}e^{-i\omega t} = (\varepsilon - \varepsilon_0)\,E_0 e^{-i\omega t}. \tag{10.32}$$

Assume that Nf_i oscillators have the eigenfrequency ω_j. Here f_i is known as the oscillator strength. Generalisation of the previous result one determines that the dielectric function of the group of oscillators is [Li (2017)]

$$\varepsilon = \varepsilon_0 + \frac{Nq^2}{m}\sum_j \frac{f_j\omega_j^2}{\omega_j^2 - \omega^2 - i\omega\Gamma_j} \tag{10.33}$$

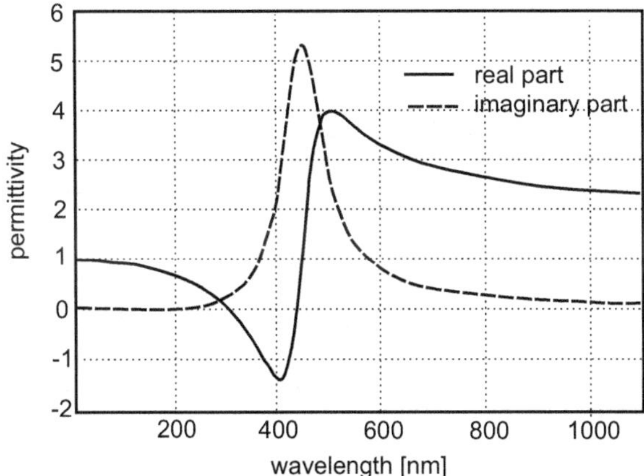

Figure 10.3: Permittivity of gold, adopted from Li book [Li (2017)].

where j refers to the resonant modes, ω_j corresponds to the resonance frequencies, f_j indicates the weighting coefficients and Γ_j is describes damping.

The Lorentz model is often used to describe frequency response of semiconductors. It works well for materials that have bound electrons. It shows strong dispersion around the resonant frequency. It is valid for photon's energies smaller than the bandgap of semiconductors. In Fig. 10.3 we plotted dielectric function for gold considering only the interband contribution of bound electrons.

10.2 Plasmons Across a Single Interface

We summarise here the analysis of SPP assuming single interface. Double-interface will be summarised later. Analysis for general geometries was analysed and summarised by [Breukelaar (2004)].

For the following discussion we assume that the system consists of two layers with z-axis perpendicular to those layers, see Fig. 10.4.

Layers are labeled using index $i = 1, 2$ where indices 1 and 2 denote dielectric and metal regions, respectively. Propagation is assumed to be along x-axis which is oriented parallel to the layers.

We start with Maxwell equations (10.1) and (10.2) in the harmonic approximation where all time dependencies are of the form $\exp(-i\omega t)$

$$\nabla \times \mathbf{E} = -i\omega\mu_0\mathbf{H} \tag{10.34}$$

$$\nabla \times \mathbf{H} = \sigma\mathbf{E} + i\omega\varepsilon_0\varepsilon\mathbf{E}. \tag{10.35}$$

In the above ε is the relative dielectric constant of a given material. In the following we analyse only non-magnetic systems with $\mu = \mu_0$. In the next steps we summarise a general analysis of the above equations for the case of a single interface. In the

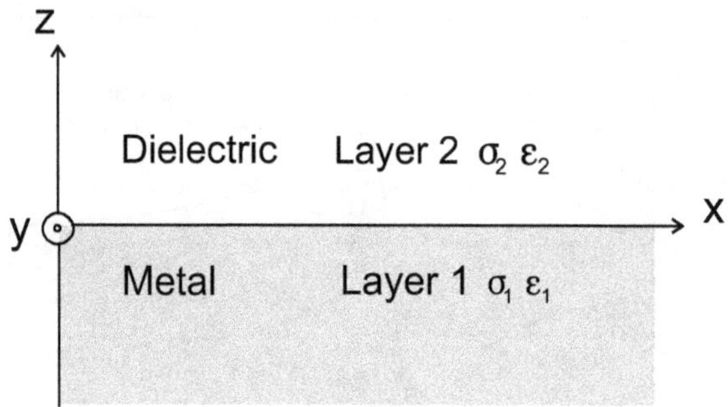

Figure 10.4: Geometry of the layers for plasmonic propagation.

explicit form, Maxwell's equations are

$$\begin{vmatrix} \mathbf{a}_x & \mathbf{a}_y & \mathbf{a}_z \\ \frac{\partial}{\partial x} & \frac{\partial}{\partial y} & \frac{\partial}{\partial z} \\ E_x & E_y & E_z \end{vmatrix} = -i\omega\mu_0\left(\mathbf{a}_x H_x + \mathbf{a}_y H_y + \mathbf{a}_z H_z\right) \tag{10.36}$$

where $\mathbf{a}_x, \mathbf{a}_y, \mathbf{a}_z$ are unit vectors along x, y, z axes, respectively and

$$\begin{bmatrix} \mathbf{a}_x & \mathbf{a}_y & \mathbf{a}_z \\ \frac{\partial}{\partial x} & \frac{\partial}{\partial y} & \frac{\partial}{\partial z} \\ H_x & H_y & H_z \end{bmatrix} = (\sigma + i\omega\varepsilon_0\varepsilon)\left(\mathbf{a}_x E_x + \mathbf{a}_y E_y + \mathbf{a}_z E_z\right). \tag{10.37}$$

Expending determinants gives

$$\frac{\partial E_z}{\partial y} - \frac{\partial E_y}{\partial z} = i\omega\mu_0 H_x \tag{10.38}$$

$$\frac{\partial E_x}{\partial z} - \frac{\partial E_z}{\partial x} = i\omega\mu_0 H_y \tag{10.39}$$

$$\frac{\partial E_y}{\partial x} - \frac{\partial E_x}{\partial y} = i\omega\mu_0 H_z \tag{10.40}$$

and

$$\frac{\partial H_z}{\partial y} - \frac{\partial H_y}{\partial z} = -(\sigma + i\omega\varepsilon_0\varepsilon)\, E_x \tag{10.41}$$

$$\frac{\partial H_x}{\partial z} - \frac{\partial H_z}{\partial x} = -(\sigma + i\omega\varepsilon_0\varepsilon)\, E_y \tag{10.42}$$

$$\frac{\partial H_y}{\partial x} - \frac{\partial H_x}{\partial y} = -(\sigma + i\omega\varepsilon_0\varepsilon)\, E_z. \tag{10.43}$$

We assume here that waveguide is extended to infinity along y-direction and therefore $\frac{\partial}{\partial y} \longrightarrow 0$. Propagation takes place along x-axis and therefore $\frac{\partial}{\partial x} \longrightarrow i\beta$. Using the above facts one obtains

$$\frac{\partial E_y}{\partial z} = i\omega\mu_0 H_x \tag{10.44}$$

$$\frac{\partial E_x}{\partial z} - i\beta E_z = i\omega\mu_0 H_y \tag{10.45}$$

$$i\beta E_y = i\omega\mu_0 H_z \tag{10.46}$$

and

$$\frac{\partial H_y}{\partial z} = (\sigma + i\omega\varepsilon_0\varepsilon)\, E_x \tag{10.47}$$

$$\frac{\partial H_x}{\partial z} - i\beta H_z = -(\sigma + i\omega\varepsilon_0\varepsilon)\, E_y \tag{10.48}$$

$$\frac{\partial H_y}{\partial x} = -(\sigma + i\omega\varepsilon_0\varepsilon)\, E_z. \tag{10.49}$$

The above equations can be separated into two groups which represent two different polarisations: transverse electric (TE) involving (H_x, H_z, E_y) and transverse magnetic (TM) involving (E_x, E_z, H_y). In the following we discuss only TM modes.

10.2.1 *TM Polarised Modes*

For TM mode [Maier (2007)] the following orientation of fields exist

$$\mathbf{E}_j = [E_x, 0, E_{z,j}]\, e^{-i\omega t} e^{i(k_x x + k_{z,j} z)} \tag{10.50}$$

$$\mathbf{H}_j = [0, H_y, 0]\, e^{-i\omega t} e^{i(k_x x + k_{z,j} z)}. \tag{10.51}$$

Using the above solutions in Maxwell's equations along with boundary conditions one gets

$$\frac{\varepsilon_1}{k_{z,1}} = \frac{\varepsilon_2}{k_{z,2}} \tag{10.52}$$

$$k_x^2 + k_{z,j}^2 = \varepsilon_j k_0^2, \qquad j = 1, 2 \tag{10.53}$$

$$E_x = \frac{k_{z,1}}{\omega\varepsilon_j} H_y, \qquad E_{z,j} = -\frac{k_x}{\omega\varepsilon_j} H_y \tag{10.54}$$

where $k_0 = \frac{\omega}{c}$. From above one derives dispersion relations for SPP as

$$k_x^2 = k_0^2 \frac{\varepsilon_1 \varepsilon_2}{\varepsilon_1 + \varepsilon_2} \tag{10.55}$$

$$k_{z,j}^2 = k_0^2 \frac{\varepsilon_j^2}{\varepsilon_1 + \varepsilon_2}. \tag{10.56}$$

In the above dispersion relation the imaginary part of k_x determines losses of SPP propagating along the interface.

Propagation constant k_x must be real for the case of the ideal metal without losses. Therefore, from dispersion relation (10.55) one obtains the requirement for

SPP, as $\varepsilon_1 + \varepsilon_2 < 0$. The condition $\varepsilon_1 + \varepsilon_2 = 0$ along with the dielectric constant for an ideal metal

$$\varepsilon(\omega) = \varepsilon_r - \frac{\omega_p^2}{\omega^2} \tag{10.57}$$

gives an expression for the surface plasmon frequency

$$\omega_{sp} = \frac{\omega_p}{\sqrt{1 + \varepsilon_2}}. \tag{10.58}$$

Comparison of dispersion relations for a single silver-air interface for an ideal metal (no losses, $Im(\varepsilon_1) = 0$) and one with losses is shown in Fig. 10.5 [Wong (2011)]. For ideal conductors any bound SPP must have frequencies obeying the relation $\omega < \omega_{sp}$ [Wong (2011)] and the propagation of any EM modes is forbidden in the region of frequencies between ω_{sp} and ω_p.

In the case of a real metal, i.e. one with losses the imaginary part of the complex propagation constant k_x represents the propagation loss. This implies that the frequency region between ω_{sp} and ω_p in the SPP dispersion relation is no longer forbidden, as shown in Fig. 10.5.

However, the notion of the surface plasmon frequency ω_{sp} is still useful, as it separates the regions of bound SPP modes for $\omega < \omega_{sp}$ and quasi-bound leaky SPP modes for $\omega > \omega_{sp}$ [Maier (2007)].

10.3 Plasmons Across Infinite Double Interface

One of the methods to reduce attenuation of SPP is to use a thin metal or stripe bounded on all sides by the same dielectric. Such structure supports long-range SPP (LRSPP) mode [Berini (2009)].

LRSPP are TM-polarised optical surface waves that propagate with low loss over long ranges (centimeters) along a symmetrically clad thin metal slab or stripe. Typically, range extension factors of 10 to 1000 are achievable with LRSPPs, relative to conventional single-interface SPPs.

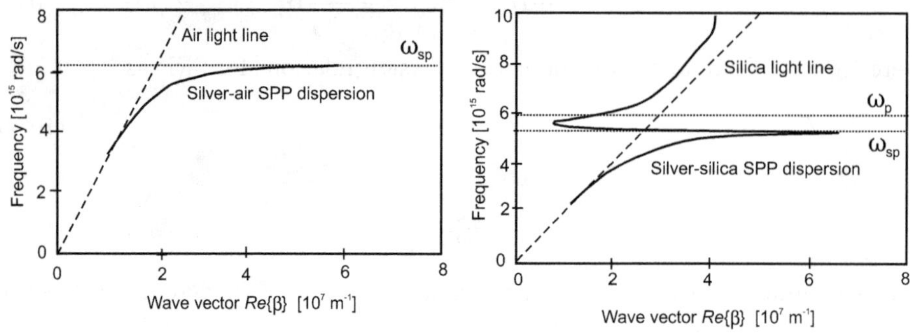

Figure 10.5: Dispersion relation for SPP at a silver-air interface without metal losses (left). Dispersion relation for SPP at a silver-silica interface considering metal losses (right).

LRSPP attenuation is at least a factor 2–3 lower compared to a single-interface SPP [Berini (2009)]. LRSPP can be considered for transmission of signals over long distances, say millimiter scale as opposed to a few microns in the case of SPP propagating at the single interface. The LRSPP mode exists on a thin metal film bounded by symmetric dielectric. In this section we provide some inside into those modes. We start with the discussion of double interface.

10.3.1 General

Typical structure consists of a thin metal layer of thicknedd $d = 2a$ sandwiched between two dielectric media, see Fig. 10.6.

We have assumed a symmetric structure. We only consider bound modes (only TM polarisation is discussed here). Introduce the following notation $k_i^2 = \beta^2 - \omega^2\mu_i\varepsilon_i$ for $i = 1, 2, 3$. The situation is described by the following equations ($a = d/2$) [Maier (2007)], [Rosenzveig (2011)]

$$
\begin{aligned}
B_y &= A_1 e^{-k_1(z-a)} e^{i\beta x} & z &> a \\
B_y &= A_2 e^{k_1(z+a)} e^{i\beta x} & z &< -a \\
B_y &= A_3 e^{k_2(z-a)} e^{i\beta x} + A_4 e^{-k_2(z+a)} e^{i\beta x} & -a &< z < a.
\end{aligned}
\tag{10.59}
$$

Substitute the above into Maxwell equations and obtain

for $z > a$

$$
B_y = A_1 e^{-k_1(z-a)} e^{i\beta x}
\tag{10.60}
$$

$$
E_x = iA_1 \frac{k_1}{\omega\mu_1\varepsilon_1} e^{-k_1(z-a)} e^{i\beta x}
\tag{10.61}
$$

$$
E_z = -A_2 \frac{\beta}{\omega\mu_1\varepsilon_1} e^{-k_1(z-a)} e^{i\beta x}
\tag{10.62}
$$

for $z < -a$

$$
B_y = A_2 e^{k_1(z+a)} e^{i\beta x}
\tag{10.63}
$$

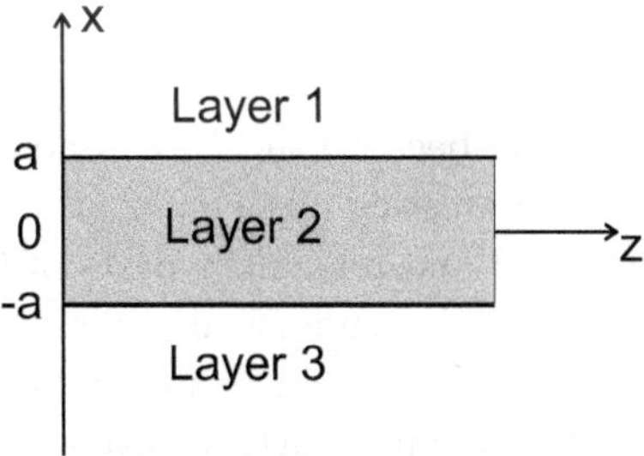

Figure 10.6: Schematic of a three layers structure.

$$E_x = iA_2 \frac{k_1}{\omega\mu_1\varepsilon_1} e^{k_1(z+a)} e^{i\beta x} \tag{10.64}$$

$$E_z = -A_2 \frac{\beta}{\omega\mu_1\varepsilon_1} e^{k_1(z+a)} e^{i\beta x} \tag{10.65}$$

for $-a < z < a$

$$B_y = \left(A_3 e^{k_2(z-a)} + A_4 e^{-k_2(z+a)} \right) e^{i\beta x} \tag{10.66}$$

$$E_x = -i\frac{1}{\omega\mu_2\varepsilon_2} \left(A_3 k_2 e^{k_2(z-a)} + A_4 k_2 e^{-k_2(z+a)} \right) e^{-i\beta x} \tag{10.67}$$

$$E_z = -\frac{\beta}{\omega\mu_2\varepsilon_2} \left(A_3 e^{k_2(z-a)} + A_4 e^{-k_2(z+a)} \right) e^{-i\beta x}. \tag{10.68}$$

The above set of equations can be expressed in a matrix form as

$$\begin{bmatrix} 1 & 0 & -1 & e^{-k_2 d} \\ 0 & 1 & -e^{-k_2 d} & -1 \\ \frac{k_1}{\varepsilon_1} & 0 & \frac{k_2}{\varepsilon_2} & -\frac{k_2}{\varepsilon_2} e^{-k_2 d} \\ 0 & \frac{k_1}{\varepsilon_1} & -\frac{k_2}{\varepsilon_2} e^{-k_2 d} & \frac{k_2}{\varepsilon_2} \end{bmatrix} \begin{bmatrix} A_1 \\ A_2 \\ A_3 \\ A_4 \end{bmatrix} = 0. \tag{10.69}$$

Determinant of the above system must vanish which gives the following dispersion relation

$$e^{-k_2 d} = \pm \frac{\frac{k_1}{\varepsilon_1} + \frac{k_2}{\varepsilon_2}}{\frac{k_1}{\varepsilon_1} - \frac{k_2}{\varepsilon_2}}. \tag{10.70}$$

Lets now discuss the above solutions.

10.3.2   *Symmetric Plasmon Modes*

Consider negative solution (10.70). Substitute it into Eq. (10.69) and assume $A_1 = 1$. Magnetic field in all three media becomes

$$B_y = e^{k_1(z+d/2)} e^{i\beta x} \qquad\qquad z < -d/2$$

$$B_y = -\frac{\sqrt{\left(\frac{k_2}{\varepsilon_2}\right)^2 - \left(\frac{k_1}{\varepsilon_1}\right)^2}}{\frac{k_2}{\varepsilon_2}} \cosh\left(k_2 z\right) e^{i\beta x} \qquad -d/2 < z < d/2 \tag{10.71}$$

$$B_y = e^{-k_1(z-d/2)} e^{i\beta x} \qquad\qquad z > d/2.$$

This is called a symmetric mode because the E_z component of the electric field is symmetric. This mode has an interesting property, that when thickness of the metallic film becomes thinner the real and imaginary parts of the propagation constant decrease [Rosenzveig (2011)]. At the same time evanescent decay length increases and the the plasmon becomes less confined to metal-dielectric interface. This in-turn results in smaller absorption and longer propagation length. Therefore this mode is called a long-range surface plasmon polariton (LRSPP). They propagate over distances of the order of millimeters.

Dispersion relation for LRSPP cannot be obtained in an explicit form and it is determined by the following relation [Bozhevolnyi (2009)]

$$\tanh\left(\frac{k_z^{(m)} t}{2}\right) = -\frac{\varepsilon_m k_z^{(d)}}{\varepsilon_d k_z^{(m)}} \tag{10.72}$$

where

$$k_z^{(m,d)} = \sqrt{k_{LRSPP}^2 - \varepsilon_{m,d} k_0^2}. \tag{10.73}$$

For thin metal films, i.e. $t \longrightarrow 0$ one can approximate $\tanh x \approx x$ and obtain

$$k_{LRSPP} \approx k_0 \sqrt{\varepsilon_d + \left(\frac{t k_0 \varepsilon_d}{2}\right)^2 \left(1 - \frac{\varepsilon_d}{\varepsilon_m}\right)^2}. \tag{10.74}$$

From above, one observes that in the limit of very thin metal film, i.e. for $t \longrightarrow 0$, $k_{LRSPP} \longrightarrow k_0 \varepsilon_d$, i.e it approaches light line of the dielectric.

The field of this mode is expelled from the metal into dielectric [Bozhevolnyi (2009)]. At the same time, from Eq. (10.74) one observes that both the real and imaginary parts of the difference between the LR-SPP propagation constant and that of light in the dielectric decreases quadratically with the decrease in the film thickness. This means that the losses of LR-SPP mode decreases to zero and the its propagation length increases.

Long propagation length of LRSPP is very useful for various practical applications such as a plasmonic waveguide. This extremely long propagation, however, leads to low confinement of SPP fields.

10.3.3 *Antisymmetric Plasmon Modes*

Here we consider positive solution of (10.70). Substitute it into Eq. (10.69) and obtain

$$
\begin{aligned}
B_y &= -e^{k_1(z+d/2)} e^{i\beta x} & z < -d/2 \\
B_y &= -\frac{\sqrt{\left(\frac{k_1}{\varepsilon_1}\right)^2 - \left(\frac{k_2}{\varepsilon_2}\right)^2}}{\frac{k_2}{\varepsilon_2}} \sinh\left(k_2 z\right) e^{i\beta x} & -d/2 < z < d/2 \\
B_y &= e^{-k_1(z-d/2)} e^{i\beta x} & z > d/2.
\end{aligned}
\tag{10.75}
$$

For this mode E_z and B_y are antisymmetric. With decreasing metal thickness both real and imaginary parts increase, plasmon becomes more confined to the interface which results in higher losses. As a result, propagation length is significantly shorter. This mode is therefore known as a short-range surface plasmon polariton (SRSPP).

Next, boundary conditions are combined with the above solutions. Across the interfaces tangential fields are continuous. This condition produces dispersion relation.

Dispersion relation for SRSPP is determined by the following relation [Bozhevolnyi (2009)]

$$\tanh\left(\frac{k_z^{(m)}t}{2}\right) = -\frac{\varepsilon_d k_z^{(m)}}{\varepsilon_m k_z^{(d)}} \tag{10.76}$$

where

$$k_z^{(m,d)} = \sqrt{k_{SRSPP}^2 - \varepsilon_{m,d}k_0^2}. \tag{10.77}$$

For thin metal films, i.e. $t \longrightarrow 0$ one obtains

$$k_{SRSPP} \approx k_0\sqrt{\varepsilon_d + \left(\frac{2\varepsilon_d}{tk_0\varepsilon_m}\right)^2}. \tag{10.78}$$

SRSPP mode behaves differently. Its propagation constant increases when the metal film thickness decreases to zero [Bozhevolnyi (2009)]. This means that the propagation distance of this mode decreases approaching zero for infinitely thin films.

In 2009 P. Berini published an extensive review on physics and applications of LRSPP [Berini (2009)]. His review covered a very broad topics including modal characteristics, excitation, field enhancement, nonlinear interactions, molecular scattering, fluorescence, to just name a few. Subsequent summary was published in 2019 [Berini (2019)].

Early studies of LRSP has been reported by Higuchi *et al.* [Higuchi *et al.* (2014)]. They theoretically predicted that the propagation length of LRSP significantly increases as the lateral width of a metal film decreases. They also fabricated silver slab waveguides. Comparison of theoretical and experimental results is shown in Fig. 10.7.

Long-range hybrid plasmonic slot waveguide was proposed and designed by Xiang and Wang [Xiang and Wang (2013)]. The mode of their design can propagate for more then 10 mm.

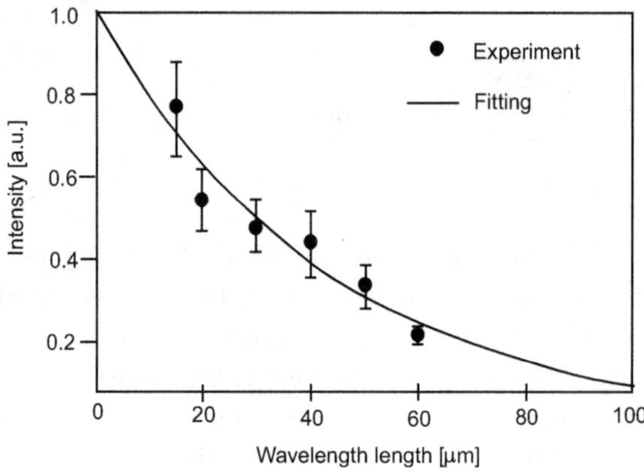

Figure 10.7: Intensity vs length, adopted from [Higuchi *et al.* (2014)].

10.4 Stripe Waveguides

Basic structure is illustrated in Fig. 10.8. Stripe waveguides (waveguides of finite width) were discussed by many groups [Berini (2001)], [Keshmarzi (2017)]. Different types of two-dimensional plasmonic waveguides are shown in Fig. 10.9, adopted from Gwo and Shih [Gwo and Shih (2016)].

Accurate modeling of stripe SPP waveguides were conducted using advanced numerical approaches by Zia *et al.* [Zia *et al.* (2005a)], [Zia *et al.* (2005b)], [Zia *et al.* (2006)]. Stripe mode were analysed by effective-index method (EIM) (described earlier), see summary by Bozhevolnyi [Bozhevolnyi (2009)]. The EIM approximates reasonably well some of the modes.

There exist four fundamental modes supported by the metal stripe, notation being used is aa_b^0, as_b^0, sa_b^0 and ss_b^0. Here letters a and s refer to asymmetric and symmetric modes, respectively. First position is associated with a horizontal dimension x and the second with vertical one y.

Index b tells that modes are purely bound, i.e. nonradiative and the superscript 0 counts the number of extrema in the horizontal distribution of E_y, not counting the corner peaks.

For $w/t >> 1$ the E_y component dominates for all modes but modes are not pure TM type because all field components, including H_z are always nonzero [Berini (2009)]. Modes were classified by Berini [Berini (2001)], [Berini (2009)].

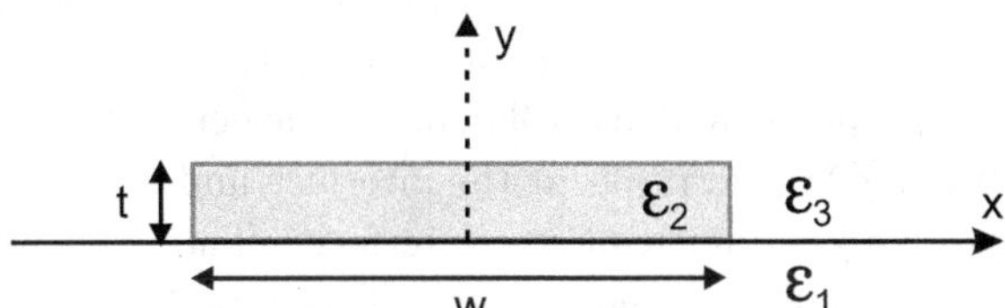

Figure 10.8: Stripe-waveguide structure. The core is consists of a lossy metal film of thickness t, width w, and permittivity ε_2. The metal film is embedded in a semi-infinite dielectric substrate of permittivity ε_1 and the semi-infinite dielectric cover of permittivity ε_3.

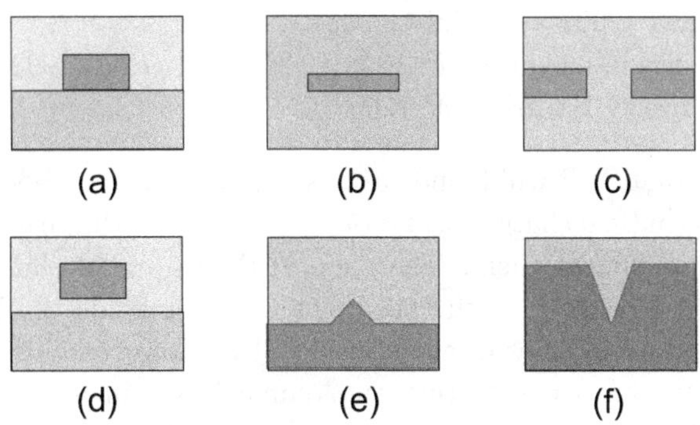

Figure 10.9: Different designs of plasmonic waveguides: (a) dielectric-loaded metal (b) insulator–metal–insulator, (c) metal slot, (d) MIS, (e) wedge, and (f) channel plasmonic waveguides.

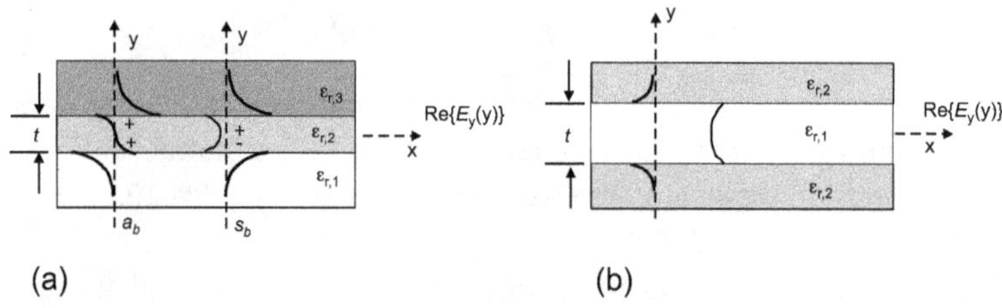

Figure 10.10: (a) Two bound SPP modes (a_b, s_b) which propagate in a metal slab having ϵ_{r2} and thickness t and bounded by semi-infinite dielectrics with ϵ_{r1} and ϵ_{r3}. (b) A symmetric bound mode which propagates in a dielectric slab with ϵ_{r1} and thickness t and bounded by semi-infinite metals with ϵ_{r2}. Adapted from Zhang [Zhang *et al.* (2012)].

The LRSPP mode is also termed the ss_b^0 mode since it provides a symmetric profile along both lateral dimensions; it is bound to the metal stripe and it is the lowest order SPP mode guided in this structure. A single LRSPP mode can be ensured in such structures by thoughtful design of stripe dimensions (width and thickness) at a given wavelength. The modes are shown schematically in Fig. 10.10 [Zhang *et al.* (2012)], [Berini and Leon (2012)].

10.5 Summary of Properties

We provide now more detailed discussion of SPP. They are a transverse magnetic (TM) polarised optical surface waves. It is an excitation which involves a charge density wave in a metal, or semiconductor and electromagnetic field (EM). In a typical geometry, it propagates along a flat metal-dielectric interface (at visible or infrared wavelengths). EM field peaks at the interface and decay exponentially into both media. It is highly confined at the interface. They can be guided beyond the diffraction limit [Takahara *et al.* (1997)], [Takahara (2009)], [Gramotnev and Bozhevolnyi (2010)].

Partial list of various structures which support SPP include:

- metal–dielectric structures including planar arrangements of metal and dielectric films [Maier (2007)], [Berini (2009)].
- metallic gratings [Ghaemi *et al.* (1998)], [Ebbesen *et al.* (1998)].
- corrugated surface [Barnes *et al.* (2003)].

It is expected that SPP will found various applications in nano-photonic circuit. They involve longitudinal charge oscillations in metals and thus can propagate much faster than electric current there. Therefore SPP can provide link between optics and electronics and result in high speed signal transmission of electrical signals. Those properties created a lot of attention in SPP. However, SPP experience also a high attenuation due to scattering and ohmic losses in a metal, which limits the propagation length, resulting in the limitation of the scope of applications (see Fig. 10.11).

10.6 Localized Surface Plasmons

Localized surface plasmons (LSP) are local oscillations in the nanostructures, see Fig. 10.12. They are different from SPP which are propagating oscillations along the metal-dielectric interface. LSP are often referrd to as localized surface plasmon resonances (LSPR).

Important systems where LSPR were observed are semiconductor nanocrystals [Agrawal *et al.* (2018)]. Various properties, like resonant absorption, scattering, and near field enhancement around nanocrystals can be tuned across a wide optical spectral range from visible to far-infrared. This can be done by synthetically varying doping levels, and via chemical oxidation and reduction.

Decay length of the electromagnetic field in localized surface plasmons is in the order of 6 nm [Sarid and Challener (2010)]. The small lenght of decay observed for LSPR reduces the sensitivity to interference from fluctuations of refractive index and results in the increased sensitivity to refractive index changes on the surface which is important for sensing applications.

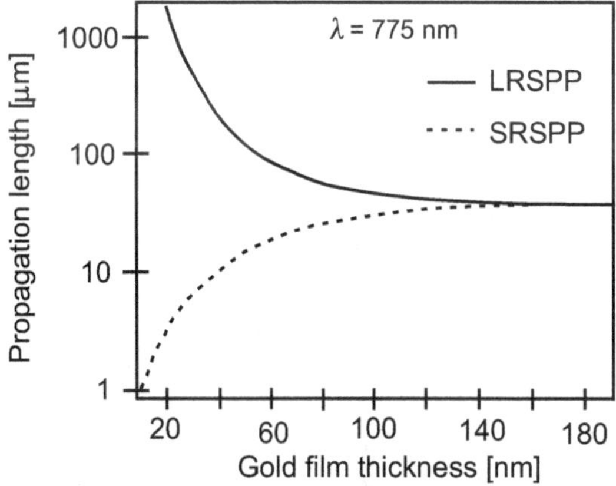

Figure 10.11: The effective indexes of short- and long-range surface plasmon-polariton modes and their propagation lengths. Adapted from Bozhevolnyi [Bozhevolnyi (2009)].

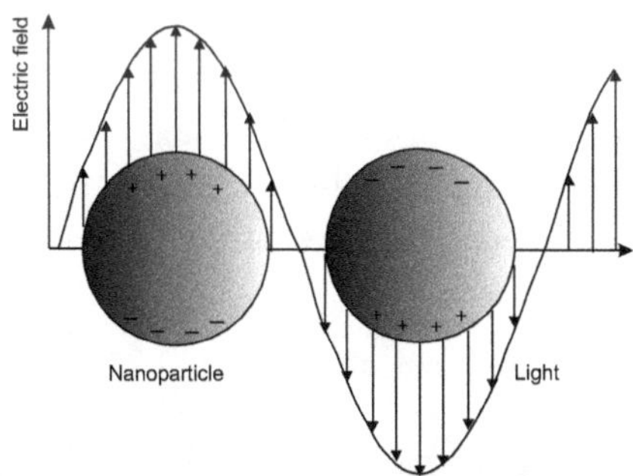

Figure 10.12: Localized surface plasmons.

10.7   Origin of Losses

Losses in surface plasmons are extensively discussed in [Boriskina *et al.* (2017)]. In general, they originate due to elastic and inelastic scattering processes with other electrons, phonons, crystal defects etc. They can be divided into three groups [Ghodsi and Kaatuzian (2020b)]

- bulk decay characterised by bulk decay rate γ_b

$$\gamma_b = \gamma_{e-e} + \gamma_{e-ph} + \gamma_{e-defect} + \cdots \tag{10.79}$$

  where the first term is due to electron-electron scattering, the second term is due to electron-phonon scattering mechanism and the third term is the electron-defect scattering. More scattering mechanisms can be introduced if needed. For example, electron-electron decay rate can be expressed as

$$\gamma_{e-e} \simeq 10^{15} \left(\frac{\hbar\omega}{E_F}\right)^2 [Hz]. \tag{10.80}$$

- surface scattering, which plays the role in plasmonic structures with dimensions of the order of several nanometers which is shorter than the mean-free path of electrons.
- Landau damping. It exists when electrons move with the velocity close to the phase velocity of the surface plasmons. At the appropriate conditions (acceleration) electrons can absorb energy from the surface plasmons. Landau damping can be expressed as [Ghodsi and Kaatuzian (2020b)]

$$\gamma_L \approx \omega_p \left(\frac{k_D}{k}\right)^3 e^{-k_D^3/(2k^2)} \tag{10.81}$$

  where $k_D = \omega_p/v_F$ and v_F is the Fermi velocity of electrons.

As mentioned earlier losses limit practical applications of SPP. Propagation length for travelling SPP is given by $L = [2Im(k_x)]^{-1}$. At visible frequencies it is of the order of 10–100 μm [Maier (2007)].

As mentioned earlier this creates a problem with the application of plasmonic in, say integrated circuits. In conventional plasmonic materials, like gold and silver the magnitude of the real part of the permittivity is very large [Markel and Sarychev (2007)].

Typical characteristics of SPP propagating along $Ag - SiO_2$ single-interface at three wavelengths are summarised in Table 10.1 from [Berini and Leon (2012)].

To finish this Section lets summarise some mechanisms of losses, namely scattering and ohmic losses in metals.

10.7.1   *Scattering*

SPP scattering process depends mainly on size, geometrical shape and dielectric constant of the surface. Defects occur when the surface is not perfectly smooth as

Table 10.1: Typical characteristics of SPP.

λ (nm)	n_{eff}	δ_w (nm)	2α (cm^{-1})	2α (dBμm^{-1})	L (μm)
360	2.537	44	5×10^5	218	022
633	1.565	176	1.6×10^3	0.71	6
1550	1.457	1269	1×10^2	0.044	100

anticipated. Three major results can occur due to defects [Zayats *et al.* (2005a)], [Tatel and Wartak (2020)]:

- Scattering of SPP into SPP in another direction (SPP reflection)
- Propagation of SPP through the defect region in the same direction as the incoming SPP (SPP transmission)
- Scattering of SPP into light.

10.7.2 *Ohmic Loss*

Ohmic losses exist due to attenuation of electromagnetic waves in metals. They are responsible for reducing propagation distances of SPPs, which, in turn reduces usefulness of plasmonic waveguides [Chang *et al.* (2011)]. Ohmic losses and SPP decay can be calculated from optical constants [Nguyen and Nguyen (2014)].

10.8 Methods for Compensating Losses

To compensate for losses we replace dielectric (medium 1) by the material with gain which is characterised by the complex permittivity $\varepsilon_1 = \varepsilon_1' + i\varepsilon_1''$ (negative value of ε_1'' represents gain). One can then analyse the conditions for a bound wave to propagate at the interface. After some algebra, one finds [Nezhad *et al.* (2004)]

$$\varepsilon_1'' = -\frac{\left(\varepsilon_1'\right)^2 \varepsilon_2''}{\left|\varepsilon_2\right|^2}. \tag{10.82}$$

The above can be related to the actual optical power gain using $\gamma = -k_0\varepsilon_1''/\left(\varepsilon_1'\right)^{1/2}$ where γ is the power gain coefficient. One obtains gain coefficient required for lossless SPP propagation

$$\gamma_0 = \frac{2\pi}{\lambda_0} \frac{\varepsilon_2''\left(\varepsilon_1'\right)^{3/2}}{\left(\varepsilon_2'\right)^2 + \left(\varepsilon_2''\right)^2}. \tag{10.83}$$

For $\gamma < \gamma_0$ the SPP propagation will still be lossy but the propagation length will increase.

Choosing an appropriate material to create gain is a challenge [Oulton (2012)]. Semiconductor materials will be a good selection as an active medium but they also introduce additional losses. Other considered solutions involved: erbium doped glass [Ambati *et al.* (2008)], gain molecules [Noginov *et al.* (2008a)], [Leon and Berini (2010)], [Gather *et al.* (2010)] and quantum dots [Grandidier *et al.* (2009a)].

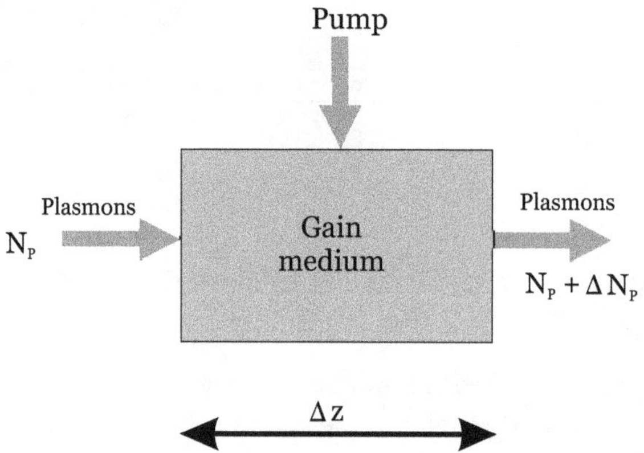

Figure 10.13: Schematic illustration of the definition of gain.

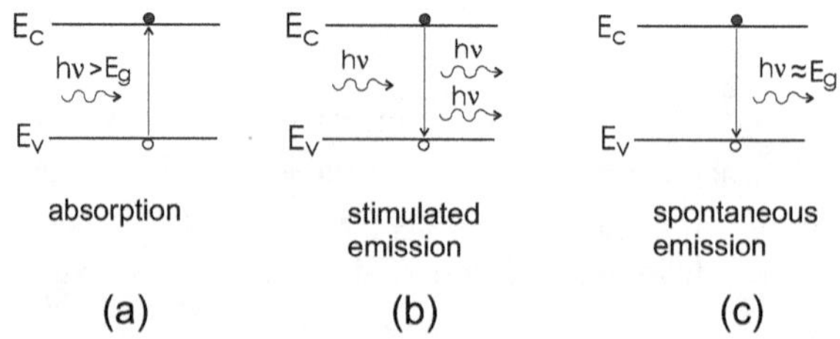

Figure 10.14: Fundamental processes in two-level system.

General picture of applification is illustrated in Fig. 10.13 in analogy to amplification of light in e.g. semiconductor lasers.

To describe details of gain process consider growth of plasmon density over a distance Δz where there exist an active region (gain) as

$$N_p + \Delta N_p = N_p e^{g\Delta z}.$$

If $g\Delta z \ll 1$ one can expand $e^{g\Delta z} \approx 1 + g\Delta z$. Using $\Delta z = v_g \Delta t$, where v_g is the group velocity and Δt is time to travel distance Δz, one finds $\Delta N_p = v_g g N_p \Delta t$. Thus generation rate of plasmons is

$$\left(\frac{dN_p}{dt}\right)_{gen} = R_{st} = \frac{\Delta N_p}{\Delta t} = v_g g N_p. \tag{10.84}$$

Active region which contains material gain in its simplest form can be represented as two level system (TLS), see Fig. 10.14 where we illustrated fundamental processes taking place there. (Original approach has been invented for optical systems and involves photons, but it can also be applied to plasmons).

Fundamental processes in TLS are:

(a) absorption
(b) spontaneous emission
(c) stimulated emission

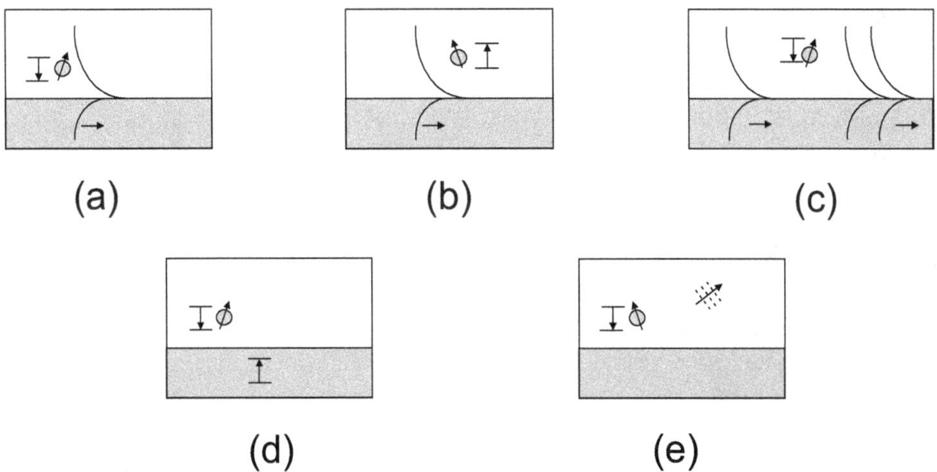

Figure 10.15: Optical processes occurring for dipoles near a single metal–dielectric interface. Adapted from Berini *et al.* [Berini and Leon (2012)].

More specifically, for plasmons some quantum processes are summarised in Fig. 10.15 [Berini and Leon (2012)]. The processes are: a. Spontaneous emission of SPPs. b. Absorption of SPPs. c. Stimulated emission of SPPs. d. Creation of electron–hole pairs. e. Spontaneous emission of radiation.

10.9 Practical Approaches to Compensate Losses in SPP

Here we summarise several methods of compensating losses in SPP propagation which were reported in literature. This part is based on our recent review [Tatel and Wartak (2020)].

One of the first reports was the paper by Nezhad *et al.* [Nezhad *et al.* (2004)] who analysed propagation of SPP along an infinite planar interface between a metal and a gain medium. They determined that material gain of the value 1260 cm^{-1} was needed to obtain lossless propagation of SPP for an Ag-InGaAsP interface at $\lambda_0 = 1550$ nm.

Another candidate to acquire gain to compensate the loss of SPs is by using multiple quantum wells. This gain medium should provide a large enough amplification to realize appropriate propagation ([Alam *et al.* (2007)]). It was reported that a gain around 1800/cm from AlGaInAs quantum wells at 1.55 μm wavelength shows that a lossless SP propagation is within current technological limits.

Another suggested method was through electrical pumping based on a Schottky-barrier diode. It shows that it is possible to fully compensate SPP propagation losses and provide net SPP gain [Fedyanin *et al.* (2012)]. SPP amplification through electrical injection is a complex phenomena incorporating electrical and optical processes. At 77K (yields a small bandgap E_g^{InAs} of InAs) the Schottky-barrier height is greater than E_g^{InAs}. Since the barrier is greater, an inversion layer is established;

this is where in equilibrium the Fermi level is higher than conduction band edge of InAs, resulting in the concentration of electrons being substantially greater that holes. Through applying a positive bias voltage, the population inversion is created in InAs by injecting both electrons and holes into the active InAs area. The electrons injected from the inversion layer allow for a shift in the quasi-Fermi level towards the conduction band. At the same instance, $p - InAs/p - AlAs_{0.16}Sb_{0.84}$ heterojunction acts as an ohmic contact for holes which are capable of penetrating from $AlAs_{0.16}Sb_{0.84}$ into $InAs$ while maintaining the hole quasi-Femi level of approximately $2k_BT$ below the valence band edge. As the energy separation between quasi-Fermi level rises over the SPP energy, the stimulated emission into SPP mode commences. The bias voltage can increase to a point where it is capable to compensate and overcome the ohmic and radiation losses of the TM_{00} mode, and even amplify it.

Having nano quantum dots as a gain medium would allow for tunable spontaneous wavelength and do not suffer from excited state quenching and photobleaching. The energy levels and transition mechanism is explored by M. Stockman ([Stockman (2008)]). To begin the surface plasmon amplification by stimulated emission of radiation (spaser), the radiation from the light pumped excites an electron hole pair in the gain medium (chromophores). The electron hole pair then relaxes into an exciton state; with the chromophores coupled to the resonator, the energy is transferred to the SP modes. From this, local fields are created, which then establishes a process which consists of the local field further exciting the gain medium. This creates somewhat of a feedback loop. If this loop creates a sufficient amount and the SP mode quality factor is high enough, then spasing will occur with that SP mode.

We finish this section with outlining structure suggested by Keshmarzi [Keshmarzi (2017)]. It is shown in Fig. 10.16.

It consists of a 20 nm thick and 1 μm wide Au stripe deposited on a 15 μm thick SIO_2 film on a Si substrate. The structure was covered with a polymer to guide a ss_b^0 mode at around 882 nm.

As a gain medium it was suggested a thin film of PMMA (polymethylmethacrylate) doped with organic dye molecules of IR-140 and was optically pumped using 8 ns laser pulses at 810 nm. The induced stimulated emission from excited dye molecules to LRSPP was at 880 nm.

The structure was modelled using commercial software COMSOL Multiphysics. Rate equations approach for four-level system had been used to model gain medium.

The refractive indices of Au and the surrounding media (SiO_2 and polymer) at $\lambda_0 = 882$ nm were assumed as $n_{Au} = 0.22138 - j5.3142$ and $n_{polymer} = n_{SiO_2} = 1.452$.

The mode effective index of ss_b^0 mode was determined from

$$n_{eff} = \frac{\beta\lambda_0}{2\pi} \tag{10.85}$$

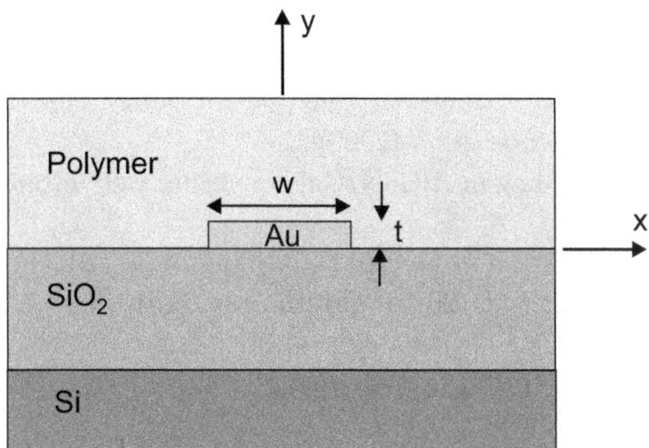

Figure 10.16: Cross-section of LRSPP waveguide, with $w = 1$ μm, $t = 20$ nm, supporting a single ss_b^0 mode at 882 nm.

to be $n_{eff} = 1.4527$. The mode power attenuation (MPA) defined as

$$MPA = \alpha \frac{20}{1000} \log_{10} e \quad [\text{dB/mm}] \tag{10.86}$$

was $MPA = 2.6$ dB/mm. The propagation length of this mode was found to be 192 μm.

10.10　Some Applications of LRSPP

The lower attenuation enables many applications of SPPs, such as waveguides, sources, near-field optics, surface-enhanced Raman spectroscopy, data storage, solar cells, chemical sensors and biosensors, interconnected and passive integrated optical structures, and the exploitation of a broad range of materials effects (e.g., electro-optic, thermo-optic, and optical gain). Some reported studies:

- Bai *et al.* [Bai *et al.* (2009)] conducted numerical studies of metallic structures with 1D periodic nanoridges attached to a thin-film amorphous Si solar cells. The excitation of SPP at the interfaces between a-Si and metal materials results in absorption enhancements (around 3.32) of light in the near-IR range. The effect can increase the efficiency of solar cell by around 17.12%.
- study the low-loss propagation of hybrid long-range surface plasmon polariton (LRSPP) modes in 2D metal–dielectric configurations [Bian *et al.* (2009)], [Chen *et al.* (2009)], [Bian and Gong (2013)].

Some of the specific applications include:

- passive integrated optics elements based on LRSPP [Charbonneau *et al.* (2006)],
- directional couplers [Boltasseva and Bozhevolnyi (2006)],
- low-loss polymer-based long-range surface plasmon polariton waveguide [Kim *et al.* (2007)],

- surface plasmon polariton based modulators and switches operating at telecom wavelengths [Nikolajsen *et al.* (2004)],
- thermally activated variable attenuation of long-range surface plasmon-polariton waves [Gagnon *et al.* (2006)],
- efficient optical coupling in AlGaN/GaN quantum well infrared photodetectors [Wang *et al.* (2016)],
- light emitting diodes [Lu *et al.* (2011)], [Zhang *et al.* (2011)],
- photodetectors [Wu *et al.* (2010)], [Muth *et al.* (2012)].

10.11 Plasmons in a DC Electric Field

Over the years there were several attempts to analyse plasmonic effects in semiconductor plasma in a constant electric field and interacting with traveling waves [Bass *et al.* (1966)], [Sydoruk (2014)], and [Cada and Pistora (2016)]. In the summary here, we concentrate on the results obtained by the last group.

One possibility to amplify SPs is by allowing them to interact with drifting electrons in an external DC electric field. Interaction and possible amplification of EM waves with moving electrons have been discussed in classic papers [Hahn and Metcalf (1939)], [Ramo (1939)]. Over the years, many groups [Bass *et al.* (1966)], [Sumi (1966)], [Yariv and Armstrong (1973)], [Gover (1976)] analysed those interactions in various materials and configurations. The specific applications to plasmonics were suggested by M. Cada [Cada and Pistora (2008)] and is extensively analysed in [Cada and Pistora (2016)].

Electrons moving under external DC field exchange their energies with the EM wave, and under certain conditions of phase matching, an EM wave is amplified. To facilitate the transfer of energy from electron stream to EM wave, their velocities should be close to each other. The electrons are accelerated and decelerated by the propagating EM wave while being moved under an external DC field.

To discuss the interaction we consider a two-dimensional configuration with the boundary lying along the z-axis and x-axis as the transversal one. Previous equations are rewritten to explicitly refer to a dielectric or a semiconductor.

A TM polarisation is only considered, which results in the following assumptions:

$$\mathbf{H}_{S,D} = [0, H_{y,S,D}, 0] \tag{10.87}$$

$$\mathbf{E}_{S,D} = [E_{x,S,D}, 0, E_{z,S,D}] \tag{10.88}$$

$$\mathbf{J} = \left[J_{x,S}, 0, J_{z,S} \right] \cdot \delta_{S,D} \tag{10.89}$$

where subscripts S, D refer to the semiconductor and dielectric materials, respectively, and

$$\delta_{S,D} = e^{i\omega t} e^{i\gamma_{S,D} x} e^{i\beta z}. \tag{10.90}$$

Here ω is the angular frequency, $\gamma_{S,D}$ and β are the transversal and longitudinal propagation constants, respectively

The remaining equations characterise electrons and are summarised below [Cada and Pistora (2016)]

$$\nabla \times \mathbf{H}_S = \mathbf{J} + \varepsilon_S \frac{\partial \mathbf{E}_S}{\partial t}, \quad \nabla \cdot \mathbf{D}_D = 0 \tag{10.91}$$

$$m \frac{d\mathbf{v}}{dt} = -e\left(\mathbf{E} + \mu_0 \mathbf{v} \times \mathbf{H}\right) \tag{10.92}$$

$$\nabla \times \mathbf{H}_D = \varepsilon_D \frac{\partial \mathbf{E}_D}{\partial t}, \quad \nabla \cdot \mathbf{H}_{S,D} = 0, \quad \varepsilon_{S,D} = \varepsilon_0 \varepsilon_{\infty,d} \tag{10.93}$$

$$\nabla \times \mathbf{E}_{S,D} = \mu_0 \frac{\partial \mathbf{H}_{S,D}}{\partial t}, \quad \nabla \cdot \mathbf{J} = -\frac{\partial \rho}{\partial t}, \quad \rho_0 = -eN \tag{10.94}$$

$$\nabla \cdot \mathbf{D}_S = \rho_S, \quad \mathbf{J} = \rho \mathbf{v}, \quad \omega_p = e\sqrt{\frac{N}{m\varepsilon_0}}. \tag{10.95}$$

Here $\mathbf{E}, \mathbf{D}, \mathbf{H}$ are the electric, displacement, and magnetic field vectors, respectively, $\mathbf{J}, \rho, N$ are the current density, charge density, and charge current density in the semiconductor, respectively, $\mathbf{v}$ is the velocity of the moving electrons, $m = m_0 m^*$ with m^* being the effective mass, e is the electron charge, ε_0 and μ_0 are the vacuum permittivity and permeability, ε_∞ and ε_d are the relative permittivity of the semiconductor and dielectric, and ω_p is the plasma frequency.

Harmonic fields are considered and the derivatives are replaced as $\frac{\partial}{\partial t} \longrightarrow -i\omega, \frac{\partial}{\partial x} \longrightarrow i\gamma_{S,D}, \frac{\partial}{\partial z} \longrightarrow i\beta$. The velocity, charge density, and current density are assumed to be modulated around constant values, which results in the following replacements

$$\mathbf{v} \longrightarrow \mathbf{v}_0 + \mathbf{v}_S \delta_S = [v_{x,0} + v_{x,S}\delta_S, 0, v_{z,0} + v_{z,S}\delta_S] \tag{10.96}$$

$$\rho \longrightarrow \rho_0 + \rho_S \delta_S \tag{10.97}$$

$$\mathbf{J} = \rho \mathbf{v} = \mathbf{J}_0 + \mathbf{J}_S \delta_S. \tag{10.98}$$

Normalize all parameters to the plasma frequency, ω_p, and the speed of light, c, like

$$\Omega = \frac{\omega}{\omega_p}, B = \frac{c\beta}{\omega_p}, \Gamma = \frac{\gamma}{\omega_p}, \Delta = \frac{v_d}{c}, \Gamma_{D,S} = \frac{c\gamma_{D,S}}{\omega_p}. \tag{10.99}$$

After long manipulations (described in [Cada and Pistora (2016)]), results in the following dispersion relation in a normalized form [Cada and Pistora (2017)] are as

$$\sqrt{\varepsilon_\infty}\left(\Omega - BA\right)\left(\Omega - BA + i\Gamma\right)\left[(\Omega + iA\Delta_0)^2 - B^2\Delta_0^2\right] - \Omega\Delta_0\left(\Omega + iA\Delta_0\right) = 0. \tag{10.100}$$

In the above Ω is the frequency, B is the complex propagation constant, Δ is the drift velocity, Γ is the electron collision loss, A is the transmission loss of the waveguide

Table 10.2: Silicon parameters used.

Symbol	Quantity	Value
ε_∞	Background permittivity	11.7
m^*	Effective electron mass	$0.26\ m_0$
α	Absorption coefficient	10^4 cm^{-1}
γ	Damping coefficient	5.6×10^{12} s^{-1}
μ_e	Electron mobility	1,200 cm V^{-1} s^{-1}

Table 10.3: Parameters of the amplifier structure.

Symbol	Quantity	Value
$\omega_p/2\pi$	Plasma frequency	3 THz
$\omega_{op}/2\pi$	Operating frequency	2 THz
α_T	Transmission loss	$10^2 - 10^3$ cm^{-1}
N	Electron concentration	3×10^{16} cm^{-3}
V_w	Waveguide volume	$1 \times 1 \times 1 \times \mu$m^3
P_{in}	Incident optical power	5 dBm

Table 10.4: Achievable performance of a terahertz amplifier structure: Transmission loss.

Symbol	Quantity	0	10^2 cm^{-1}	10^3 cm^{-1}
l_p(nm)	Plasma frequency	48.7	48.6	48.4
g_{lp}(dB	Operating frequency	0.6	0.48	0.39
g_{lw}(dB)	Transmission loss	12.3	10.2	8

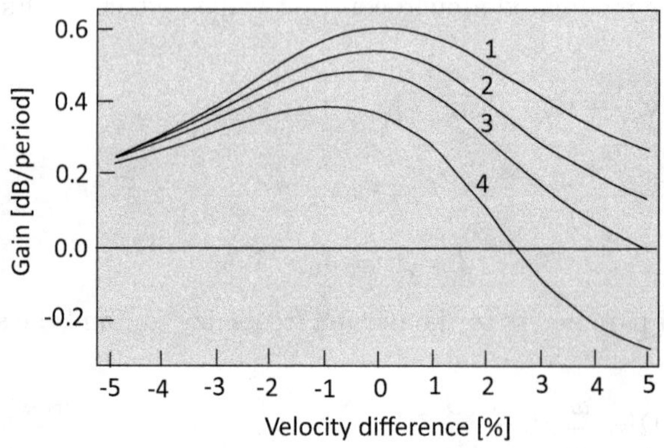

Figure 10.17: Gain per period as a function of velocity difference. Adapted from [Cada and Pistora (2017)].

mode, Δ_0 is the velocity of a natural waveguide mode, and ε_∞ is the background permittivity of the semiconductor.

For the analysis of Eq. (10.100), the parameters summarised in Tables 10.2, 10.3 and 10.4 were selected. Results of numerical analysis of Eq. (10.100) are shown in Fig. 10.17.

As expected the electrons moving along the interface affect the speed and the attenuation of the SPPs. Electrons moving in the same direction as EM wave give the energy to the wave which results in the decrease of attenuation.

During studies based on analysis of Eq. (10.100) it was observed that SPP move too fast compared to the drift velocities of electrons under dc electric fields. Further studies are needed to design structure where both velocities are comparable. Materials with large mobilities will be good candidates as a gain medium.

Chapter 11

Quantum Plasmonics

One considers SPP as an electromagnetic wave interacting with electrons. As a wave it should exhibit wave-like behavior such as, interference, diffraction etc. Quantum plasmonics deals with effects associated with the quantum nature of both electrons and electromagnetic fields.

We start this chapter with a review of several experiments which show wave nature of SPP. We concentrate on why quantum effects are necessary for a description of small size nanoparticles. Theoretical aspects will be discussed later-on. Some general literature on quantum plasmonics is [Fitzgerald *et al.* (2016)], [Jacob (2012)], [de Leon *et al.* (2012)], [Fitzgerald (2018)], [Stockman *et al.* (2018)], [Jacak (2020)]. We start with summarising early evidence of quantum effects in plasmonics. More advanced experiments will be discussed later.

11.1 Early Evidence of Quantum Effects in Plasmonics

For small metal nanostructures, say around 5 nm the electrons will start experiencing quantum effects. This has been observed experimentally [Knight *et al.* (1984)], [Zheng *et al.* (2007)]. Below we provide a short summary.

- Knight *et al.* [Knight *et al.* (1984)] obtained mass spectra for sodium clusters of N atoms per cluster (for values of N = 4 to 100). The spectra show large peaks or steps at N = 8, 20, 40, 58, and 92. The experimental results were explained using a one-electron shell model in which independent delocalized atomic 3s electrons are bound in a spherically symmetric potential well.
- Zheng *et al.* [Zheng *et al.* (2007)] reported on excitation and emission spectra of different Au nanoclusters. They observed fluorescence arising from intraband transitions of the free electrons, see Fig. 11.1. Emission maxima shift to longer wavelengths with increasing initial Au concentrations, suggesting that increasing nanocluster size leads to lower-energy emission. In Fig. 11.1 we showed emission spectra under long-wavelength ultraviolet (UV)-lamp irradiation (366 nm).

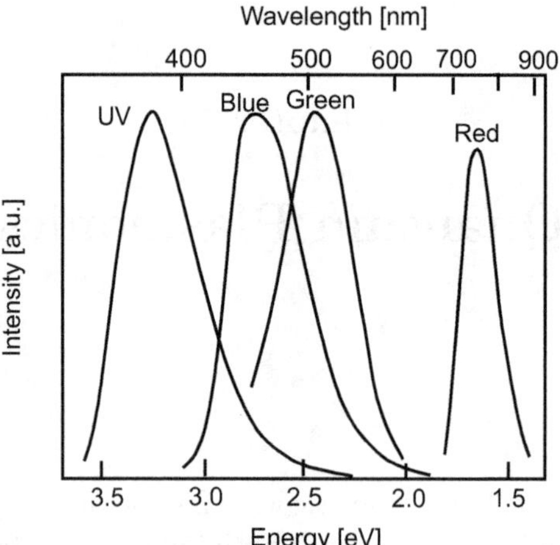

Figure 11.1: Emission spectra of different Au nanoclusters. Adapted from Zheng [Zheng *et al.* (2007)].

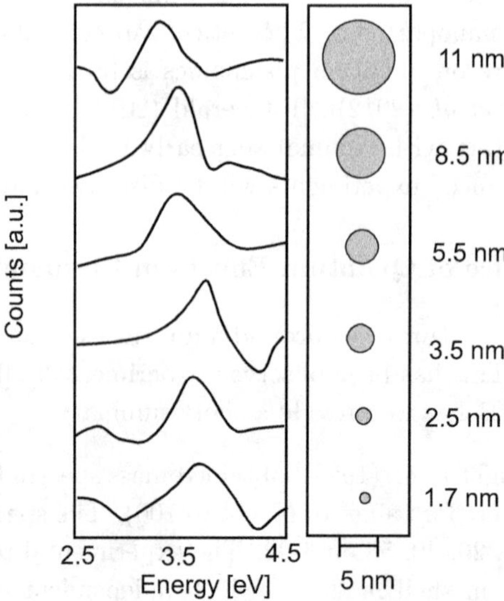

Figure 11.2: Correlating Ag nanoparticle geometry with plasmonic EELS data. Adapted from [Scholl *et al.* (2012)]. EELS data from particles ranging from 11 nm to 1.7 nm in diameter and the corresponding STEM image of each specimen. The electron beam was directed onto the edge of the particles so that only the surface resonance is shown.

- Scholl *et al.* [Scholl *et al.* (2012)] (see also [de Abajo (2012)]) studied individual ligand-free silver nanoparticles for sizes ranging from 20 nanometers to about 2 nanometers, using electron energy-loss spectroscopy (EELS). They found typical nonlocal effects as the particle size is decreased: a blueshift and a broadening of the SP peak. Some of their experimental data are shown in Fig. 11.2.

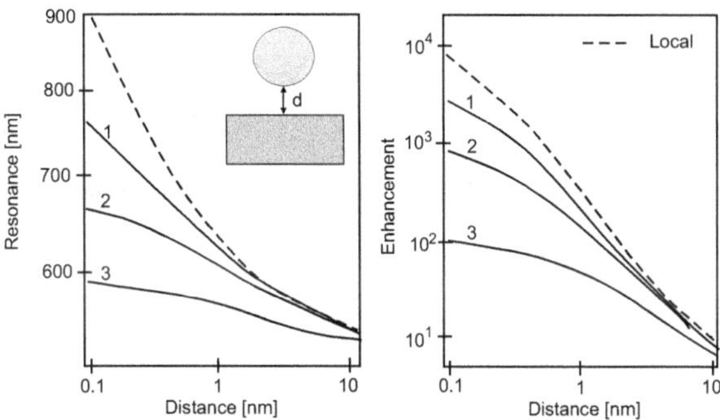

Figure 11.3: Behavior of the film-coupled nanosphere, assuming a local model and the nonlocal model with various values of β, as a function of separation distance. Adapted from [Ciraci *et al.* (2012)]. (A) Position of the peak scattering intensity as a function of gap size. (B) The corresponding field enhancement ratio. Lines 1, 2, 3 correspond to the values of β as follows $\beta_1 = 1.0 \times 10^6$ m/s, $\beta_2 = 3.0 \times 10^6$ m/s, $\beta_3 = 1.0 \times 10^7$ m/s.

- Ciraci *et al.* [Ciraci *et al.* (2012)] used chemically deposited sub-nanometre molecular layers to precisely control the separation between gold spheres and a gold film. This allowed a detailed study of the effects of nonlocality. It was found that the hydrodynamic model (HM) can give excellent predictions down to gap-sizes of 1 nm. Behavior of the film-coupled nanosphere assuming a local model and the nonlocal model as a function of separation distance is shown in Fig. 11.3. Calculations were performed assuming a gold nanosphere of radius r = 30 nm on a film 300 nm thick. Parameter β which appears in the thermodynamic model (to be discussed in the theory Section) is approximately equal to the speed of sound in the degenerate Fermi plasma of conduction electrons.

11.1.1 *Comparison of Classical and Quantum Plasmonics*

In typical plasmonics experiments conducted with metals, due to high density of electrons the discrete energy levels can be ignored as the spacing between energy levels is very small. Deviations from classical behaviour for the electrical (such as the Coulomb blockade) and optical properties are observed when the energy level spacing are comparable to the thermal energy.

In Fig. 11.4 we compared classical and quantum effects in plasmonics. Electrons occupy states up to Fermi energy. In classical limit plasmons are visualized as oscillations within an electron gas, whereas in the quantum limit the main role play transitions between occupied and empty states within electron gas.

In particles larger than about 10 nanometres, plasmons exist as collective oscillations of a gas of conduction electrons. In particles smaller than 10 nanometres, plasmons are associated with quantum electron transitions between occupied and unoccupied energy levels.

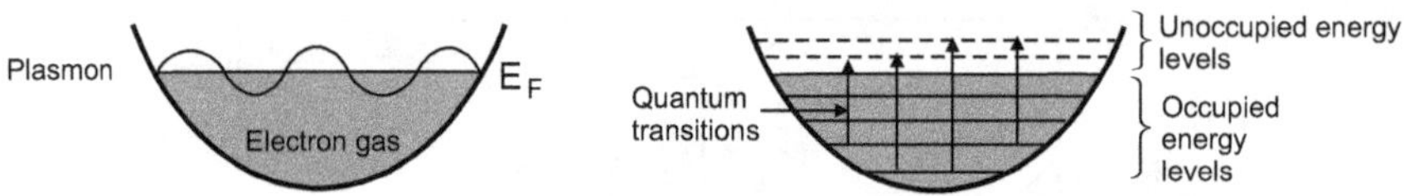

Figure 11.4: Comparison of classical (left) and quantum (right) plasmons.

Now we will discuss some experiments showing typical quantum-mechanical nature of plasmons.

11.2 Young's Double-Slit Experiments

Young's double slit experiment can be considered as a fundamental one in quantum mechanics. Lets start with its optical version.

11.2.1 *Basics — Optics*

Principle of double-slit experiment is illustrated in Fig. 11.5. It consists of a source, screen with two small slits a and b and an observation screen. The experiment has been performed with photons as well as with more massive particles, like electrons or neutrons. It can be conducted with the "normal" (classical) source of light and also with sources which emit individual photons when the intensity of the source is keep very low so there is only one particle at a given time in the apparatus which interferes with itself. In both cases the results are identical. As a result, the interference pattern is produced which is shown in Fig. 11.5. The state of the particle is in the coherent superposition [Bouwmeester and Zeilinger (2000)]

$$|\psi\rangle = \frac{1}{\sqrt{2}} \left(|\psi_a\rangle + |\psi_b\rangle \right) \tag{11.1}$$

where $|\psi_a\rangle$ and $|\psi_b\rangle$ describe the quantum state with only slit a or slit b open.

The interpretation of the experiment (when conducted with classical light with many photons simultaneously) within the framework of classical wave theory of light is as follows: the light wave from point source upon reaching the screen with slits converts both slits into sources of new light waves (Huygens' principle). Slits a and b act as new light sources. Emitted light waves interfere which results in the characteristic interference pattern of intensity distribution.

11.2.2 *Young-Type Experiments with Plasmons*

11.2.2.1 *Schouten (2005) experiment*

Schouten *et al.* [Schouten *et al.* (2005)] reported on experimental and theoretical studies of the optical transmission of a thin metal layer perforated by two parallel subwavelength slits. The slits were separated by many optical wavelengths.

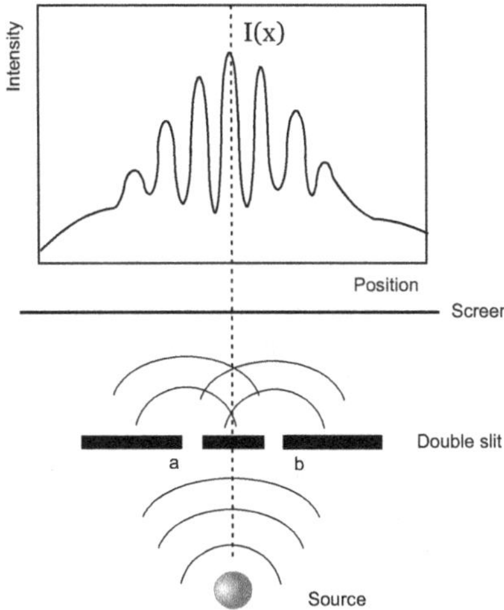

Figure 11.5: Principle of the double-slit-experiment.

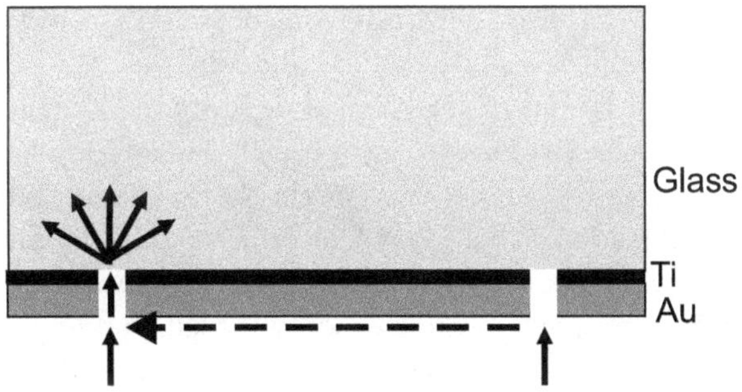

Figure 11.6: Two interfering paths leading to light emission from the slit at the left. The dashed line indicates the propagating surface plasmon. Adapted from [Schouten *et al.* (2005)].

Sample used in the experiment was formed of a 200 nm thick gold film, evaporated on top of a 0.5 mm thick fused quartz substrate. Two-slits created were 50 μm long and 0.2 μm wide and were separated by a distance of 4.9, 9.9, 14.8, 19.8, or 24.5 μm, respectively. The schematic of slits is shown in Fig. 11.6.

The structure was illuminated at normal incidence by a narrow-band cw Ti:sapphire laser, tunable between 740 and 830 nm. The slits transmit part of the incident radiation which creates Young-type interference pattern. Also, each slit scatters part of the incident wave into a plasmonic channel. Slits also provide mechanism for backconverting a surface plasmons into free-space radiation.

Incident TM light excites surface plasmons at one of the slits which subsequently propagate and eventually interfere with the light, see Fig. 11.6. In their experiment the far-field interference pattern results due to interference of four paths, two of which are partially plasmonic, while the other two are photonic all the way.

11.2.2.2 *Zia 2007 paper*

In Fig. 11.7, from Zia *et al.* paper [Zia and Brongersma (2007)] we showed a schematic of their experimental configuration. The SPPs are created and propagate along two metal stripe waveguides. The stripes play also the role of the two holes in the traditional Young experiment. A photon scanning tunnelling microscope (PSTM) is used to visualize propagation of SPP.

The detected signal provides information of the local field intensity just under the tip and the propagation, diffraction and interference of the SPPs.

In Fig. 11.8 we reproduced experimental results of Zia *et al.* [Zia and Brongersma (2007)] which demonstrate propagation of guided polaritons and also their diffraction and interference. Plots are offset by -0.25 increments along the y-axis. Orientation of axis is shown in the figure.

11.2.2.3 *Alam (2013)*

Alam [Alam (2013)] considered two parallel waveguides which produced interference patterns with SPPs. SPP interference was studied using SPP tomography.

His sample consists of a glass substrate coated with a layer of chromium. The thickness of the chromium is maintained at approximately 2 nm. On top of this layer, a layer of gold of 50 nm in thickness is deposited. Chromium layer acts as an adhesive layer between the glass substrate and the gold layer. A thin layer of polymethylmethacrylate (PMMA), a transparent thermoplastic, is deposited on top of the gold layer. Thickness of this PMMA layer is roughly 100 nm.

The sample was pattern using electron beam lithography technique. First, three scattering feature 50 μm in length are etched by focusing an electron beam on the sample. The periods of these lines are roughly 600 nm, and each line is about 300 nm wide. Next, two parallel single mode Dielectric Loaded Surface Plasmon Polariton Waveguide (DLSPPW) of 3 μm long and a width of 600 nm each are etched into the PMMA layer. Figure 11.9 shows top view of the sample.

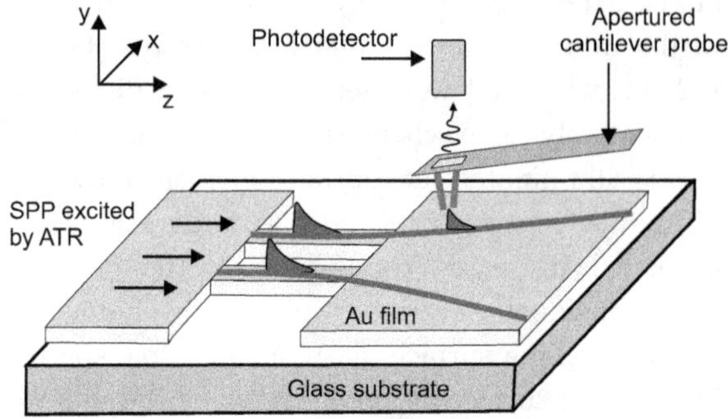

Figure 11.7: Schematic of a double-slit experiment for surface plasmon polaritons (SPPs). It must be noted here that ATR stands for Attenuated Total Reflection. [In this book ATR usually stands for "absorbed, transmitted and reflected" — Ed.] Adapted from [Zia and Brongersma (2007)].

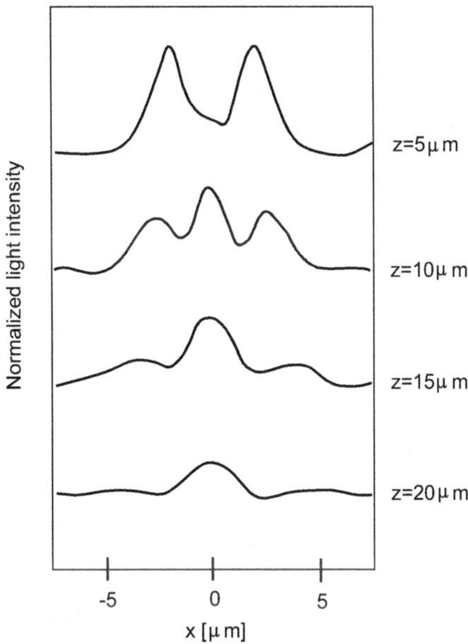

Figure 11.8: Experimental results of double-slit experiment. Adapted from [Zia and Brongersma (2007)].

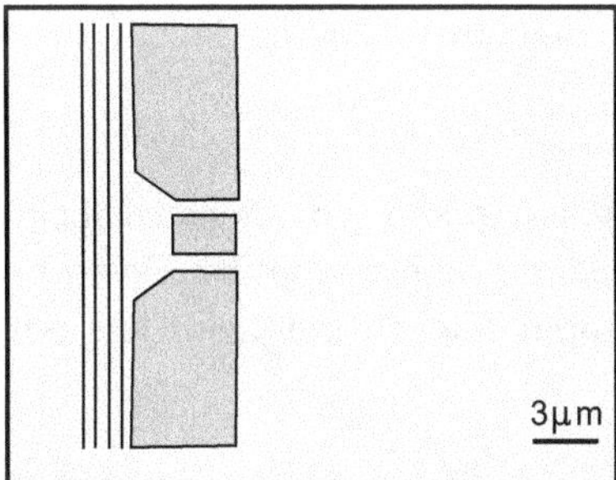

Figure 11.9: Schematic diagram of the top view of the sample. Adapted from [Alam (2013)].

Results show typical interference patterns as depicted in Fig. 11.5. SPPs propagates through the waveguides and at the end they interfere to produce diffraction pattern. Zeroth and first-order interference maxima were clearly identified in the experiment. Separation between the wave guides was 0.75 μm.

11.3 Quantum Information at the Nanoscale. HOM Interference

The Hong–Ou–Mandel (HOM) effect is a two-photon interference effect in quantum optics which was demonstrated in 1987 by three physicists from the University of Rochester: Chung Ki Hong, Zhe Yu Ou and Leonard Mandel, [Hong *et al.* (1987)].

The effect occurs when two identical single-photon waves enter a 1:1 beam splitter, one in each input port. When the photons are identical, they will extinguish each other. If they become more distinguishable, the probability of detection will increase. In this way the interferometer can accurately measure bandwidth, path lengths and timing

11.3.1 *Optical Hong–Ou–Mandel Effect*

In Fig. 11.10 we show a schematic view of Hong–Ou–Mandel interference experiment with photons. The effect can be directly observed using single photons. Photons (modes) are generated on the left and (typically) travel via optical fiber, denoted as a' and b'. They interfere at a beam splitter (BS); the output photons from the beam splitter are denoted a and b. They are collected by two photodetectors, amplified and analysed.

The coincidence rate of the detectors will drop to zero when the identical input photons overlap perfectly in time. This is called the HOM dip. When the two photons are perfectly distinguishable, the dip completely disappears. The precise shape of the dip is directly related to the power spectrum of the single-photon wave packet and is therefore determined by the physical process of the source. Analysis of the effect is conducted in a few steps.

First, consider classical light of amplitude E which hits beam splitter and it is transmitted and reflected as follows, see Fig. 11.11 [Agarwal (2013)]

$$E_t = t_1 E, \qquad E_r = r_1 E \tag{11.2}$$

where t_1 and r_1 are transmission and reflections coefficients (amplitude). In quantum theory field amplitudes are replaced by Heisenberg operators, as $E \longrightarrow a, E_t \longrightarrow c, E_r \longrightarrow d$. Input and output fields are related as [Agarwal (2013)], see Fig. 11.12

$$\begin{pmatrix} c \\ d \end{pmatrix} = U \begin{pmatrix} a \\ b \end{pmatrix}. \tag{11.3}$$

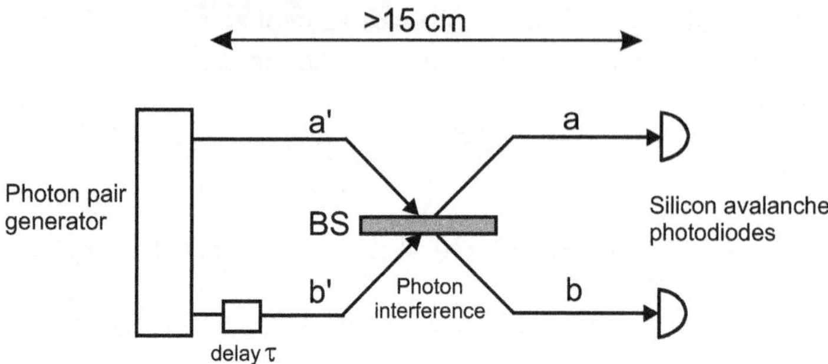

Figure 11.10: Conventional (optical) two-photon quantum interference.

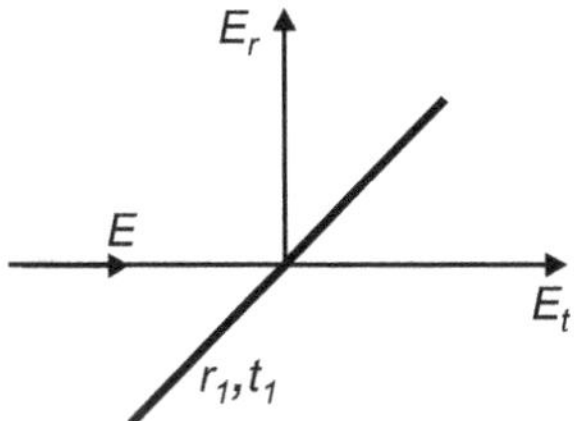

Figure 11.11: Schematic illustration of classical beam splitter.

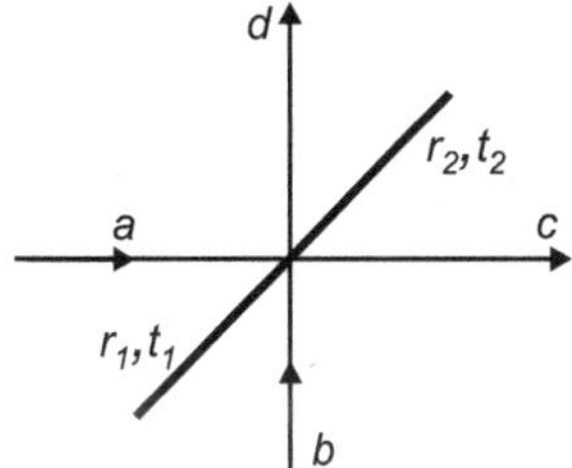

Figure 11.12: Schematic illustration of quantum beam splitter.

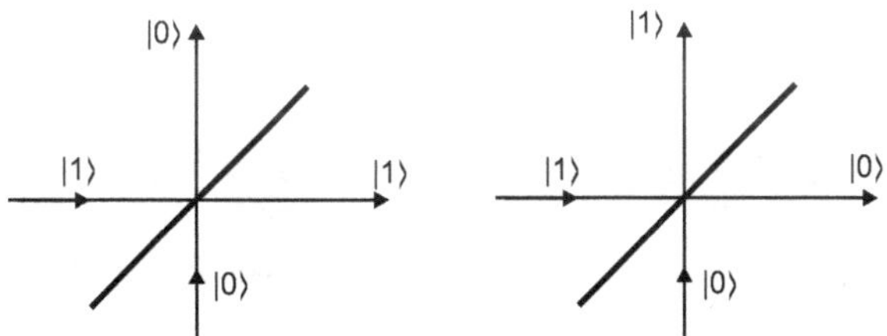

Figure 11.13: Possible paths for a single photon at beam splitter.

The transformation matrix U must be unitary and can be expressed as

$$U = \begin{pmatrix} \cos\theta & e^{i\phi}\sin\theta \\ -e^{-i\phi}\sin\theta & \cos\theta \end{pmatrix} \tag{11.4}$$

For a 50–50 beam splitter popular form of matrix U is

$$U = \frac{1}{\sqrt{2}} \begin{pmatrix} 1 & i \\ i & 1 \end{pmatrix}. \tag{11.5}$$

Interesting physics can be performed when fixed number photons (usually small) enter beam splitter. We will use Dirac notation. For that introduce state $|n, m\rangle$ where n and m designate number of photons. Let us consider the case when no light enters port b and a single photon enters port a. The input state is $|1, 0\rangle$. The output state is

$$|1, 0\rangle_{out} = \cos\theta \, |1, 0\rangle + i \sin\theta \, |0, 1\rangle. \tag{11.6}$$

Creation of this state is illustrated in Fig. 11.13.

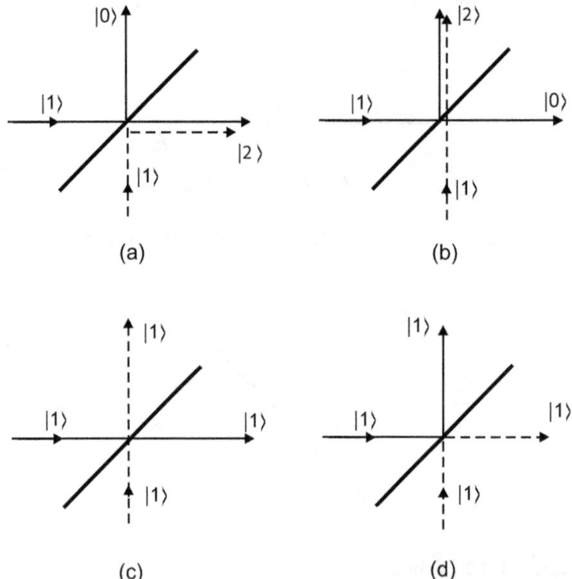

Figure 11.14: Possible paths when two photons enter beam splitter.

Single photon can take two paths, it can be transmitted or reflected.
In the case of two photons incident on beam splitter the output state is

$$|1,1\rangle_{out} = \left(\cos^2\theta - \sin^2\theta\right)|1,1\rangle + \sqrt{2}i\sin\theta\cos\theta\,|2,0\rangle + |0,2\rangle\,. \qquad (11.7)$$

Creation of the above state involves four paths as shown in Fig. 11.14.

We stress the possibility of quantum interference associated with paths (iii) and (iv). Important case of interference exists for 50-50 beam splitter for $\theta = \pi/4$. In that case Eq. (11.7) reduces to

$$|1,1\rangle_{out} = \frac{i}{\sqrt{2}}\left(|2,0\rangle + |0,2\rangle\right)\,. \qquad (11.8)$$

One can observe complete cancellation of contributions from paths (iii) and (iv) which is known as Hong, Ou, Mandel effect [Hong *et al.* (1987)].

Recent review [Bouchard *et al.* (2021)]. Typical result with two photons is the same as for plasmons, see Fig. 11.16.

11.3.2 *HOM Effect with Plasmons*

One of the first questions about SPP is the investigation of the physical nature of single surface plasmon polaritons. Important evidence was obtained by observing HOM effect with plasmons which will be discussed in this section. The experiments provided also direct evidence for the bosonic nature of SPP.

As an important step in analyzing quantum properties of plasmons, in 2012 Maier's group [Martino *et al.* (2012)] showed that SPP are not negatively affected by high optical losses. They showed that the high optical losses of a plasmon waveguides do not affect quantum statistics of single photons when converted into single SPPs. Now, we will discuss in detail experiments.

11.3.2.1 *Experiment by Reinier W. Heeres et al. (2013)*

Plasmon interference device and experimental set-up is shown in Fig. 11.15. Photon pairs are generated by generator at 1064 nm and enter plasmonic directional coupler which works as beamsplitter. Detection is performed by superconducting single-photon detectors (SSPDs) which operate at 4 K. They are fabricated from niobium nitrade and consist of a meandering line (100 nm width) of a thin (5 nm). They are able to absorb single phonons. After absorption the detector switches locally to the normal, resistive state. Created current goes to an amplifier and after it a voltage pulse is measured. Reported coincidence results are typical as shown in Fig. 11.16.

11.3.2.2 *G. Di Martino et al. (2014)*

Observation of quantum interference in the plasmonic Hong–Ou–Mandel effect was also reported by DiMartino *et al.* [Martino *et al.* (2014)]. They demonstrated a direct bosonic nature of single SPPs. Their experimental configuration was similar to that of Heeres *et al.* and involved the following steps.

(a) Photon pair generation. Photon pairs are generated via spontaneous parametric down-conversion by using a pump laser focused onto a BBO crystal and filtered by using interference filters.
(b) Microscopy. The photons from single-mode fiber are collimated and half-wave plates are used to optimize SPP excitation. A time delay is introduced on one path.
(c) Plasmonic beam splitter. The photons are focused onto separate spots on the input gratings by using a microscope objective.
(d) Detection and analysis. The outputs of the multimode fibers are sent to avalanche photodiodes.

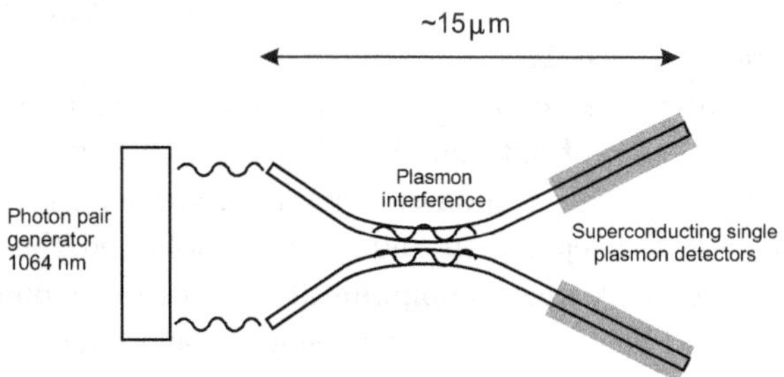

Figure 11.15: HOM interference of plasmons. Adapted from [Heeres *et al.* (2013)].

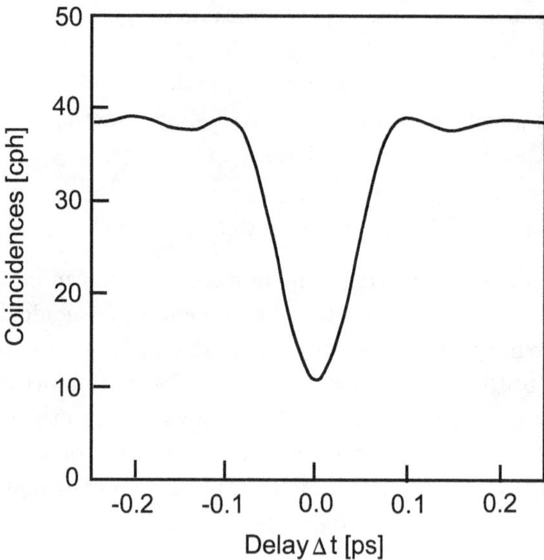

Figure 11.16: HOM coincidence count versus time delay. Adapted from [Martino *et al.* (2014)].

The interference is observed by counting coincidences of photon detection at two output ports. For zero time delay a dip in coincidences is observed which clearly indicates quantum interference. Typical experimental results are shown in Fig. 11.16.

11.4 Tunneling of Plasmonics

Tunneling is generally considered as a strong manifestation of quantum effects. Savage *et al.* [Savage *et al.* (2012)] reveal the quantum regime of tunnelling plasmonics.

The experimental scheme is illustrated in Fig. 11.17. Central role is played by two atomic force microscope (AFM) tips terminated by two gold nanoparticles separated by a distance d. They form nano cavity which supports plasmonic resonances created by strong coupling between localized plasmons on each tip. Tips are illuminated from the side in a dark field configuration. Tips can move with a nanometre precision due to three-axis piezoelectric stages. A supercontinuum laser with polarization parallel to the tip-tip axis is used as an excitation source over a wide wavelength range of 450–1700 nm.

Savage demonstrated that quantum effects play critical role when separation between tips is of the order of 0.3 nm. At this scale spatially non-local tunnelling plasmonics controls the optical response. They used two gold nanostructures with controllable subnanometre separation and carefully followed the evolution of the plasmonic modes. They were able to pinpoint at what distance quantum tunnelling sets in, and establish a quantum limit for plasmonic-field enhancement. Their observations are in good agreement with quantum based models of plasmonic systems [Esteban *et al.* (2012)].

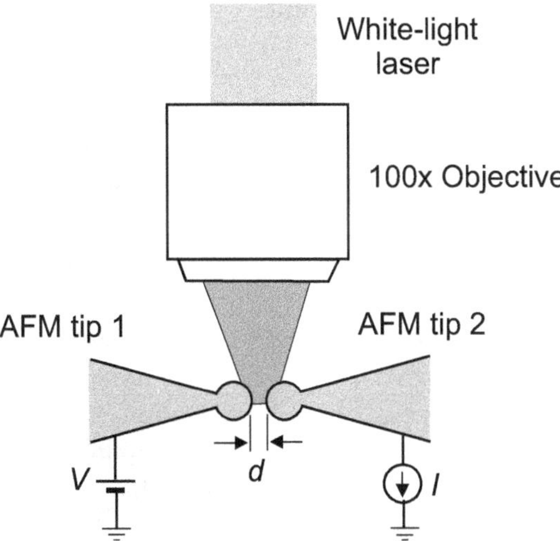

Figure 11.17: Scheme for simultaneous optical and electrical measurements of plasmonic cavity formed between two Au-coated tips. Adapted from [Savage *et al.* (2012)].

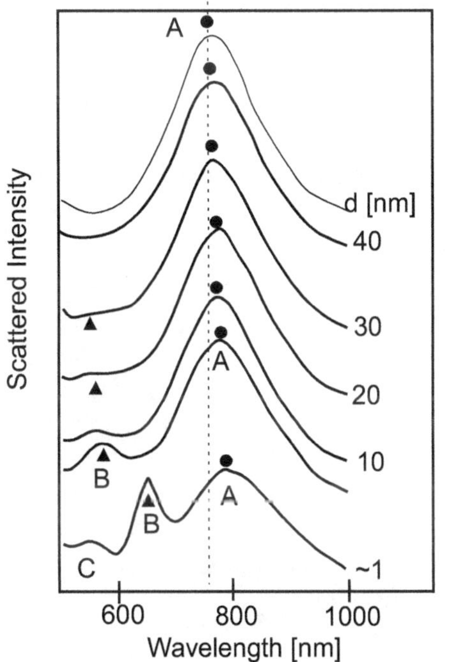

Figure 11.18: Measured dark field scattering spectra (vertically offset for clarity) from the NP dimer at different separations *d*. The LSPP resonances are labelled A-C. Adapted from [Savage *et al.* (2012)].

They simultaneously measured electrical conductance and dark-field optical back-scattering with increasing force applied to the inter-tip cavity. Experimental results are shown in Fig. 11.18.

11.5　Quantum Models of Plasmonics

As mentioned earlier using plasmonic effects it is possible to confine and enhance light at nano-meter scales. At those dimensions it becomes necessary to include quantum effects for a full description.

One of the subsystems forming plasmons, namely electrons in metals were long described using quantum-mechanical tools. Extensive history of developments was summarised by Fitzgerald [Fitzgerald (2018)]. Here, we summarise various approaches to quantize surface plasmons.

The quantum size effects start to appear when the de Broglie wavelength of the valence electrons in the nanoparticle is of the same order as its size.

It has been predicted, see e.g. [Alivisatos (1996)] that spherical nanoparticles with diameters of the order of 1–10 nm, i.e. intermediate between small molecules and bulk metals will show-up underlying electronic band structure of nanoparticle which own its origin due to quantum mechanics.

We start with a simple illustration and in the next steps summarise more advanced developments.

11.5.1　*Cordaro (2017). Quantum Drude Model (Au Nanowires)*

We provide a simple illustration of incorporation of quantum effects to the Drude model for Au nanowires [Cordaro *et al.* (2017)]. Following Genzel *et al.* [Genzel *et al.* (1975)] they determined dielectric function for Au nanowire using a simple model of noninteracting electrons constrained by the physical boundaries of the nanowire. Electrons perform quantum mechanical transitions from occupied to final states. The resulting dielectric constant is

$$\varepsilon^{m,n}(\omega) = \varepsilon_\infty + \omega_p^2 \sum_i \sum_f \frac{S_{if}}{\omega_{if}^2 - \omega^2 - i\Gamma'\omega} \qquad (11.9)$$

where

$$S_{if} = \frac{2M\omega_{if}}{N\hbar} \left| \langle\psi_i| \, x \, |\psi_f\rangle \right|^2 \qquad (11.10)$$

$$\omega_{if} = \frac{E_f - E_i}{\hbar} \qquad (11.11)$$

where E_f and E_i are the energies of the final and initial states, respectively. The damping coefficient Γ' is given by

$$\Gamma' = \Gamma + v_F/R_{eff} \qquad (11.12)$$

where v_F is the Fermi velocity and R_{eff} is an effective radius parameter.

Quantum size effects affect ε^{xx} which in-turn affects dispersion relation of SPP. A blue-shift in surface plasmon resonance energy appears which can be evaluated. The condition to have surface plasmon resonance is

$$Re\varepsilon^{xx}(\omega_{spQ}) = -\varepsilon_2 \qquad (11.13)$$

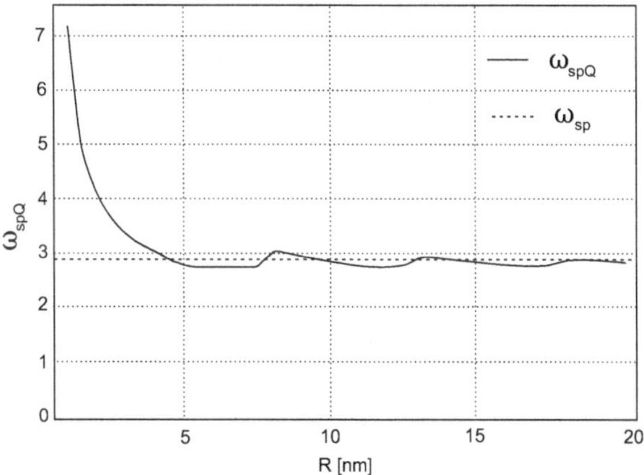

Figure 11.19: Dependence of ω_{spQ} vs radius of nanowire. Adapted from [Cordaro *et al.* (2017)].

The above was solved numerically. Some of the results are illustrated in Fig.11.19. We show the change of ω_{spQ} as a function of R. There is a consistent blue-shift as the radius of the wire decreases below 5 nm. For larger radius values $\omega_{spQ} \to \omega_{sp} = \frac{\omega_p}{\sqrt{1+\varepsilon_\infty}}$, i.e. it approaches classical limit.

11.5.2 *Dasgupta (1977)*

Dasgupta [Dasgupta (1977)] derived a pair of matrix equations describing the surface plasmon dispersion for a very small spherical metal particle using the quantum mechanical random-phase approximation (RPA). He derived dispersion relation as the condition that a solution of the RPA integral equation for charge fluctuation would exist for the particle.

The resulting equations are complicated and will not be discussed here. Please consult original paper for details.

11.5.3 *Nakamura (1983)*

Nakamura [Nakamura (1983)] developed an approach to quantize non-radiative surface plasma oscillations in a semi-infinite metal using a hydrodynamic jellium model for electrons in the metal. He considered a standard system where electrons occupy a half-space $z < 0$ and the metal-vacuum boundary surface is at $z = 0$.

Basic equations consists of Maxwell's equations and the hydrodynamic Bloch equation for the electron gas in the linearized forms (observe system of units)

for metal:

$$\nabla \cdot \mathbf{H} = 0 \tag{11.14}$$

$$\nabla \cdot \mathbf{E} = 4\pi n e \tag{11.15}$$

$$\nabla \times \mathbf{E} = -\frac{1}{c}\frac{\partial \mathbf{H}}{\partial t} \tag{11.16}$$

$$\nabla \times \mathbf{H} = \frac{1}{c}\frac{\partial \mathbf{E}}{\partial t} + \frac{4\pi e n_0}{c}\mathbf{v} \tag{11.17}$$

$$m n_0 \frac{\partial \mathbf{v}}{\partial t} = e n_0 \mathbf{E} - m\beta^2 \nabla n \tag{11.18}$$

for vacuum:

$$\nabla \cdot \mathbf{H} = 0 \tag{11.19}$$

$$\nabla \cdot \mathbf{E} = 0 \tag{11.20}$$

$$\nabla \times \mathbf{E} = -\frac{1}{c}\frac{\partial \mathbf{H}}{\partial t} \tag{11.21}$$

$$\nabla \times \mathbf{H} = \frac{1}{c}\frac{\partial \mathbf{E}}{\partial t}. \tag{11.22}$$

In the above equations $\mathbf{E}$, $\mathbf{H}$, n and $\mathbf{v}$ are fluctuations of the electric field, magnetic field, electron density and average velocity of electrons in the metal respectively; e and m are the charge and mass of an electron; n_0 is the electron density in the undisturbed state of the electron gas, $\beta^2 = \frac{3}{5}v_F^2$ and v_F is the Fermi velocity of electrons.

For the following we introduce electromagnetic scalar and vector potentials ϕ and $\mathbf{A}$ and assume Coulomb gauge, i.e. $\nabla \cdot \mathbf{A} = 0$. The Lagrangian of the system which produces the above equations is

$$L = \iiint d^3x \left\{ \frac{1}{2}m n_0 \left(\frac{\partial \mathbf{u}}{\partial t}\right)^2 + \frac{e n_0}{c}\frac{\partial \mathbf{u}}{\partial t}\cdot \mathbf{A} + e n_0 \phi \nabla \cdot \mathbf{u} + m\beta^2 n\,(\nabla \cdot \mathbf{u}) \right.$$
$$\left. \frac{m\beta^2}{2n_0}n^2 + \frac{1}{8\pi}\left[\left(-\frac{1}{c}\frac{\partial \mathbf{A}}{\partial t} - \nabla\phi\right)^2 - (\nabla \times \mathbf{A})^2\right] \right\} \tag{11.23}$$

where $\mathbf{u}$ is the displacement vector of the electrons and $\mathbf{v} = \frac{\partial \mathbf{u}}{\partial t}$.

Based on the above Lagrangian, in the long wavelength limit he predicted behavior of the surface loss intensity of scattered electrons passing through the metal foil.

11.5.4　*Nga (2015) — Second Quantization*

Nga *et al.* [Nga *et al.* (2015)] formulated the following model for SPP

$$H = \sum_k H_k = \sum_k \left\{ \omega_{\gamma k} a_k^\dagger a_k + E_{pk} c_k^\dagger c_k + g_k \left(a_k^\dagger c_k + c_k^\dagger a_k \right) \right\}. \tag{11.24}$$

Here $a_k^\dagger$ and a_k are creation and annihilation operators for photons with momentum k and $c_k^\dagger$ and c_k are creation and annihilation operators for surface plasmons; $\omega_{\gamma k} = ck$ is the photon energy, $E_{pk} = \omega_p/\sqrt{2}$ is the energy of the surface plasmon and g_k is the plasmon-photon coupling constant.

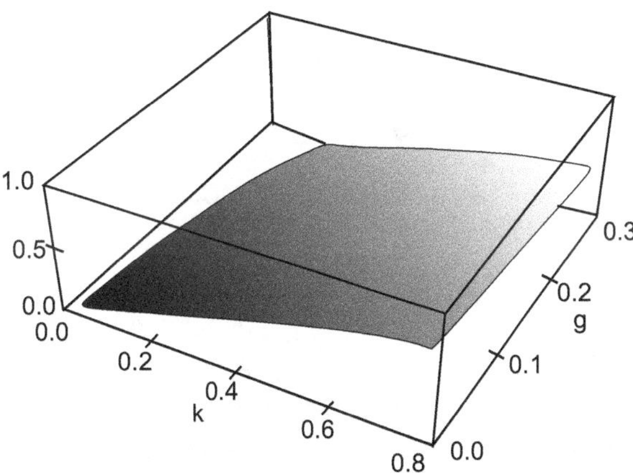

Figure 11.20: Dispersion relation from Nga 2015.

Performing Bogoliubov transformation they obtained dispersion relation for surface plasmons

$$\omega_{SP}(k) = \frac{1}{2}\left[\left(ck - \frac{\omega_p}{\sqrt{2}}\right) - \sqrt{\left(ck - \frac{\omega_p}{\sqrt{2}}\right)^2 + 4g_k^2}\,\right].\qquad(11.25)$$

The surface plasmon polariton dispersion relation $\omega_{SP}(k)$ depends on wave vector k and coupling constant g_k is presented in the Fig. 11.20.

Dispersion relation of SPP depends on wave vector k and coupling constant g_k.

11.5.5 *Fundamental Hydrodynamic Approach*

As indicate above local model of the free electrons breaks down at distances around a few nanometers. In such cases the introduction of quantum effects is necessary.

Approach based on hydrodynamic model is probably the first step in introducing quantum effects into plasmonics. Basic step is that interactions and Pauli principle result in a repulsive force between charge carriers. This force is introduced as a pressure in an electron gas and it enters into hydrodynamic model. The pressure for a three-dimensional gas is [Ciraci *et al.* (2012)]

$$p(\mathbf{r},t) = p_0\left[\frac{n(\mathbf{r},t)}{n_0}\right]^{5/3}\qquad(11.26)$$

where $n(\mathbf{r},t)$ is the electron fluid density, n_0 is the equilibrium charge density and $p_0 \sim n_0 E_F$ with E_F being the Fermi energy.

The nonlocal electromagnetic response of metal is described by the Euler's equation

$$mn(\mathbf{r},t)\left[\frac{\partial \mathbf{v}}{\partial t} + (\mathbf{v}\cdot\nabla)\,\mathbf{v}\right] + \gamma mn(\mathbf{r},t)\,\mathbf{v} = en(\mathbf{r},t)\,\mathbf{v}\times\mathbf{B} - \nabla p\qquad(11.27)$$

where m is the free electron mass, γ is the electron collision rate and $\mathbf{v}$ is the electron velocity. The current density $\mathbf{J} = en\left(\mathbf{r}, t\right)\mathbf{v}$ satisfies the continuity equation

$$\nabla \cdot \mathbf{J} = -e\frac{\partial}{\partial t}n\left(\mathbf{r}, t\right). \tag{11.28}$$

Combining the above equations one determines the following equation obeyed by current [Ciraci *et al.* (2012)]

$$-\beta^2 \nabla\left(\nabla \cdot \mathbf{J}\right) + \frac{\partial^2}{\partial t^2}\mathbf{J} + \gamma\frac{\partial}{\partial t}\mathbf{J} = \frac{e^2 n_0}{m}\frac{\partial}{\partial t}\mathbf{E} \tag{11.29}$$

where $\beta \sim \sqrt{E_F/m}$ is approximately the speed of sound in the Fermi plasma. Assuming harmonic propagation, from Eq. (11.29) and Maxwell's equations one obtains two coupled equations which describe electromagnetic response of a metal

$$\nabla \times \nabla \times \mathbf{E} - k_0^2 \mathbf{E} = i\omega\mu_0\mathbf{J} \tag{11.30}$$

$$\beta^2 \nabla\left(\nabla \cdot \mathbf{J}\right) + \left(\omega^2 + i\gamma\omega\right)\mathbf{J} = i\omega\omega_p^2\varepsilon_0\mathbf{E} \tag{11.31}$$

where $\omega_p = \sqrt{n_0 e^2/\left(\varepsilon_0 m\right)}$ is the plasma frequency.

Using the above relations one can evaluate the longitudinal dielectric function ε_L which become nonlocal and depends on the propagation vector $\mathbf{k}$

$$\varepsilon_L\left(\mathbf{k}, \omega\right) = 1 - \frac{\omega_p^2}{\omega^2 + i\gamma\omega - \beta^2\left|\mathbf{k}\right|}. \tag{11.32}$$

11.5.6 *Field-Theoretical Approaches*

Nguyen *et al.* [Nguyen and Nguyen (2015)] presented a review of the theoretical research on the quantum theory of plasmons and plasmon–photon interaction. It is based on advanced field-theoretical approaches. In this approach plasmons are defined as the quanta of the quantized plasmonic field. The electron–electron Coulomb interaction is taken into account.

In the model it was assumed that electrons of density $n\left(\mathbf{x}, t\right)$ oscillate in the field of positive ions and that n_0 is the average density of electrons. Let $\delta\mathbf{r}\left(\mathbf{x}, t\right)$ be the displacement vector of the electron of mass m at position $\mathbf{x}$ at time t. We Fourier transform it as

$$\delta\mathbf{r}\left(\mathbf{x}, t\right) = \frac{1}{\sqrt{V}}\sum_{\mathbf{k}} e^{i\mathbf{kx}}\sum_{i=1}^{3}\mathbf{e}_{\mathbf{k}}^{(i)}q_{\mathbf{k}}(t) \tag{11.33}$$

where $\mathbf{e}_{\mathbf{k}}^{(i)}$ are unit vectors satisfying the condition

$$\mathbf{ke}_{\mathbf{k}}^{(i)} = \begin{cases} 0, & i = 1, 2 \\ k, & i = 3 \end{cases} \tag{11.34}$$

where $k = \left|\mathbf{k}\right|$. The terms with $i = 1, 2$ describe transverse displacements and term with $i = 3$ describe the longitudinal displacement along the direction of the wave vector $\mathbf{k}$.

Introduce $\widetilde{\sigma}_{\mathbf{k}}(t)$ as

$$\sqrt{n_0 m}\, q_{\mathbf{k}}(t) = \widetilde{\sigma}_{\mathbf{k}}(t). \tag{11.35}$$

Finally, consider Fourier transform

$$\sigma(\mathbf{x}, t) = \sum_{\mathbf{k}} e^{i\mathbf{k}\mathbf{x}} \widetilde{\sigma}_{\mathbf{k}}(t). \tag{11.36}$$

The scalar field $\sigma(\mathbf{x}, t)$ can be quantized and its quanta are plasmons. In the harmonic approximation the plasmons are dispersionless.

In the canonical quantum mechanics the plasmon-photon interaction Lagrangian is expressed as

$$L_{int}(t) = \sum_{n=1}^{\infty} L_{int}^{(n)}(t). \tag{11.37}$$

Terms up to second order are

$$L_{int}^{(1)}(t) = \frac{e n_0}{c} \int d\mathbf{x} \int d\mathbf{y} \sum_{i} A_i(\mathbf{x}, t) F_i(\mathbf{x} - \mathbf{y}) \dot{\sigma}_i(\mathbf{y}, t) \tag{11.38}$$

$$L_{int}^{(2)}(t) = -\frac{e n_0}{c} \int d\mathbf{x} \int d\mathbf{y}_1 \int d\mathbf{y}_2 \sum_{i} A_i(\mathbf{x}, t) F_i(\mathbf{x} - \mathbf{y}_1) \sum_{j}$$
$$\times \nabla_j F_i(\mathbf{x} - \mathbf{y}_2) \dot{\sigma}_i(\mathbf{y}_1, t) \sigma_i(\mathbf{y}_2, t) \tag{11.39}$$

where

$$F_i(\mathbf{x} - \mathbf{y}) = \frac{1}{\sqrt{n_0 m}} \frac{1}{\sqrt{V}} \sum_{\mathbf{k}} \frac{k_i}{k} e^{i\mathbf{k}(\mathbf{x} - \mathbf{y})}. \tag{11.40}$$

The above results show that the interaction processes are nonlocal ones. The physical origin of the nonlocality is the complex structure of plasmons as composite quasiparticles: they cannot be considered as point particles, as it is assumed in all phenomenological theories.

11.6 Time-Dependent DFT

As indicated earlier there has been several recent experiments which indicate the need to go beyond the classical models to explain and predict the plasmonic response at the nanoscale. One of approaches which handle this problem is a fully *ab initio* time-dependent density functional theory (TDDFT) [Ullrich and Yang (2014)], [Varas *et al.* (2016)]. It was used to determine optical response of a prototypical plasmonic system, namely nanoparticle dimer. We start with explaining theoretical approach.

TDDFT is an extension of Hohenberg–Kohn–Sham density functional theory (DFT) [Hohenberg and Kohn (1964)], [Kohn and Sham (1965)]. It describes many-body N-electron system in an external potential with the following assumptions:

(a) ground-state electron density, $n_0(\mathbf{r})$ unambigously defines the many-electron ground state $|\Psi_0\rangle$,
(b) there exists an energy functional $E[n]$ whose minimum equals the ground-state energy and it is reached at $n(\mathbf{r}) = n_0(\mathbf{r})$,
(c) to find a minimum of this functional one defines a fictitious system of independent fermions (the so-called Kohn–Sham (KS) system) whose ground-state density is also $n_0(\mathbf{r})$.

11.6.1   *Density Functional Theory. General*

One starts with a many-body problem for N interacting electron system in an external potential $V_{ext}(\mathbf{r}_i)$. It is described by the following Hamiltonian

$$H = -\left[\sum_{i=1}^{N}\frac{\hbar^2}{2m}\nabla_i^2 + \sum_{i=1}^{N}eV_{ext}(\mathbf{r}_i) - \frac{1}{2}\sum_{i\neq j}^{N}\frac{e^2}{|\mathbf{r}_i - \mathbf{r}_j|}\right]. \tag{11.41}$$

The presence of the electron-electron interaction (last term) makes the solution very difficult (even numerically). The solution is described by time-independent nonrelativistic Schrödinger equation for i-th eigenstate

$$H\Psi_i(\mathbf{r}_1, \mathbf{r}_2, ...\mathbf{r}_N) = E_i\Psi_i(\mathbf{r}_1, \mathbf{r}_2, ...\mathbf{r}_N) \tag{11.42}$$

where $i = 0, 1, 2, \ldots$ (Case with $i = 0$ corresponds to the ground state).

The Hohenberg–Kohn theorem [Hohenberg and Kohn (1964)] guarantees the existence of an effective, position dependent single-electron potential V_{eff} which replaces all the interactions of a given electron and it reproduces the exact ground-state density of the interacting system.

The above problem can thus be replaced by a system of non-interacting electrons described by

$$-\left[\sum_{i=1}^{N}\frac{\hbar^2}{2m}\nabla_i^2 + \sum_{i=1}^{N}eV_{eff}(\mathbf{r}_i)\right]\Psi = E\Psi. \tag{11.43}$$

To solve this problem we apply separation of variables method and write the many-electron wave function as a product of single-electron wave functions

$$\Psi(\mathbf{r}_1, \mathbf{r}_2, \ldots, \mathbf{r}_N) = \psi(\mathbf{r}_1)\psi(\mathbf{r}_2)\ldots\psi(\mathbf{r}_N). \tag{11.44}$$

The energy of the total system becomes sum of single-electron energies ε_i as

$$E = \varepsilon_1 + \varepsilon_2 + \ldots \varepsilon_N. \tag{11.45}$$

Thus the many-electron Schrödinger equation is reduced into single-electron wave equation of the form

$$\left[-\frac{\hbar^2}{2m}\nabla_i^2 - eV_{eff}(\mathbf{r}_i)\right]\psi(\mathbf{r}_i) = \varepsilon_i\psi(\mathbf{r}_i). \tag{11.46}$$

11.6.1.1 *The Kohn–Sham approach*

The main problem is how to determine an effective potential V_{eff}. This is done within the so-called Kohn–Sham formalism. One defines a noninteracting system which, be definition reproduces the exact ground-state density of the interacting system.

To follow main literature, in the following we will work in the atomic system of units where $\hbar = m = e = 1$. The many-body Hamiltonian is thus written as

$$H = \widehat{T} + \widehat{V} + \widehat{W} \tag{11.47}$$

the kinetic energy and scalar potential operator are

$$\widehat{T} = -\sum_{i=1}^{N} \frac{1}{2} \nabla_i^2, \quad \widehat{V} = \sum_{i=1}^{N} v\left(\mathbf{r}_i\right). \tag{11.48}$$

The electron-electron interaction operator (assuming Coulomb interaction) is

$$\widehat{W} = \sum_{i \neq j}^{N} w\left(\left|\mathbf{r}_i - \mathbf{r}_j\right|\right) = -\frac{1}{2} \sum_{i \neq j}^{N} \frac{1}{\left|\mathbf{r}_i - \mathbf{r}_j\right|} \tag{11.49}$$

The exact ground-state density $n_0(\mathbf{r})$ is expressed in terms of a single-electron wave functions as

$$n_0(\mathbf{r}) = \sum_{i=1}^{N} \left|\varphi_i\left(\mathbf{r}\right)\right|^2. \tag{11.50}$$

The above wave functions $\varphi_i\left(\mathbf{r}\right)$ obey

$$\left[-\frac{1}{2} \nabla_i^2 - +v_s\left[n\right]\left(\mathbf{r}\right)\right] \varphi_i\left(\mathbf{r}\right) = \varepsilon_i \varphi_i\left(\mathbf{r}\right). \tag{11.51}$$

Potential v_s is defined in such a way that calculated wave functions φ_i give the exact ground-state density of the interacting system. This potential acts on a single electron and it is a functional of the density, $v_s\left[n\right]\left(\mathbf{r}\right)$. Equation (11.51) is known as Kohn–Sham equation.

Effective potential v_s is written as

$$v_s\left[n\right]\left(\mathbf{r}\right) = v_0\left(\mathbf{r}\right) + v_H\left[n\left(\mathbf{r}\right)\right] + v_{xc}\left[n\left(\mathbf{r}\right)\right] \tag{11.52}$$

where $v_0\left(\mathbf{r}\right)$ is a external background potential. The remaining terms, $v_H\left[n\left(\mathbf{r}\right)\right] + v_{xc}\left[n\left(\mathbf{r}\right)\right]$ account for electronic many-body effects. A main part of it is the classical Coulomb potential due to a given density. It is also known as a Hartree potential is

$$v_H\left[n\left(\mathbf{r}\right)\right] = \int \frac{n\left(\mathbf{r}'\right)}{\left|\mathbf{r} - \mathbf{r}'\right|} d\mathbf{r}'. \tag{11.53}$$

The last term, $v_{xc}\left[n\left(\mathbf{r}\right)\right]$ is called the exchange-correlation (xc) potential.

At this point we are ready to introduce the total energy functional as [Ullrich and Yang (2014)]

$$E[n] = T[n] + \int v_0(\mathbf{r}) n(\mathbf{r}) d\mathbf{r} + W[n]$$

$$= T_s[n] + \int v_0(\mathbf{r}) n(\mathbf{r}) d\mathbf{r} + (T[n] - T_s[n] + W[n])$$

$$\equiv T_s[n] + \int v_0(\mathbf{r}) n(\mathbf{r}) d\mathbf{r} + E_H[n] + E_{xc}[n] \tag{11.54}$$

where $T[n]$ is the kinetic energy functional of an interacting system and $T_s[n]$ is the kinetic energy functional of a noninteracting system which is expressed as

$$T_s[n] = -\frac{1}{2} \sum_{i=1}^{N} \varphi_i^*(\mathbf{r}) \nabla_i^2 \varphi_i(\mathbf{r}). \tag{11.55}$$

The Hartree energy is

$$E_H[n] = \frac{1}{2} \int d\mathbf{r} \int \frac{n(\mathbf{r}) n(\mathbf{r}')}{|\mathbf{r} - \mathbf{r}'|} d\mathbf{r}'. \tag{11.56}$$

The exchange-correlation energy is

$$E_{xc}[n] = T[n] - T_s[n] + W[n] - E_H[n]. \tag{11.57}$$

Exchange-correlation potential is thus expressed as the functional derivative

$$v_{xc}[n(\mathbf{r})] = \frac{\delta E_{xc}[n]}{\delta n(\mathbf{r})}. \tag{11.58}$$

Combining the above results allows to express total energy functional as

$$E[n] = \sum_{i=1}^{N} \varepsilon_i - E_H[n] - \int v_{xc}(\mathbf{r}) n(\mathbf{r}) d\mathbf{r} + E_{xc}[n]. \tag{11.59}$$

11.6.1.2 *LDA*

In the local density approximation (LDA) the exchange-correlation energy of a nonuniform electron gas is the following function of a slowly varying density of electrons

$$E_{sc}[n(\mathbf{r})] = \int \varepsilon_{sc}[n(\mathbf{r}')] n(\mathbf{r}') d\mathbf{r}' \tag{11.60}$$

where ε_{sc} is the exchange-correlation energy per particle in the homogeneous system having density $n(\mathbf{r})$. Within this approximation the LDA exchange-correlation potential is [Perdew and Zunger (1981)], [Zuloaga (2011)]

$$V_{sc}(r_s) = \begin{cases} -\frac{0.611}{r_s} + 0.03109 \log r_s - 0.0584 \\ \quad + 0.0013 r_s \log r_s - 0.007 r_s & r_s < 1 \\ -\frac{0.611}{r_s} - \frac{0.1423(1 + 1.23\sqrt{r_s} + 0.445 r_s)}{(1 + 1.1592\sqrt{r_s} + 0.3334 r_s)^2} & r_s > 1. \end{cases} \tag{11.61}$$

11.6.2 *TDDFT*

Time evolution of many-body system is described by time-dependent Schrödinger equation

$$i\frac{\partial}{\partial t}\Psi_i\left(\mathbf{r}_1,\mathbf{r}_2,\ldots,\mathbf{r}_N\right) = H(t)\Psi_i\left(\mathbf{r}_1,\mathbf{r}_2,\ldots,\mathbf{r}_N\right) \tag{11.62}$$

where time-dependent Hamiltonian is

$$H(t) = \widehat{T} + \widehat{V}(t) + \widehat{W}. \tag{11.63}$$

External potential operator is explicitly time-dependent

$$\widehat{V}(t) = \sum_{i=1}^{N} v\left(\mathbf{r}_i,t\right). \tag{11.64}$$

This is an initial value problem where external perturbation starts working on the ground state $|\Psi_0\rangle$ at time t_0. The exact time-dependent density $n\left(\mathbf{r},t\right)$ is expressed in terms of noninteracting wave functions as

$$n\left(\mathbf{r},t\right) = \sum_{i=1}^{N} \left|\varphi_i\left(\mathbf{r},t\right)\right|^2 \tag{11.65}$$

which satisfy the time-dependent Kohn–Sham equation

$$\left[-\frac{1}{2}\nabla_i^2 - +v_s\left[n\right]\left(\mathbf{r},t\right)\right]\varphi_i\left(\mathbf{r},t\right) = i\frac{\partial}{\partial t}\varphi_i\left(\mathbf{r},t\right). \tag{11.66}$$

In the above the time-dependent effective potential is

$$v_s\left[n,\Psi_0,\Phi_0\right]\left(\mathbf{r},t\right) = v\left(\mathbf{r},t\right) + v_H\left(\mathbf{r},t\right) + v_{xc}\left[n,\Psi_0,\Phi_0\right]\left(\mathbf{r},t\right) \tag{11.67}$$

where $v\left(\mathbf{r},t\right)$ is the time-dependent external potential, Ψ_0 is the initial many-body state of the exact interacting system and Φ_0 is the initial state of the Kohn–Sham system. The time-dependent Hartree potential is

$$v_H\left(\mathbf{r},t\right) = \int \frac{n\left(\mathbf{r}',t\right)}{\left|\mathbf{r}-\mathbf{r}'\right|}d\mathbf{r}'. \tag{11.68}$$

11.6.3 *Results*

First, we present some results for a single particle. In Fig. 11.21 we show the local electromagnetic field enhancements calculated at the peak resonant frequency using classical electrodynamics and quantum mechanical (using TDDFT) approaches [Zuloaga (2011)] as a function of separation d from the particle surface. For values of d larger than 10 Bohr the results are similar and start to deviate for smaller values of d. The TDDFT results show a peak at approximately $d = 3$ Bohr from the surface. The difference between classical and quantum approaches becomes maximum for $d = 0$.

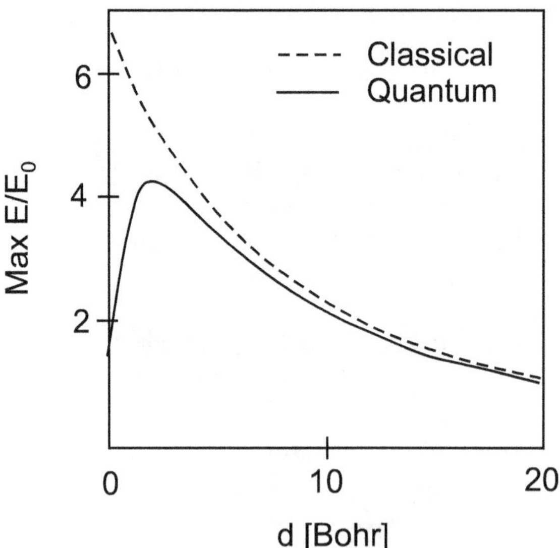

Figure 11.21: Electric field enhancements as a function of distance d from the particle tip for an aspect ratio of (=1 (top) and (=3 (bottom). The classical calculations are shown with red lines and the TDDFT results are shown with black lines. Adapted from [Zuloaga (2011)].

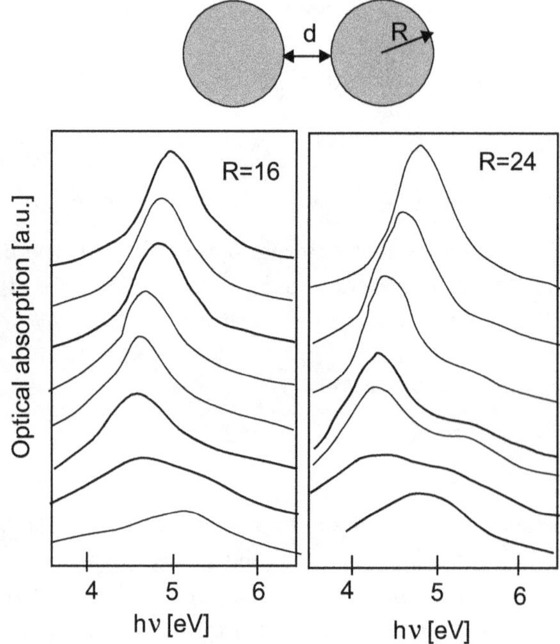

Figure 11.22: TDLDA absorption spectra for the $R = 16b$ (left panel) and $R = 24b$ (right panel) for different separations d equal to 0, 1, 2, 3, 4, 5, 6, 7, 8, 10, 12, 14, 16, 20, and 24b (from bottom curve and up), $b = 0.0529$ nm (bohr). The spectra were calculated using an energy broadening of 0.27 eV. Adapted from Zuloaga *et al.* [Zuloaga *et al.* (2009)]

Some important results for a dimer were reported by Zuloaga *et al.* [Zuloaga *et al.* (2009)]. They used TDDFT to analyze plasmonic resonances in a nanoparticle dimer as a function of the separation between particles, see Fig. 11.22. They observed quantum mechanical effects, like tunnelling and screening for separations smaller then 1 nm. Those effects significantly modify the optical response and drastically reduce the electromagnetic field enhancements relative to predictions by classical theories.

For larger separations, the dimer plasmons are well described by classical electromagnetic theory.

Similar results were also reported by Alrasheed *et al.* [Alrasheed and Fabrizio (2017)] and Fitzgerald [Fitzgerald (2018)].

Chapter 12

Nanolasers

A minimum volume in which photons can be localised is about the same as cube with dimensions equal to the photon's wavelength. However, in recent years it was shown that it is possible to beat this restriction by replacing photons with surface plasmon-polaritons (SPP).

SPP are electromagnetic waves, but at the same frequency as photons they are much better localised. Using SPP instead of photons makes it possible to "compress" light and thus overcome the diffraction limit.

The above observation opened the door to create plasmonic laser. It is an example of a new class of lasers known as nanolasers which are summarised here.

12.1 Definition of Nanolaser

Making photonics (also known as optoelectronics) compatible with modern electronics is a long dream of researchers. On the electronic side, in the very large-scale integration (VLSI) process it is possible to create an integrated circuit by combining millions of MOS transistors onto a single chip. VLSI are approaching gate dimensions of the order of a few nm whereas active photonic devices based on semiconductors utilizing dielectric cavity resonators are limited by the diffraction limit. Thus, conventional semiconductor lasers are limited to a minimum three-dimensional (3D) volume of $(\lambda/2n)^3$, where λ is the free-space wavelength (1.3–1.55 μm for optical communication over fiber) and n is the refractive index of the dielectric resonator (typical vales for semiconductors are 2.5–3.5). Reduction of the volume of the active region is also the potential factor in lowering threshold current.

As the dimensions of active optical nano devices scale down, we are approaching the situation where there is effectively only one optical emission mode. One can call such structures *nanolasers*. Alternatively, nanolasers can be defined as structures having dimensions smaller than the wavelength of light in all three dimensions [Gourley (1998)]. Recent book on the subject of nanolasers is by Qing Gu and Yeshaiahu Fainman [Gu and Fainman (2017)].

The smallest conventional semiconductor lasers available commercially today are VCSELs. In Fig. 12.1 we showed in-plane laser where light propagates in the plane

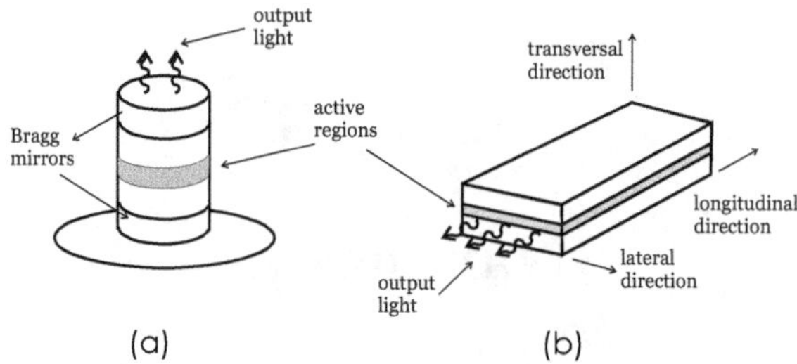

Figure 12.1: Basic types of semiconductor lasers.

[Agrawal and Dutta (2000)] and vertical cavity surface emitting laser (VCSEL) [Coldren *et al.* (1999)]. The largest dimension of in-plane structures is typically in the range of 250 microns (longitudinal direction). The structure of interest to us is the one where light propagates perpendicularly to wafer's surface which is known as VCSEL. Typical diameter of VCSEL cylinder is about 10 µm.

12.1.1 *Types of Nanolasers*

However, in the past two decades new type of even smaller devices are emerging. Examples include microcavity lasers [McCall *et al.* (1992)], photonic crystal lasers [Strauf and Jahnke (2011)], [Tandaechanurat *et al.* (2011)], [Nozaki *et al.* (2007)], [Painter *et al.* (1999)], [Park *et al.* (2004)], [Altug *et al.* (2006)], polariton lasers [Schneider *et al.* (2013)], semiconductor wire lasers [Huang *et al.* (2001)], [Dasgupta *et al.* (2014)], metal-cladded semiconductor microlasers [Hill *et al.* (2007)], [Nezhad *et al.* (2010)], [Ding *et al.* (2012)], [Khajavikhan *et al.* (2012)], and plasmonic nanolasers [Oulton *et al.* (2009)], [Lu *et al.* (2012)], [Lu *et al.* (2014)], [Sidiropoulos *et al.* (2014)], [Zhang *et al.* (2014)].

Reported progress made it possible to reduce cavity size, reduce threshold current and also increase power conversion efficiency. Also, new mechanisms allowing better confinement of light were invented. With this in mind, nanolasers were classified as [Jeong *et al.* (2020)]

- photonic bandgap (photonic crystal (PC)),
- surface plasmon (SP) polaritons,
- parity-time (PT) symmetry,
- photonic topological insulators,
- bound states in the continuum (BIC).

Main properties of the above devices are summarised in a Table 12.1 [Jeong *et al.* (2020)].

In the following we will discuss some of them and will finish with summarising spaser devices.

Table 12.1: Main characteristics of nanolasers.

Type	Principle of operation	Mode volume	Q factor
PC	Photonic bandgap	$\approx \left(\frac{\lambda}{n}\right)^3$	> 1000
Plasmonic	Confinement by SP	$<< \left(\frac{\lambda}{n}\right)^3$	< 1000
PT	Gain/loss manipulation	$\approx \left(\frac{\lambda}{n}\right)^3$	> 1000
Topological	Topological properties	$\approx \left(\frac{\lambda}{n}\right)^3$	> 1000
BIC	Destrictive interference of leaky modes	$\approx \left(\frac{\lambda}{n}\right)^3$	∞

12.2 Progress in Nanolasers

Metallic cavities can guide light at high concentrations and in volumes with dimensions significantly smaller than the wavelenght of light (at subwavelength regime). This property strongly suggest that metallic cavities can be used as lasers and amplifiers. In short, the operation of nanolasers is possible due to the strong confinement in wavelength scale by cavities. Here we provide a short outline on the progress with nanolasers research.

Main steps (subjective) in the development of nanolasers are summarised in a Table 12.2. Note. The designs 1–5 did not exceeded diffraction limit.

As mentioned earlier the very small (by volume) lasers were vertical cavity surface emitting lasers which were invented in the 1980s. Since then a significant progress has been made in reducing cavity size due to invention of a new mechanisms of strong light confinement.

In the following we will briefly summarise some of the early designs.

12.2.1 *Microdisk Lasers*

We briefly summarise results obtained for microdisk lasers. Those are lasers in the form of a small disk with quantum wells forming active region. As an example, for operation at $\lambda = 1550$ nm the disk radius is in the range of 0.5 μm–10 μm and

Table 12.2: Main steps in nanolasers development.

	Milestone	Year	Reference
1	Microdisk laser	1992	[McCall *et al.* (1992)]
2		1995	[Frateschi and Levi (1995)]
3	Photonic crystal laser	1999	[Painter *et al.* (1999)]
4		2002	[Loncar *et al.* (2002)]
5	Photoluminescence of bulk ZnO	2001	[Huang *et al.* (2001)]
6	Spaser (theory)	2003	[Bergman and Stockman (2003)]
7	Spaser (experiment)	2009	[Noginov *et al.* (2009)]
8		2009	[Oulton *et al.* (2009)]

thickness 0.05–0.3 µm [Frateschi and Levi (1995)]. Analysis of the device is based on two-dimensional Helmholtz equation for optical field $E(r,\theta) = R(r)e^{iZ\theta}$ which is separable in r and θ. Optical field can therefore be separated as

$$r^2\frac{d^2R(r)}{dr^2} + r\frac{dR(r)}{dr} - \left(k^2r + Z\right)R(r) = 0 \tag{12.1}$$

and

$$\frac{d^2\Theta(\theta)}{d\theta^2} - Z^2\Theta(\theta) = 0 \tag{12.2}$$

where $k = n_{eff}\omega/c$ and Z is complex constant.

For the analysis it was assumed that optical resonances are appoximated by whispering gallery modes (WGM). The modes consist of light traveling around the perimeter of the disk, see Fig. 12.2. Spectrum of the lasing line of a practical device is shown in Fig. 12.3, from [McCall *et al.* (1992)]. Laser consists of a 5 µm in diameter disk with six 100 Å InGaAs quantum wells.

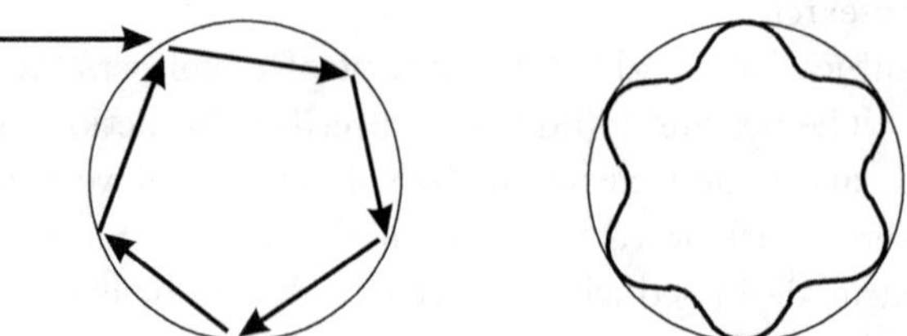

Figure 12.2: Illustration of whispering gallery modes.

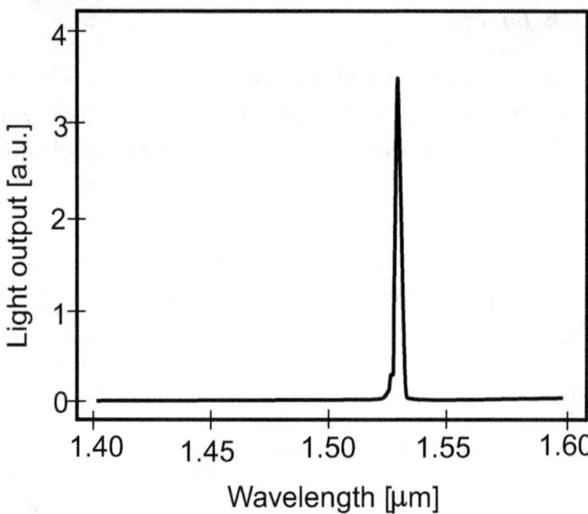

Figure 12.3: Indicative spectrum of a microdisc laser.

12.2.2 *Photonic Crystal Laser*

A photonic crystal (PC) structure consists of a drilled repeating pattern of holes through the laser material. This pattern is called a photonic crystal. One can deliberately introduce an irregularity, or defect, into the crystal pattern, for example by slightly shifting the positions of two holes. The photonic crystal structure and the defect prevent light of most frequencies from existing in the structure, with the exception of a small band of frequencies that can exist in the region near the defect.

Photonic-crystal defect microcavities can provide extremely small mode volume, [Painter *et al.* (1999)], [Vahala (2003)]. Schematic of the laser proposed by Painter *et al.* [Painter *et al.* (1999)] is shown in Fig. 12.4. The interhole separation is 515 nm. It was fabricated from InGaAsP grown by Metal-Organic Chemical Vapour Deposition (MOCVD) on an InP substrate. The active region consists of four (here only two are shown) 9 nm 0.85% compressively strained InGaAsP quantum wells. Two-dimensional photonic crystal hexagonal lattice was formed by etching, resulting in air holes that penetrate through the active region and into an underlying sacrificial InP layer.

Over the last few years various devices have been fabricated and characterised [Park *et al.* (2008)], [Nozaki *et al.* (2008)], [Yoshie *et al.* (2004)], [Nozaki *et al.* (2007)]. Such structures are associated with single-photon sources which are required for quantum computing and quantum communications. Single-photon sources are recent applications of the Purcell effect in quantum-dot microcavities.

Scientists from the Yokohama National University in Japan [Nozaki *et al.* (2007)] have reported interesting results concerning PC nanolasers. They demonstrated high-performing room-temperature nanolaser in the form of PC slab. The laser is made of a GaInAsP. This ultrasmall laser has a modal volume close to the diffraction limit. When operating in a high-Q mode (about 20,000) it will be useful for optical devices in optical integrated circuits. In a moderate-Q (1500) configuration the nanolaser needs only an extremely small amount of external power to bring the device to the threshold of producing laser light. In this near-thresholdless operation, it might permit the emission of very low light levels, even single photons.

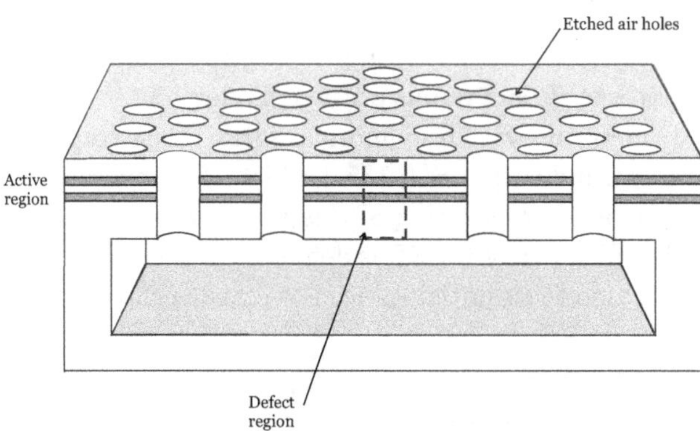

Figure 12.4: Cross-section through the middle of the photonic crystal microcavity. A defect is formed (shown in the middle) by removing a single hole. Adapted from [Painter *et al.* (1999)].

12.3   Summary of Analysis of Nanowire Nanolasers

Those devices will be discussed in more details. We start with summarising wave-guiding properties in nanowires.

12.3.1   *General Theory*

Main types nanolasers are in the form of a long cylinders which can be approximated as quasi-one-dimensional systems (quantum wires). Therefore, before discussing nanolasers we will summarise properties of quasi one-dimensional plasmons. We follow Sarid and Challener [Sarid and Challener (2010)]. In cylindrical coordinate systems Maxwell's equations are

$$\nabla \times \mathbf{H} = \left(\frac{1}{\rho}\frac{\partial H_z}{\partial \phi} - \frac{\partial H_\phi}{\partial z}\right)\widehat{\rho} + \left(\frac{\partial H_\rho}{\partial z} - \frac{\partial H_z}{\partial \rho}\right)\widehat{\phi} + \frac{1}{\rho}\left(\frac{\partial (\rho H_\phi)}{\partial \rho} - \frac{\partial H_\rho}{\partial \phi}\right)\widehat{\mathbf{z}}$$

$$= \frac{\partial D_\rho}{\partial t}\widehat{\rho} + \frac{\partial D_\phi}{\partial t}\widehat{\phi} + \frac{\partial D_z}{\partial t}\widehat{\mathbf{z}} = \frac{\partial \mathbf{D}}{\partial t} \tag{12.3}$$

and

$$\nabla \times \mathbf{E} = \left(\frac{1}{\rho}\frac{\partial E_z}{\partial \phi} - \frac{\partial E_\phi}{\partial z}\right)\widehat{\rho} + \left(\frac{\partial E_\rho}{\partial z} - \frac{\partial E_z}{\partial \rho}\right)\widehat{\phi} + \frac{1}{\rho}\left(\frac{\partial (\rho E_\phi)}{\partial \rho} - \frac{\partial E_\rho}{\partial \phi}\right)\widehat{\mathbf{z}}$$

$$= -\frac{\partial B_\rho}{\partial t}\widehat{\rho} - \frac{\partial B_\phi}{\partial t}\widehat{\phi} - \frac{\partial B_z}{\partial t}\widehat{\mathbf{z}} = -\frac{\partial \mathbf{B}}{\partial t}. \tag{12.4}$$

The general solution can be expressed in tems of Bessel and Hankel functions and their derivatives with respect to the argument. The equation used to determine the SP wavevector is

$$\left[\frac{1}{R\gamma_1}\frac{J_n'(iR\gamma_1)}{J_n(iR\gamma_1)} + \frac{1}{R\gamma_2}\frac{H_n^{(1)\prime}(iR\gamma_2)}{H_n^{(1)}(iR\gamma_2)}\right]\left[\frac{\varepsilon_1}{R\gamma_1}\frac{J_n'(iR\gamma_1)}{J_n(iR\gamma_1)} + \frac{\varepsilon_2}{R\gamma_2}\frac{H_n^{(1)\prime}(iR\gamma_2)}{H_n^{(1)}(iR\gamma_2)}\right]$$

$$+ \left(\frac{nk_z}{R^2 k_0}\right)^2 \left(\frac{1}{\gamma_1^2} - \frac{1}{\gamma_2^2}\right)^2 = 0 \tag{12.5}$$

where $\gamma_i^2 = k_z^2 - k_0^2 \varepsilon_{ri}, i = 1, 2$. ε_r represents the relative permittivity of the medium, n is an integer, R is a radius of the cylinder.

The analysed nanowire was in cylindrical form made from silver with radius 50 nm with refractive index $n_{Ag} = 0.135 + i3.99$. The operating wavelength was $\lambda = 633$ nm. The fields of propagating SP (assuming $n = 1$) are shown in Fig. 12.5.

The propagation distance of the SP is defined as a distance at which the electric field drops $1/e$ compared to its initial value. For planar propagation the propagation distance is

$$d_{plane} = \frac{1}{Im(k_{SP})} = \frac{1}{k_0 Im(\beta)} \approx \frac{2(\varepsilon_m')^2}{k_0 \varepsilon_d^{3/2} \varepsilon_m''} \tag{12.6}$$

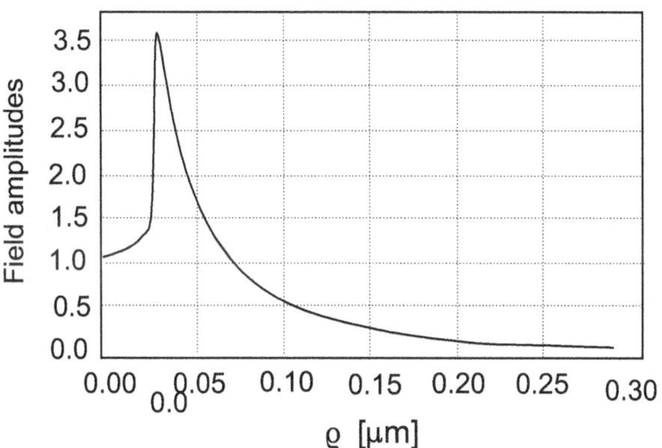

Figure 12.5: Amplitude of the electric field of a cylindrical nanowire as a function of the radial distance. Adapted from [Sarid and Challener (2010)].

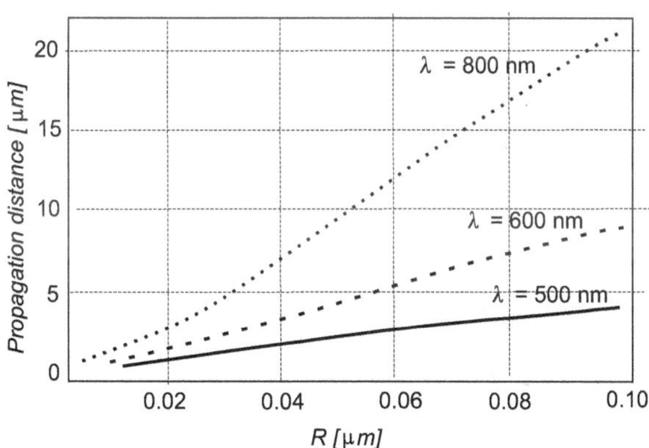

Figure 12.6: Propagation distance of a SP along a cylindrical Ag nanowire, for three wavelengths (500 nm, 600 nm and 800 nm) as a function of the radius of the cylinder. Adapted from [Sarid and Challener (2010)].

where ε_d is the relative permittivity of the surrounding dielectric, ε_m is the complex relative permittivity of the metal and k_0 is the free space wavevector. Propagation distance as a function of the radius of the cylinder is shown in Fig. 12.6. It increases as the radius increases and also it increases with increasing wavelength as the metal's conductivity increases.

12.3.2 *Dispersion Relation of the Lowest Order*

From the above general results one can obtain specific cases. The characteristic equation of the lowest order (0th) TM mode is [Song (2012)]

$$\frac{\gamma_2 I_2\left(\gamma_1 a\right) K_0\left(\gamma_2 a\right)}{\gamma_1 I_0\left(\gamma_1 a\right) K_1\left(\gamma_2 a\right)} = -\frac{\varepsilon_2}{\varepsilon_1} \tag{12.7}$$

where I_j and K_j ($j = 1, 2$) are the jth-order modified Bessel functions and γ_j is defined as

$$\gamma_j = \left(\beta^2 - \varepsilon_j \mu_0 \omega^2\right)^{1/2} \tag{12.8}$$

The SPP dispersion relation is very sensitive to the size and shape of nanowires and nanowires with different forms of cross-sections experience different plasmon resonances. For example, when the diameter increases the plasmon resonance frequency also increases [Lim *et al.* (2005)].

12.3.3   *General Properties*

Important types of nanolasers are based on semiconductor wires. Typical wires have dimensions of tens to hundreds of nm in diameter and tens to hundreds of μm in length. They can be considered as 1D structures. Popular material used is a zinc oxide (ZnO) [van Vugt *et al.* (2006)].

In Fig. 12.7 we shown schematically of a semiconductor nanowire laser in air. It behaves as a Fabry–Perot cavity due to large difference in refractive indices between the wire and the surrounding media (air or quartz).

Optical feedback needed for lasing is realized by reflecting guided photons back and forth along the nanowire longitudinal direction.

The conditions for lasing (some were already mention) are [Gwo and Shih (2016)]

    1. optical confinement (diffraction limit)

$$d > \frac{\lambda_0}{2n_{eff}} \tag{12.9}$$

where d is the wire's diameter,

    2. phase (coherence) limit

$$L = \frac{\lambda_0}{2n_{eff}}m, \quad m = 1, 2, 3, \ldots \tag{12.10}$$

This condition determines a minimum cavity length which is associated with the diffraction limit along the longitudinal direction,

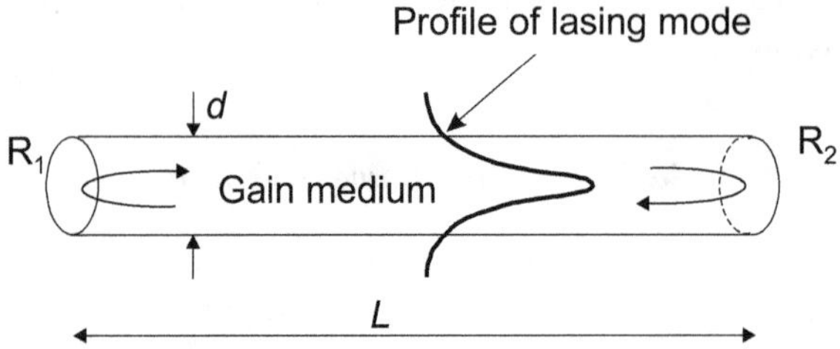

Figure 12.7: Schematic of a semiconductor wire laser. The semiconductor wire serves as the photonic Fabry–Pérot cavity and the optical gain medium. Adapted from [Gwo and Shih (2016)]

3. the lasing threshold condition

$$g = \alpha_m + \frac{1}{L_{gain}} \ln \frac{1}{\sqrt{R_1 R_2}} \qquad (12.11)$$

where α_m is the modal loss, L_{gain} is the gainlength and R_1 and R_2 are reflectivities.

12.4 Some Examples

We discuss here some of more interesting examples of nanolasers as reported in the literature. For recent reviews, see [Couteau *et al.* (2015)], [Balykin (2018)], [Hill (2018)], [Zhang *et al.* (2019)].

12.4.1 *First Demonstrations of Plasmonic Nanolasers*

First experimental demonstrations of plasmonic nanolasers has been reported in 2009 by three independent groups: Hill *et al.* [Hill *et al.* (2009)] fabricated nanolaser based on a metal-insulator semiconductor-insulator-metal plasmonic gap mode, where vertical confinement was achieved by means of a double heterostructure, Oulton *et al.* [Oulton *et al.* (2009)] had plasmonic nanolaser based on a two dimensionally confined nanowire plasmonic mode and Noginov *et al.* [Noginov *et al.* (2009)] demonstrated spaser based on a three-dimensionally confined metal nanoparticle mode. Noginov's device consisted of a gold nanosphere of radius 7 nm surrounded by a dielectric shell of a 21 nm outer radius containing immobilised dye molecules. It was optically pumped in the absorption band of the dye. It developed a relatively narrow spectrum, see Figs. 12.8 and 12.9.

12.4.2 *More Demonstrations*

As mentioned earlier, Hill *et al.* [Hill *et al.* (2007)], see Fig. 12.10 developed electrically injected plasmon lasers. For the first time they reported laser operation in an electrically pumped metallic-coated nanocavity formed by a semiconductor heterostructure encapsulated in a thin gold film. The demonstrated lasers show a low threshold current and their dimensions are smaller than the smallest electrically pumped lasers reported so far.

Marell *et al.* [Marell *et al.* (2011)] investigated electrically pumped, distributed feedback (DFB) lasers, based on gap-plasmon mode metallic waveguides. The waveguides have nano-scale widths below the diffraction limit and incorporate vertical groove Bragg gratings. In their structure semiconductor core consists of an InP/InGaAs/InP heterojunction. It is shielded from the silver cladding by a SiN layer. The authors demonstrated strong reduction of he linewidth and also superlinear light-current characteristics.

In the same year Ding *et al.* [Ding *et al.* (2011)] reported continuous operation of two metallic cavity lasers under electric injection operation at $T = 260$ K with semiconductor core encapsulated in silver. Volumes of physical cavities of their

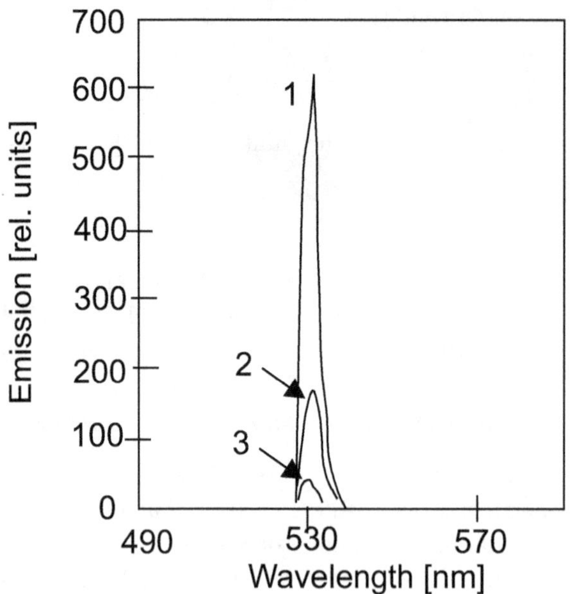

Figure 12.8: Stimulated emission spectra of the nanoparticle sample pumped with 22.5 mJ (1), 9 mJ (2), 4.5 mJ (3) 5-ns optical parametric oscillator pulses at 15488 nm. Adapted from [Noginov *et al.* (2009)].

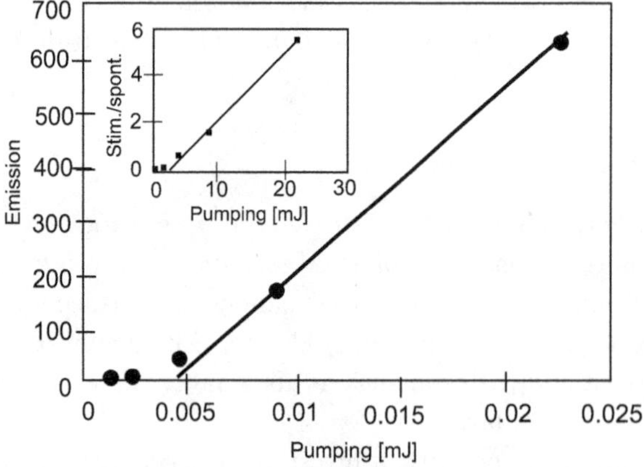

Figure 12.9: Input-output curve. Inset shows the ratio of the stimulated emission intensity (integrated between 526 nm and 537 nm) to the spontaneous emission background (integrated at < 526 nm and > 537 nm).

devices were: $0.96\lambda^3$ (at $\lambda = 1563.4$ nm) and $0.78\lambda^3$ (at $\lambda = 1488.7$ nm). The main problem of their devices was insufficient dissipation of heat.

Some more experimental developments of plasmon lasers reported at that time were: Ma *et al.* in 2010 [Ma *et al.* (2010)], Kumar [Kumar *et al.* (2008)] and MIT team [Bourzac (2009)].

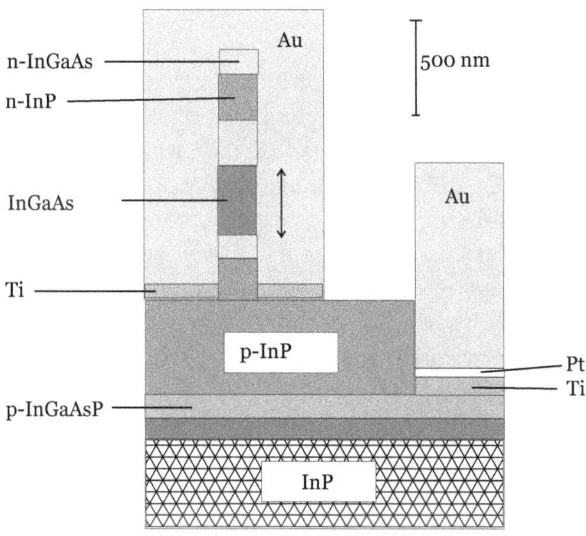

Figure 12.10: Schematic of the nano laser. Adapted from [Hill *et al.* (2007)].

We finish this part with a discussion of the recently suggested a new type of electrically driven nanolaser. As it is known, most of the current generation of nanolasers are optically pumped, making them unsuitable for use on electronic chips. A chip intended for mass production and real-life applications must incorporate hundreds of nanolasers and operate on an ordinary printed circuit board. A practical laser needs to be both electrically pumped and operate at room temperature.

Fedyanin *et al.* [Fedyanin *et al.* (2020)] recently discussed the possibility of electrically pumped nanolasers with a potential ability to operate at room temperature and with the predicted output power of up to 100 µW which is comparable to much larger photonic lasers. In their approach the estimated mode volume is less then $\lambda^3/30$ where λ is the operational free-space wavelength (according to authors, their nanolasers could be made even smaller). Therefore the volume occupied by the SPPs in the proposed nanolaser is 30 smaller than the equivalent volume occupied by photons.

Fedyanin *et al.* conducted extensive computer simulations of their proposed structure. The laser cavity consists of a subwavelength ring-resonator design with an InGaAs active layer; the laser emits at a 1.95 µm. Electrical pumping is based on a Au/InAsP/InGaAs)/AlInAs double heterostructure with a tunneling Schottky contact. The pumping is done along the direction where SPP propagate, i.e. across the interface between the plasmonic metal and semiconductor.

In closing, we mention recent suggestion of an electrically pumped GaAs quantum dot plasmonic nanolaser [Ghodsi and Kaatuzian (2020a)]. They used a semiclassical rate equations approach and determined that their structure can generate 1 mW of output power for a 23.6 mA injected current.

12.5 Basic Principles of Spaser

The concept of stimulated emission to sustain plasmonic oscillations in a resonator was first described by David Bergman and Mark Stockman in 2003 [Bergman and Stockman (2003)]. They suggested name SPASER which is an acronym for surface plasmon amplification by stimulated emission of radiation. A spaser is effectively a nanoscale laser with subwavelength dimensions and a low-Q plasmonic resonator, which sustains its oscillations using stimulated emission of surface plasmons. It is anticipated that they can be integrated with electronics on a chip.

As explained recently by Ning [Ning (2021)] there is no difference in the essential physics between surface plasmon laser regime and spaser regime. The differences and similarities are illustrated in Fig. 12.11.

12.5.1 *General*

Spaser is based on a single metallic nanoparticle resonator with the optical feedback provided by the localised surface plasmon resonance. Spasers are the nanoplasmonic analogs of lasers: instead of photons, spasers generate coherent surface plasmons (collective electron oscillations at the surface of a metal) in a resonant nanoparticle. They are considered as ideal sources of coherent optical fields at the nanoscale.

Typical spaser operates under optical pumping where light acting on a metal nanoparticles produces surface plasmons with the same frequency as the light. However, as mentioned before, light cannot be focused to spots smaller than about half of its wavelength (diffraction limit). On the other hand, plasmons are spread over much shorter distances and so can beat diffraction limit. In short, spaser amplifies surface plasmons, which, in turn, produce light waves.

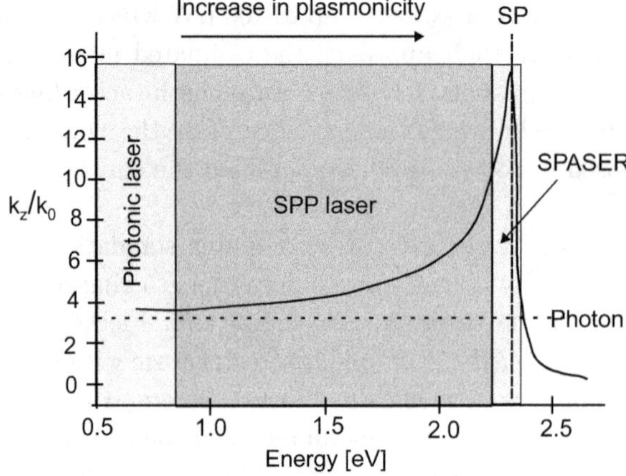

Figure 12.11: The normalised propagation wavevector versus energy plot for a semiconductor-silver interface; the range of energy is divided roughly into three regions depending on the "plasmonicity", or the degree of proximity to the surface plasmon resonance (SP). We call the three regions: photonic laser (conventional laser), SPP-Laser, and SPASER.

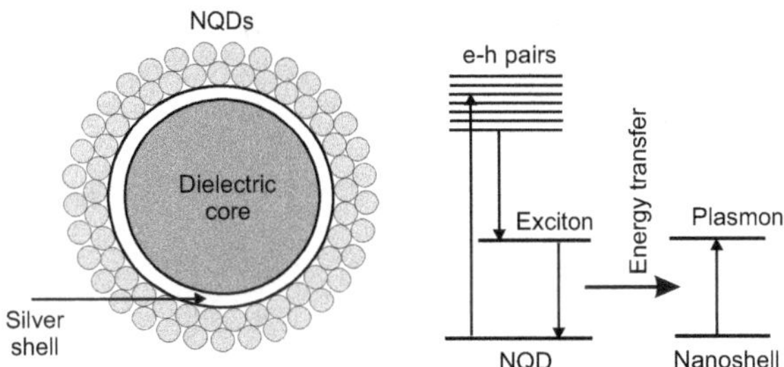

Figure 12.12: Schematic of a spaser and energy transitions, from [Stockman (2008)].

We explain basic principles following [Stockman (2008)], [Stockman (2010)]. In Fig. 12.12 we show a schematic of a spaser which consists of a silver nanoshell on a dielectric core (with a radius of 10–20 nm), and surrounded by two dense monolayers of nanocrystal quantum dots (NQDs). In the figure we also show energy levels and transitions in a spaser.

In short, spaser operates by optical pumping which forces transitions into electron–hole (e–h) pairs (vertical black arrow). The e–h pairs relax to excitonic levels. The exciton recombines and its energy is transferred (without radiation) to the plasmon excitation of the metal nanoparticle (nanoshell) through resonant coupled transitions.

As mentioned earlier, experimental demonstration of nanoparticle spaser has been reported by Noginov *et al.* [Noginov *et al.* (2009)]. Comprehensive summary of recent developments of the theory and technology of spasers has been reported by Malin Premaratne and Mark I. Stockman [Premaratne and Stockman (2017)]. Quantum theory of a spaser-based nanolaser, under the bad-cavity approximation was reported in [Parfenyev and Vergeles (2014)].

12.5.2 *Quantum Theory of Spaser*

Here we present summary of quantum theory of spaser as summarised by Stockman and collaborators [Stockman (2010)], [Stockman (2013)], [Premaratne and Stockman (2017)], [Stockman (2020)]. It is splitted into several parts.

12.5.2.1 *SP eigenmodes and their quantisation*

The resonator of a spaser can be a metalic nanoparticle with a size R which is much smaller than the wavelength λ,i.e. $R \ll \lambda$ and also $R \preceq l_s$, where l_s is a skin depth

$$l_s = \frac{\lambda}{2\pi}\left[Re\left(\frac{-\varepsilon_m^2}{\varepsilon_m + \varepsilon_d}\right)^{1/2}\right]^{-1} \tag{12.12}$$

where ε_m is the dielectric function of the metal and ε_d is that of the embedding dielectric. For metals like silver, gold, copper $l_s \approx 25$ nm in the entire optical region. For such small systems one can use the so-called quasistatic approximation.

The equation for electrostatic potential derived from Maxwell equations is

$$\frac{\partial}{\partial \mathbf{r}} \varepsilon\left(\mathbf{r}\right) \frac{\partial}{\partial \mathbf{r}} \varphi\left(\mathbf{r}\right) = 0. \tag{12.13}$$

Dielectric function has the following space and frequency dependence

$$\varepsilon\left(\mathbf{r}, \omega\right) = \varepsilon_m\left(\omega\right) \Theta\left(\mathbf{r}\right) + \varepsilon_d\left[1 - \Theta\left(\mathbf{r}\right)\right] = \varepsilon_d \left[1 - \frac{\Theta\left(\mathbf{r}\right)}{s\left(\omega\right)}\right] \tag{12.14}$$

where $s\left(\omega\right)$ is

$$s\left(\omega\right) = \frac{\varepsilon_d}{\varepsilon_d - \varepsilon_m\left(\omega\right)} \tag{12.15}$$

and $\Theta\left(\mathbf{r}\right)$ is the so-called characteristic function of the nanosystem which is equal to 1 when $\mathbf{r}$ belongs to the metal and 0 otherwise.

From above one formulates the eigenvalue problem

$$\frac{\partial}{\partial \mathbf{r}} \Theta\left(\mathbf{r}\right) \frac{\partial}{\partial \mathbf{r}} \varphi_n\left(\mathbf{r}\right) - s_n \frac{\partial^2}{\partial \mathbf{r}^2} \varphi_n\left(\mathbf{r}\right) = 0 \tag{12.16}$$

where ω_n is the corresponding eigenfrequency and $s_n = s\left(\omega_n\right)$ is the corresponding eigenvalue.

The eigenfunctions $\varphi_n\left(\mathbf{r}\right)$ satisfy the homogeneous Dirichlet–Neumann boundary conditions on the surface S surrounding the system. One obtains

$$\int_V \varepsilon\left(\mathbf{r}, \omega\right) \left|\nabla \varphi_n\left(\mathbf{r}\right)\right|^2 dV = \varepsilon_d \left[1 - \frac{s_n}{s\left(\omega\right)}\right] \tag{12.17}$$

where V is the volume of the system and

$$\varphi_1\left(\mathbf{r}\right)\big|_{\mathbf{r}\in S} = 0, \;\; \text{or} \;\; \mathbf{n}\left(\mathbf{r}\right) \frac{\partial}{\partial \mathbf{r}} \varphi_1\left(\mathbf{r}\right)\big|_{\mathbf{r}\in S} = 0 \tag{12.18}$$

where $\mathbf{n}\left(\mathbf{r}\right)$ is a normal vector to the surface S at a point $\mathbf{r}$.

Quantization of SP is performed by neglecting losses. It is done by expressing electric field as a sum over the eigenmodes

$$\mathbf{E}\left(\mathbf{r}\right) = -\sum_n A_n \nabla \varphi_n\left(\mathbf{r}\right) \left(a_n^\dagger + a_m\right) \tag{12.19}$$

where $a_n^\dagger$ and a_m are creation and annihilation operators which obey the following commutation relations

$$\left[a_n^\dagger, a_m\right] = \delta_{nm}. \tag{12.20}$$

Additionally $-\nabla \varphi_n\left(\mathbf{r}\right) = \mathbf{E}_n\left(\mathbf{r}\right)$ is the modal field of an n-th mode and A_n is a normalization constant

$$A_n = \left(\frac{4\pi \hbar s_n}{\varepsilon_d s_n'}\right), \;\; s_n' = Re \left.\frac{ds\left(\omega\right)}{d\omega}\right|_{\omega=\omega_n}. \tag{12.21}$$

Normalization constant A_n is obtained by evaluating average energy of SP as

$$\langle H_{SP} \rangle = \frac{1}{8\pi} \int_V \frac{\partial}{\partial \omega} \left[\omega \varepsilon\left(\mathbf{r}, \omega\right)\right] \sum_n \langle \mathbf{E}_n^\dagger\left(\mathbf{r}\right) \mathbf{E}_n\left(\mathbf{r}\right) \rangle dV \tag{12.22}$$

where

$$H_{SP} = \sum_n \hbar \omega_n \left(a_n^\dagger a_n + \frac{1}{2} \right). \qquad (12.23)$$

12.5.2.2 *Quantum density matrix equations for spasers*

Spaser Hamiltonian is

$$H = H_g + H_{SP} - \sum_p \mathbf{E}_p \left(\mathbf{r}_p \right) \cdot \mathbf{d}^{(p)} \qquad (12.24)$$

where H_g is the Hamiltonian of the gain medium and p is an index (label) of the gain medium chromophore, $\mathbf{r}_p$ is its coordinate vector and $\mathbf{d}^{(p)}$ is its dipole moment operator. In a typical approach the active medium is treated quantum mechanically and the SP quasiclassically.

Density matrix of a p-th chromophore obeys the equation

$$i\hbar \frac{\partial \rho^{(p)}}{\partial t} = \left[\rho^{(p)}, H \right] \qquad (12.25)$$

Spaser can be modelled as either two-level or three-level system. For TLS model with rotating wave approximation one obtains the following necessary condition for spasing [Premaratne and Stockman (2017)]

$$\frac{(\gamma_m + \gamma_{ph})^2 \sum_p \left| {}^{(m)}_a \Omega_p \right|^2}{\gamma_m \gamma_{ph} \left[(\omega_{21} - \omega_m)^2 + (\gamma_m + \gamma_{ph})^2 \right]} \geq 1 \qquad (12.26)$$

where ${}^{(m)}_a \Omega_p$ is the single-plasmon Rabi frequency representing the spasing transition in the p-th chromophore.

12.5.3 *Possible Applications of Nanolasers*

There are many possible applications of nanolasers [Xu *et al.* (2019)], [Ma and Oulton (2019)]. They can be used as a light source for scanning near-field optical microscopes, in nanolithography, in nanoscale circuits, to just name a few. Using nanolasers one can resolve details beyond the reach of standard light microscopy. It is also potentially possible to etch patterns much smaller than the width of a human hair. In the following we describe one which seems very interesting.

To improve conventional optical interconnects Stockman suggested to use spasers for on-chip interconnects [Stockman (2018)]. The idea has the potential to eliminate critical problem in current microprocessors and to increase clock rate which, at present does not exceed several GHz. The limit is associated with the fundamental physics, namely that the coupling between the transistors on a processor chip is electrostatic. When a transistor flops its state, the interconnect must be recharged by the current from a single transistor, which requires a long time and releases electrostatic energy as heat [Azzam *et al.* (2020)].

Stockman suggested to use SPPs to bring a signal from one transistor to another [Stockman (2018)]. In his solution processor includes a transistor pair. The first transistor is coupled to a spaser and pumps a spaser that has the same $\sim$10 nm size as the transistor. The spaser generates SPPs which are fed to a plasmonic interconnect wire. The SPPs propagate on the plasmonic interconnect wire and are detected by a phototransistor. The phototransistor generates an output current that is fed to a gate terminal of the second transistor to charge it. In an independent study Stockman showed that a single nanoscale transistor produces sufficient drive current to electrically pump a spaser.

Bibliography

Agarwal, G. S. (2013). *Quantum Optics* (Cambridge University Press, Cambridge).

Agrawal, A., Cho, S. H., Zandi, O., Ghosh, S., Johns, R. W., and Milliron, D. J. (2018). Localized surface plasmon resonance in semiconductor nanocrystals, *Chemical Reviews* **118**, pp. 3121–3207.

Agrawal, G. and Dutta, N. (2000). *Semiconductor Lasers. Second Edition* (Kluwer Academic Publishers, Boston/Dordrecht/London).

Alam, M., Meier, J., Aitchison, J., and Mojahedi, M. (2007). Gain assisted surface plasmon polariton in quantum wells structures, *Optics Express* **15**, pp. 176–182.

Alam, M. K. (2013). *Double Slit Diffraction Experiment with Surface Plasmon Polaritons. Master Thesis*, Ph.D. thesis, Texas Tech University.

Alekseyev, L. and Narimanov, E. (2006). Slow light and 3d imaging with non-magnetic negative index systems, *Opt. Exp.* **14**, p. 498.

Alexander, R. W., Kovener, G. S., and Bell, R. J. (1974). Dispersion Curves for Surface Electromagnetic Waves with Damping, *Phys. Rev. Lett.* **32**, p. 154.

Alivisatos, A. (1996). Semiconductor clusters, nanocrystals, and quantum dots, *Science* **271**, pp. 933–937.

Alrasheed, S. and Fabrizio, E. D. (2017). Efect of surface plasmon coupling to optical cavity modes on the field enhancement and spectral response of dimer-based sensors, *Scientific Reports* **7**, pp. 1–11.

Altug, H., Englund, D., and Vuckovic, J. (2006). Ultrafast photonic crystal nanocavity laser, *Nat. Phys.* **2**, p. 484-8.

Alu, A. and Engheta, N. (2002). Tuning the scattering response of optical nanoantennas with nanocircuit loads, *Nat. Photonics* **70**, p. 37.

Alzar, C. G., Martinez, M. G., and Nussenzveig, P. (2002). Classical analog of electromagnetically induced transparency, *Am. J. Phys.* **70**, p. 37.

Ambati, M. *et al.* (2008). Observation of stimulated emission of surface plasmon polaritons, *Nano Letters* **8**, pp. 3998–4001.

Ashroft, N. W. and Mermin, N. D. (1976). *Solid State Physics* (Harcourt College Publishers, Ford Worth).

Azzam, S. *et al.* (2020). Ten years of spasers and plasmonic nanolasers, *Light Science and Applications* **9**, p. 90.

Baba, T. (2008). Slow light in photonic crystals, *Nat. Photonics* **2**, p. 465.

Bai, W., Gan, Q., Bartoli, F., Zhang, J., Cai, L., Huang, Y., and Song, G. (2009). Design of plasmonic back structures for efficiency enhancement of thin-film amorphous si solar cells, *Optics Letters* **34**, pp. 3725–3727.

Balykin, V. (2018). Plasmon nanolaser: current state and prospects, *Physics-Uspekhi* **61**, pp. 846–870.

Barnes, W., Dereux, A., and Ebbesen, T. (2003). Surface plasmon subwavelength optics, *Nature* **424**, pp. 824–830.

Barron, L. D. (2004). *Molecular Light Scattering and Optical Activity*, 2nd edn. (Cambridge University Press, Cambridge).

Bass, F., Khankina, S., and Yakovenko, V. (1966). The low frequency properties of a semiconducting plasma situated in a constant electric field, *Soviet Physics JETP* **23**, pp. 70–75.

Benech, P. and Khalil, D. (1995). Rigorous spectral analysis of leaky structures: application to the prism coupling problem, *Opt. Com.* **118**, p. 220.

Berenger, J. P. (2008). A Perfectly Matched Layer for the Absorption of Electromagnetic Waves, *J. Comp. Phys.* **2**, p. 307.

Bergman, D. and Stockman, M. I. (2003). Surface plasmon amplification by stimulated emission of radiation: Quantum generation of coherent surface plasmons in nanosystems, *Phys. Rev. Lett.* **90**, p. 027402.

Berini, P. (2001). Plasmon-polariton waves guided by thin lossy metal films of finite width: Bound modes of asymmetric structures, *Phys. Rev.* **B 63**, p. 125417.

Berini, P. (2009). Long-range surface plasmon polaritons, *Advances in Optics and Photonics* **1**, pp. 484–588.

Berini, P. (2019). Highlighting recent progress in long-range surface plasmon polaritons: guest editorial, *Advances in Optics and Photonics* **11(2)**.

Berini, P. and Leon, I. D. (2012). Surface plasmon-polariton amplifiers and lasers, *Nature Photonics* **6**, pp. 16–24.

Berman, P. R. (2005). Goos-Hänchen shift in negatively refractive media, *Phys. Rev.* **E 66**, p. 067603.

Bian, Y. and Gong, Q. (2013). Multilayer metal-dielectric planar waveguides for subwavelength guiding of long-range hybrid plasmon polaritons at 1550 nm, *J Opt* **16**, p. 015001.

Bian, Y., Zheng, Z., Zhao, X., Zhu, J., and Zhou, T. (2009). Symmetric hybrid surface plasmon polariton waveguides for 3d photonic integration, *Opt Express* **17**, pp. 21320–21325.

Bigelow, M. S. *et al.* (2003a). Observation of Ultraslow Light Propagation in a Ruby Crystal at Room Temperature, *Phys. Rev. Lett.* **90**, p. 113903.

Bigelow, M. S. *et al.* (2003b). Superluminal and slow light propagation in a room-temperature solid, *Science.* **301**, p. 200.

Blumenthal, D. J. *et al.* (1994). Photonic packet switches: architectures and experimental implementations, *Proc. IEEE* **82**, p. 1650.

Boardman, A. D. *et al.* (2007). Creating stable gain in active metamaterials, *J. Opt. Soc. Am.* **B 24**, p. A53.

Boltasseva, A. and Bozhevolnyi, S. I. (2006). Directional couplers using long-range surface plasmon polariton waveguides, *IEEE J. Select. Topics Quantum Electron.* **12**, pp. 1233–1241.

Boriskina, S. V., Cooper, T. A., Zeng, L., Ni, G., Tong, J. K., Tsurimaki, Y., Huang, Y., Meroueh, L., Mahan, G., and Chen, G. (2017). Losses in plasmonics: from mitigating energy dissipation to embracing loss-enabled functionalities, *Advances in Optics and Photonics* **9**, pp. 775–827.

Born, M. and Wolf, E. (1999). *Principles of Optics*, 7th edn. (Cambridge University Press, Cambridge).

Bouchard, F., Sit, A., Zhang, Y., Fickler, R., Miatto, F. M., Yao, Y., Sciarrino, F., and Karimi, E. (2021). Two-photon interference: the Hong-Ou-Mandel effect, *Rep. Prog. Phys.* **84**, p. 012402.

Bourzac, K. (2009). The smallest laser ever made, *MIT Technology Review* **August 17**.

Bouwmeester, D. and Zeilinger, A. (2000). The physics of quantum information: Basic concepts, in D. Bouwmeester, A. Ekert, and A. Zeilinger (eds.), *The Physics of Quantum Information* (Springer, Berlin).

Boyd, R. (2008). *Nonlinear Optics (3rd ed.)* (Academic Press, San Diego).

Boyd, R. W. and Gauthier, D. J. (2006). Transparency on an optical chip, *Nature* **441**, p. 701.

Bozhevolnyi, S. and Vohnsen, B. (1995). Direct observation of surface polariton localization caused by surface roughness, *Opt. Com.* **117**, p. 417.

Bozhevolnyi, S. I. (2009). Introduction to surface plasmon-polariton waveguides, in S. I. Bozhevolnyi (ed.), *Plasmonic. Nanoguides and Circuits* (Pan Stanford, Singapore).

Breukelaar, I. G. (2004). Surface plasmon-polaritons in thin metal strips and slabs: waveguiding and mode cutoff, .

Brillouin, L. (1960). *Wave Propagation and Group Velocity* (Academic Press, New York).

Burke, J. J., Stegeman, G. I., and Tamir, T. (1986). Surface-polariton-like waves guided by thin, lossy metal films, *Phys. Rev.* **B 33**, p. 5186.

Burstein, E. and DeMartini, F. (eds.) (1974). *Proceedings of the First Taormina Research Conference on the Structure of Matter* (Pergamon Press, New York).

Cada, M. and Pistora, J. (2008). Optical plasmons in semiconductors, *The Institute of Photonic Sciences, Barcelona, Spain.*

Cada, M. and Pistora, J. (2016). Plasmon dispersion at an interface between a dielectric and a conducting medium with moving electrons, *The IEEE Journal of Quantum Electronics* **52**, p. 7200107.

Cada, M. and Pistora, J. (2017). Design of a new terahertz nanowaveguide amplifier, *IEEE Photonics Journal* **9**, p. 2200905.

Caloz, C. and Itoh, T. (2006). *Electromagnetic Metamaterials: Transmission Line theory and Microwave Applications* (Wiley, New Jersey).

Chang, W.-S. *et al.* (2011). Low absorption losses of strongly coupled surface plasmons in nanoparticle assemblies, *PNAS* **108**, pp. 19879–19884.

Charbonneau, R., Scales, C., Breukelaar, I., *et al.* (2006). Passive integrated optics elements based on long-range surface plasmon polaritons, *J. Lightwave Technol.* **24**, pp. 477–494.

Cheianov, V. V., Falko, V., and Altshuler, B. L. (2007). The focusing of electron flow and a Veselago lens in graphene p-n junctions, *Phys. Rev.* **E 315**, p. 1252.

Chen, H. *et al.* (2004). Left-handed materials composed of S-shaped resonators, *Phys. Rev.* **E 70**, p. 057605.

Chen, H. *et al.* (2005). Negative refraction of a combined double S-shaped metamaterial, *Appl. Phys. Lett.* **86**, p. 151909.

Chen, H. *et al.* (2006a). Equivalent circuit model for left-handed metamaterials, *J. Appl. Phys.* **100**, p. 024915.

Chen, H.-T. *et al.* (2006b). Negative refraction of a combined double S-shaped metamaterial, *Nature* **444**, p. 597.

Chen, J., Li, Z., Yue, S., and Gong, Q. (2009). Hybrid long-range surface plasmon-polariton modes with tight field confinement guided by asymmetrical waveguides, *Opt Express* **17**, pp. 23603–23609.

Cheng, D. K. (1993). *Fundamentals of Engineering Electromagnetics* (Prentice Hall, Upper Saddle River, New Jersey).

Chilwell, J. and Hodgkinson, I. (1984). Thin-films field-transfer matrix theory of planar multilayer waveguides and reflection from prism-loaded waveguides, *J. Opt. Soc. Am. A* **1**, p. 742.

Ciraci, C., Hill, R. T., Mock, J. J., Urzhumov, Y., Fernández-Domínguez, A. I., Maier, S. A., Pendry, J. B., Chilkoti, A., and Smith, D. R. (2012). Probing the ultimate limits of plasmonic enhancement, *Science* **337**, pp. 1072–1074.

Coldren, L., Temkin, H., and Wilmsen, C. (1999). Introduction to VCSELs, in C. Wilmsen, H. Temkin, and L. Coldren (eds.), *Vertical-Cavity Surface-Emitting Lasers* (Cambridge University Press, Cambridge).

Cole, J. P. (1995). A high accuracy FDTD algorithm to solve microwave propagation and scattering problems on a coarse grid, *IEEE Trans. Microw. Theory Tech.* **43**, p. 2053.

Cole, J. P. (2002). A high accuracy FDTD algorithm to solve microwave propagation and scattering problems on a coarse grid, *IEEE Trans. Antennas Propagat.* **50**, p. 1185.

Collin, R. E. (2001). *Foundations for Microwave Engineering* (Wiley, IEEE Press, New Jersey).

Cordaro, C., Piccitto, G., and Priolo, F. (2017). Quantum plasmonic waveguides: Au nanowires, *Eur. Phys. J. Plus* **132**, p. 453.

Couteau, C., Larrue, A., Wilhelm, C., and Soci, C. (2015). Nanowire lasers, *Nanophotonics* **4**, pp. 90–107.

Cummer, S. A. (2003). Dynamics of causal beam refraction in negative refractive index materials, *Appl. Phys. Lett.* **82**.

Darmanyanab, S. A., Nevièreb, M., and Zakhidovc, A. A. (2003). Surface modes at the interface of conventional and left-handed media, *Opt. Com.* **225**, p. 233.

Dasgupta, B. B. (1977). Surface plasmon dispersion for very small metallic spheres: A quantum mechanical formulation, *Z. Physik B* **27**, pp. 75–79.

Dasgupta, N., Sun, J., Liu, C., Brittman, S., Andrews, S., Lim, J., Gao, H., Yan, R., and Yang, P. (2014). 25th anniversary article: semiconductor nanowires synthesis, characterization, and applications, *Adv. Mater.* **26**, p. 2137-84.

de Abajo, F. J. G. (2012). Plasmons go quantum, *Nature* **483**, pp. 417–418.

de Leon, N., Lukin, M., and Park, H. (2012). Quantum plasmonic circuits, *IEEE J. Select. Topics Quantum Electron.* **18**, pp. 1781–1791.

Delves, L. M. and Lyness, J. N. (1967). A Numerical Method for Locating the Zeros of an Analytic Function, *Mathematics of Computation* **21**, p. 543.

Depine, R. A. and Lakhtakia, A. (2004). Comment I on Resonant and antiresonant frequency dependence of the effective parameters of metamaterials, *Phys. Rev. E* **70**, p. 048601.

Derkachova, A., Kolwas, K., and Demchenko, I. (2016). Dielectric function for gold in plasmonics applications: Size dependence of plasmon resonance frequencies and damping rates for nanospheres, *Plasmonics* **11**, p. 941-951.

Ding, K., Liu, Z., Yin, L., Hill, M., Marell, M., van Veldhoven, P., Nöetzel, R., and Ning, N. (2012). Room-temperature continuous wave lasing in deep-subwavelength metallic cavities under electrical injection, *Phys. Rev. B* **85**, p. 041301.

Ding, K., Liu, Z., Yin, L., *et al.* (2011). Electrical injection, continuous wave operation of subwavelength-metallic cavity lasers at 260k, *Appl. Phys. Lett.* **98**, p. 231108.

Dogariu, A., Kuzmich, A., and Wang, L. J. (2001). Transparent anomalous dispersion and superluminal light-pulse propagation at a negative group velocity, *Phys. Rev. A* **63**, p. 053806.

Dolling, G. *et al.* (2006). Low-loss negative-index metamaterial at telecommunication wavelengths, *Opt. Lett.* **31**, p. 1800.

Ebbesen, T., Lezec, H., Ghaemi, H., Thio, T., and Wolff, P. (1998). *Nature* **391**, pp. 667–669.

Economou, E. N. (1969). Surface Plasmons in Thin Films, *Phys. Rev.* **182**, p. 539.

Efros, A. L. (2004). Comment II on Resonant and antiresonant frequency dependence of the effective parameters of metamaterials, *Phys. Rev. E* **70**, p. 048602.

Eleftheriades, G. V. (2007). Analysis of bandwidth and loss in negative-refractive-index transmission-line (NRI-TL) media using coupled resonators, *IEEE Microw. Wireless Compon. Lett.* **17**, p. 412.

Eleftheriades, G. V. and Balmain, K. G. (eds.) (2005). *Negative Refraction Metamaterials: Fundamental Principles and Applications* (Wiley IEEE Press, New Jersey).

Engelen, R. *et al.* (2008). Two Regimes of Slow-Light Losses Revealed by Adiabatic Reduction of Group Velocity, *Phys. Rev. Lett.* **101**, p. 103901.

Engheta, N. (2007). Circuit Elements at Optical Frequencies: Nanoinductors, Nanocapacitors, and Nanoresistors, *Phys. Rev. Lett.* **317**, p. 1698.

Engheta, N., Salandrino, A., and Alu, A. (2005). Circuit Elements at Optical Frequencies: Nanoinductors, Nanocapacitors, and Nanoresistors, *Phys. Rev. Lett.* **95**, p. 095504.

Engheta, N. and Ziolkowski, R. W. (eds.) (2006). *Electromagnetic Metamaterials: Physics and Engineering Explorations* (Wiley IEEE Press, New Jersey).

Esteban, R., Borisov, A. G., Nordlander, P., and Aizpurua, J. (2012). Bridging quantum and classical plasmonics with a quantum-corrected model, *Nature Communications* **3**, p. 825.

Eyges, L. (1980). *The Classical Electromagnetic Field* (Dover Publications, New York).

Fang, A., Koschny, T., Wegener, M., and Soukoulis, C. (2009). Self-consistent calculations of metamaterials with gain, *Phys. Rev. B* **79**, p. 241104.

Fedyanin, D., Krasavin, A., Arsenin, A., and Zayats, A. (2012). Surface plasmon polariton amplification upon electrical injection in highly integrated plasmonic circuits, *Nano Letters* **12**, pp. 2459–2463.

Fedyanin, D. Y., Krasavin, A. V., Arsenin, A. V., and Zayats, A. V. (2020). Lasing at the nanoscale: coherent emission of surface plasmons by an electrically driven nanolaser, *Nanophotonics* **9**, pp. 3965–3975.

Feigenbaum, E., Kaminski, N., and Orenstein, M. (2008). Negative Group Velocity: Is It a Backward wave or Fast Light? *arXiv:0807.4915* **18**, p. 598.

Fitzgerald, J. M. (2018). *Electromagnetic field enhancement in classical and quantum plasmonics*, Ph.D. thesis, Imperial College, London.

Fitzgerald, J. M., Narang, P., Craster, R. V., Maier, S. A., and Giannini, V. (2016). Quantum plasmonics, *Proceedings of the IEEE* **104**, pp. 2307–2322.

Frateschi, N. C. and Levi, A. F. J. (1995). Resonant modes and laser spectrum of microdisk lasers, *Appl. Phys. Lett.* **66**, pp. 2932–2934.

Gagnon, G., Lahoud, N., Mattiussi, G. A., and Berini, P. (2006). Thermally activated variable attenuation of long-range surface plasmon-polariton waves, *J. Lightwave Technol.* **24**, pp. 4391–4402.

Gather, M., Meerholz, K., Danz, N., and Leosson, K. (2010). Net optical gain in a plasmonic waveguide embedded in a fluorescent polymer, *Nature Photonics* **4**, pp. 457–461.

Gehrig, E. *et al.* (2006). Dynamic Spatiotemporal Speed Control of Ultrashort Pulses in Quantum-Dot SOAs, *IEEE J. Quantum Electron.* **42**, p. 1047.

Genov, D. A., Ambati, M., and Zhang, X. (2007). Surface Plasmon Amplification in Planar Metal Films, *IEEE J. Quantum Electron.* **43**, p. 1104.

Genzel, L., Martin, T. P., and Kreibig, U. (1975). Dielectric function and plasma resonances of small metal particles, *Z. Physik B* **21**, pp. 339–346.

Gersen, H. *et al.* (2005). Real-Space Observation of Ultraslow Light in Photonic Crystal Waveguides, *Phys. Rev. Lett.* **94**, p. 073903.

Ghaemi, H., Thio, T., Grupp, D., Ebbesen, T., and Lezec, H. (1998). Surface plasmons enhance optical transmission through subwavelength holes, *Phys. Rev.* **B 58**, pp. 6779–6782.

Ghodsi, H. and Kaatuzian, H. (2020a). Design and analysis of an electrically pumped GaAs quantum dot plasmonic laser, *Optic* **203**, p. 164027.

Ghodsi, H. and Kaatuzian, H. (2020b). Nanoscale plasmon sources: physical principles and novel structures, in C. J. Bueno-Alejo (ed.), *Nanoplasmonics* (IntechOpen Limited, London).

Gorkunov, M., Lapine, M., Shamonina, E., and Ringhofer, K. H. (2002). Effective magnetic properties of a composite material with circular conductive elements, *Eur. Phys. J.* **B 28**, p. 263.

Gourley, P. (1998). Nanolasers, *Sci. Am.* **278**, pp. 56–61.

Gover, A. (1976). *Wave interactions in periodic structures and periodic dielectric waveguides*, Ph.D. thesis, CalTech, Physics.

Gramotnev, D. K. and Bozhevolnyi, S. I. (2010). Plasmonics beyond the diffraction limit, *Nature Photonics* **4**, pp. 83–91.

Grandidier, J., des Francs, G. C., Massenot, S., Bouhelier, A., Markey, L., Weeber, J.-C., Finot, C., and Dereux, A. (2009a). Gain-assisted propagation in a plasmonic waveguide at telecomwavelength, *Nano Letters* **9**, pp. 2935–2939.

Grandidier, J. *et al.* (2009b). Gain-Assisted Propagation in a Plasmonic Waveguide at Telecom Wavelength, *Nano Lett.*, p. 2935.

Gu, Q. and Fainman, Y. (2017). *Semiconductor Nanolasers* (Cambridge University Press, Cambridge).

Gwo, S. and Shih, C.-K. (2016). Semiconductor plasmonic nanolasers: current status and perspectives, *Rep. Prog. Phys.* **79**, pp. 1–34.

Hahn, W. and Metcalf, G. (1939). Velocity-modulated tubes, *Proceedings of the IRE* **27**, pp. 106–116.

Hau, L. V. *et al.* (1999). Light speed reduction to 17 metres per second in an ultracold atomic gas, *Nature* **297**, p. 594.

He, Y., Cao, Z., and Shen, Q. (2005). Guided optical modes in asymmetric left-handed waveguides, *Opt. Com.* **245**, p. 125.

Heeres, R. W., Kouwenhoven, L. P., and Zwiller, V. (2013). Quantum interference in plasmonic circuits, *Nature Nanotechnology* **8**, pp. 719–722.

Helszajn, J. (2000). *Ridge Waveguides and Passive Microwave Components* (IEE Press, London).

Higuchi, M., Miyata, M., and Takahara, J. (2014). Super long-range surface plasmon polaritons in a silver nano-slab waveguide, in A. D. Boardman (ed.), *Plasmonics: Metallic Nanostructures and Their Optical Properties XII*, Vol. 9163 (Proc. of SPIE), p. 91632V.

Hill, M., Oei, Y., *et al.* (2007). Lasing in metallic-coated nanocavities, *Nature Photonics* **1**, pp. 589–594.

Hill, M. T. (2018). Electrically pumped metallic and plasmonic nanolasers, *Chin. Phys. B* **27**, p. 114210.

Hill, M. T., Marell, M., Leong, E., *et al.* (2009). Lasing in metal-insulator-metal subwavelength plasmonic waveguides, *Optics Express* **17 (13)**, pp. 11107–11112.

Hohenberg, P. and Kohn, W. (1964). Inhomogeneous electron gas, *Phys. Rev.* **136**, p. B864.

Hong, C., Ou, Z., and L. Mandel (1987). Measurement of subpicosecond time intervals between two photons by interference, *Phys. Rev. Lett.* **59**, pp. 2044–2046.

Houck, A. A., Brock, J. B., and Chuang, I. L. (2003). Experimental Observations of a Left-Handed Material That Obeys Snell's Law, *Phys. Rev. Lett.* **90**, p. 137401.

Hu, J. and Menyuk, C. R. (2009). Understanding leaky modes: slab waveguide revisited, *Advances in Optics and Photonics* **1**, p. 58.

Huang, M. H., Mao, S., Feick, H., Yan, H., Wu, Y., Kind, H., Weber, E., Russo, R., and Yang, P. (2001). Room-temperature ultraviolet nanowire nanolasers, *Science* **292**, pp. 1897–1899.

Huang, W. P. (1993). Simulation of three-dimensional optical waveguides by a full-vector beam propagation method, *IEEE J. Quantum Electron.* **29**, p. 2639.

Inoue, K. (2002). Observation of small group velocity in two-dimensional AlGaAs-based photonic crystal slabs, *Phys. Rev. B* **65**, p. 121308(R).

Jacak, W. A. (2020). *Quantum nano-plasmonics* (Cambridge University Press, Cambridge).

Jackson, J. D. (1975). *Classical Electrodynamics*, 2nd edn. (Wiley, New York).

Jacob, Z. (2012). Quantum plasmonics, *MRS Bulletin* **37**, pp. 761–767.

Jeong, K.-Y. *et al.* (2020). Recent progress in nanolaser technology, *Advanced Materials* **32**, p. 2001996.

Jin, J. (2002). *The Finite Element Method in Electrodynamics* (Wiley, New York).

Johnson, P. B. and Christy, R. W. (1972). Optical constants of the noble metals, *Phys. Rev. B* **6**, pp. 4370–4379.

Karalis, A. *et al.* (2005). Surface-Plasmon-Assisted Guiding of Broadband Slow and Subwavelength Light in Air, *Phys. Rev. Lett.* **95**, p. 063901.

Kash, M. M. *et al.* (1999). Ultraslow Group Velocity and Enhanced Nonlinear Optical Effects in a Coherently Driven Hot Atomic Gas, *Phys. Rev. Lett.* **82**, p. 5229.

Kashiwa, T. *et al.* (2002). The phase velocity error and stability condition of the three-dimensional nonstandard FDTD method, *IEEE Trans. Magn.* **38**, p. 661.

Kastel, J. *et al.* (2007). Tunable Negative Refraction without Absorption via Electromagnetically Induced Chirality, *Phys. Rev. Lett.* **99**, p. 073602.

Kekatpure, R. D. *et al.* (2009). Solving dielectric and plasmonic waveguide dispersion relations on a pocket calculator, *Opt. Exp.* **17**, p. 24112.

Keshmarzi, E. K. (2017). *Long-range surface plasmon polariton active structures based on optically-pumped dye-doped polymer gain media*, Ph.D. thesis, Carleton University.

Khajavikhan, M., Simic, A., Katz, M., Lee, J., Slutsky, B., Mizrahi, A., Lomakin, V., and Fainman, Y. (2012). Thresholdless nanoscale coaxial lasers, *Nature* **482**, p. 204-7.

Kim, J. T., Park, S., Ju, J. J., Park, S. K., and Kim, M. (2007). Low-loss polymer-based long-range surface plasmon polariton waveguide, *IEEE Photonics Technology Letters* **19**, pp. 1374–1376.

Kim, K.-B. and McCall, M. W. (2003). Relationship between Surface Roughness of Indium Tin Oxide and Leakage Current of Organic Light-Emitting Diode, *Jpn. J. Appl. Phys.* **42**, p. L438.

Kinsler, P. and McCall, M. W. (2008). Criteria for negative refraction in active and passive media, *Microw. Opt. Technol. Lett.* **50**, p. 1804.

Knight, W., Clemenger, K., de Heer, W. A., Saunders, W. A., Chou, M., and Cohen, M. L. (1984). Electronic shell structure and abundances of sodium clusters, *Phys. Rev. Lett.* **52**, p. 2141.

Kock, W. E. (1948). Metallic Delay Lenses, *Bell Syst. Tech. J.* **27**, p. 58.

Kogelnik, H. and Ramaswamy, V. (1974). Scaling Rules for Thin-Film Optical Waveguides, *Appl. Opt.* **13**, p. 1857.

Kohn, W. and Sham, L. (1965). Self-consistent equations including exchange and correlation effects, *Phys. Rev.* **140**, pp. A1133–8.

Kolomenski, A. *et al.* (2009). Propagation length of surface plasmons in a metal film with roughness, *Appl. Opt.* **48**, p. 5683.

Koschny, T., Zhang, L., and Soukoulis, C. M. (2005). Isotropic three-dimensional left-handed metamaterials, *Phys. Rev.* **B 71**, p. 121103(R).

Koschny, T. *et al.* (2003). Isotropic three-dimensional left-handed metamaterials, *Phys. Rev.* **E 68**, p. 065602.

Koschny, T. *et al.* (2004). Reply to Comments on Resonant and antiresonant frequency dependence of the effective parameters of metamaterials, *Phys. Rev.* **E 70**, p. 048603.

Kovener, G. S., Alexander, R. W., and Bell, R. J. (1976). Surface electromagnetic waves with damping. I. Isotropic media, *Phys. Rev.* **B 14**, p. 1458.

Kumar, P., Tripathi, V., and Liu, C. (2008). A surface plasmon laser, *J. Appl. Phys.* **104(3)**, p. 033306.

Kwon, M. S. and Shin, S. Y. (2004). Simple and fast numerical analysis of multilayer waveguide modes, *Opt. Com.* **233**, p. 119.

Lee, H. *et al.* (2005). Realization of optical superlens imaging below the diffraction limit, *New J. Phys.* **7**, p. 255.

Leon, D. and Berini, P. (2010). Amplification of long-range surface plasmons by a dipolar gain medium, *Nature Photonics* **4**, pp. 382–387.

Leong, E. P. *et al.* (2011). Effect of Surface Morphology on the Optical Properties in Metal-Dielectric-Metal Thin Film Systems, *ACS Appl. Mater.* **3**, p. 1148.

Li, Y. (2017). *Plasmonic Optics: Theory and Applications* (SPIE Press, Bellingham, Washington, USA).

Lim, J., Imura, K., Nagahara, T., Kim, S., and Okamoto, H. (2005). Imaging and dispersion relations of surface plasmon modes in silver nanorods by near-field spectroscopy, *Chem. Phys. Lett.* **412**, pp. 41–45.

Linden, S. *et al.* (2004). Magnetic Response of Metamaterials at 100 Terahertz, *Science* **306**, p. 1351.

Liu, C. *et al.* (2001). Observation of coherent optical information storage in an atomic medium using halted light pulses, *Nature* **409**, p. 490.

Liu, R. *et al.* (2007). Description and explanation of electromagnetic behaviors in artificial metamaterials based on effective medium theory, *Phys. Rev.* **E 76**, p. 026606.

Locharoenrat, K. (2015). *Optical Properties of Solids, An Introductory Textbook* (Stanford Publishing Pvt. Ltd, Singapore).

Logeeswaran, V. J. *et al.* (2007). Ultra-smooth metal surfaces generated by pressure-induced surface deformation of thin metal films, *Appl. Phys.* **A 87**, p. 187.

Loncar, M., Yoshie, T., Scherer, A., Gogna, P., and Qiu, Y. (2002). Low-threshold photonic crystal laser, *Appl. Phys. Lett.* **81**, pp. 2680–2682.

Loudon, R. (1970). The propagation of electromagnetic energy through an absorbing dielectric, *J. Phys. A: Gen. Phys.* **3**, p. 233.

Lu, C.-H., Lan, C.-C., Lai, Y.-L., Li, Y.-L., and Liu, C.-P. (2011). Enhancement of green emission from InGaN/GaN multiple quantum wells via coupling to surface plasmons in a two-dimensional silver array, *Adv Funct Mater* **21**, pp. 4719–4723.

Lu, W. T. *et al.* (2010). Storing light in active optical waveguides with single-negative materials, *Appl. Phys. Lett.* **96**, p. 211112.

Lu, Y. *et al.* (2012). Plasmonic nanolaser using epitaxially grown silver film, *Science* **337**, pp. 450–3.

Lu, Y. *et al.* (2014). All-color plasmonic nanolasers with ultralow thresholds: autotuning mechanism for single-mode lasing, *Nano Lett.* **14**, pp. 4381–8.

Lukin, M. D. and Imamoglu, A. (2001). Controlling photons using electromagnetically induced transparency, *Nature* **413**, p. 273.

Ma, R. M. and Oulton, R. F. (2019). Applications of nanolasers, *Nature Nanotechnology* **14**, pp. 12–22.

Ma, R. M., Oulton, R. F., Sorger, V. J., *et al.* (2010). Room-temperature sub-diffraction-limited plasmon laser by total internal reflection, *Nature Materials* **10**, pp. 110–113.

Maier, S. (2007). *Plasmonics: fundamentals and applications* (Springer).

Maier, S. and Atwater, H. (2005). Plasmonics: Localization and guiding of electromagnetic energy in metal/dielectric structures, *J. Appl. Phys.* **98**, p. 011101.

Marasinghe, D. (2018). *Drude-Lorentz analysis of the optical properties of the quasiptwo-dimensional dichalcogenides* $2H - NbSe_2$ *and* $2H - TaSe_2$, Master's thesis, The University of Akron.

Marcatili, E. A. (1969). Dielectric Rectangular Waveguide and Directional Coupler for Integrated Optics, *Bell Syst. Tech. J.* **48**, p. 2071.

Marell, M. J., Smalbrugge, B., J.Geluk, E., *et al.* (2011). Plasmonic distributed feedback lasers at telecommunications wavelengths, *Optics Express* **19**, pp. 15109–15118.

Markel, V. and Sarychev, A. (2007). Propagation of surface plasmons in ordered and disordered chains of metal nanospheres, *Phys. Rev. B* **75**, p. 085426.

Markos, P. and Soukoulis, C. M. (2003). Transmission properties and effective electromagnetic parameters of double negative metamaterials, *Opt. Exp.* **11**, p. 649.

Martino, G. D., Sonnefraud, Y., Tame, M. S., Kena-Cohen, S., Dieleman, F., Ozdemir, S. K., Kim, M. S., and Maier, S. A. (2014). Observationof quantum interference in the plasmonic Hong-Ou-Mandel effect, *Physical Review Applied* **1**, p. 034004.

Martino, G. D. *et al.* (2012). Quantum statistics of surface plasmon polaritonsin metallic stripe waveguides, *Nano Letters* **12**, p. 2504.

Maxwell, J. C. (1891). *A Treatise on Electricity and Magnetism*, 3rd edn. (Clarendon Press, Oxford).

McCall, S., Levi, A., Slusher, R., Pearton, S., and Logan, R. (1992). Whispering-gallery mode microdisk lasers, *Appl. Phys. Lett.* **60**, p. 289-291.

Michelotti, F. *et al.* (2009). Thickness dependence of surface plasmon polariton dispersion in transparent conducting oxide films at 1.55 microm, *Opt. Lett.* **34**, p. 839.

Mickens, R. E. (1984). *Nonstandard Finite Difference Models of Differential Equations* (World Scientific, Singapore).

Mickens, R. E. (1989). Exact solutions to a finite-difference model of a nonlinear reaction-advection equation: Implications for numerical analysis, *Num. Methods Partial Dif. Eqn.* **5**, p. 313.

Mickens, R. E. (1993). Construction of a Finite-Difference Scheme that Exactly Conserves Energy for a Mixed Party Oscillator, *J. of Sound and Vibration* **172**, p. 142.

Mickens, R. E. (1996). Exact finite difference schemes for the wave equation with spherical symmetry, *J. Diff. Equations Appl.* **2**, p. 263.

Mickens, R. E. (1997). Exact Finite Difference Schemes for Two-Dimensional Advection Equations, *J. of Sound and Vibration* **207**, p. 426.

Milonni, P. W. (2005). *Fast Light, Slow Light and Left-Handed Light* (IoP Publishing, Bristol).

Min, C. and Veronis, G. (2010). Theoretical investigation of fabrication-related disorders on the properties of subwavelength metal-dielectric-metal plasmonic waveguides, *Opt. Exp.* **18**, p. 20939.

Muth, J., Lee, J., Shmagin, I., Kolbas, R., Jr, H. C., Keller, B., Mishra, U., and DenBaars, S. (2012). *Appl. Phys. Lett.* **100**, p. 181104.

Nagpal, P. *et al.* (2009). Ultrasmooth Patterned Metals for Plasmonics and Metamaterials, *Science* **31**, p. 594.

Nakamura, Y. (1983). Quantization of non-radiative surface plasma oscillations, *Progress of Theoretical Physics* **70**, pp. 908–918.

Nezhad, M., Simic, A., Bondarenko, O., Slutsky, B., Mizrahi, A., Feng, L., Lomakin, V., and Fainman, Y. (2010). Roomtemperature subwavelength metallo-dielectric lasers, *Nat. Photon.* **4**, p. 395-9.

Nezhad, M., Tetz, K., and Hainman, Y. (2004). Gain assisted propagation of surface plasmon polaritons on planar metallic waveguides, *Optics Express* **12**, pp. 4072–4079.

Nga, D. T. T., Lan, N. T. P., and Viet, N. A. (2015). Second quantization model of surface plasmon polariton at metal planar surface, *Journal of Physics: Conference Series* **627**, p. 012018.

Nguyen, B. H. and Nguyen, V. H. (2014). Dispersion and attenuation of surface plasmon polariton at metal dielectric interface, *Advances in Natural Sciences: Nanoscience and Nanotechnology* **5**, p. 035002.

Nguyen, V. H. and Nguyen, B. H. (2015). Basics of quantum plasmonics, *Adv. Nat. Sci.: Nanosci. Nanotechnol.* **6**, p. 023001.

Nikolajsen, T., Leosson, K., and Bozhevolnyi, S. I. (2004). Surface plasmon polariton based modulators and switches operating at telecom wavelengths, *Appl. Phys. Lett.* **85**, pp. 5833–5835.

Ning, C.-Z. (2021). Spaser or plasmonic nanolaser? — reminiscences of discussions and arguments with Mark Stockman, *Nanophotonics* **10**, pp. 3619–3622.

Nistad, B. and Skaar, J. (2008). Causality and electromagnetic properties of active media, *Phys. Rev. E* **78**, p. 036603.

Noginov, M., Zhu, G., Belgrave, M., Bakker, R., Shalaev, V., Narimanov, E., Stout, S., Herz, E., Suteewong, T., and Wiesner, U. (2009). Demonstration of a spaser-based nanolaser, *Nature* **460**, pp. 1110–1112.

Noginov, M., Zhu, G., Mayy, M., Ritzo, B. A., Noginova, N., and Podolskiy, V. A. (2008a). Stimulated emission of surface plasmon polaritons, *Phys. Rev. Lett.* **101**, p. 226806.

Noginov, M. A. *et al.* (2008b). Causality and electromagnetic properties of active media, *Opt. Exp.* **16**, p. 1385.

Noginov, M. A. *et al.* (2011). Transparent conductive oxides: Plasmonic materials for telecom wavelengths, *Appl. Phys. Lett.* **99**, p. 021101.

Notomi, M. *et al.* (2001). Extremely Large Group-Velocity Dispersion of Line-Defect Waveguides in Photonic Crystal Slabs, *Phys. Rev. Lett.* **87**, p. 253902.

Nozaki, K., Kita, S., and Baba, T. (2007). Room temperature continuous wave operation and controlled spontaneous emission in ultrasmall photonic crystal nanolaser, *Opt. Express* **15**, pp. 7506–7514.

Nozaki, K., Watanabe, H., and Baba, T. (2008). Photonic crystal nanolaser monolithically integrated with passive waveguide for effective light extraction, *Appl. Phys. Lett.* **92**, p. 021108.

O'Brien, S. and Pendry, J. B. (2002). Photonic band-gap effects and magnetic activity in dielectric composites, *J. Phys.: Condens. Matter* **14**, p. 4035.

O'Donnell, K. A., Torre, R., and West, C. S. (1997). Observations of second-harmonic generation from randomly rough metal surfaces, *Phys. Rev. B* **55**, p. 7985.

Okamoto, K. *et al.* (2004). Surface-plasmon-enhanced light emitters based on InGaN quantum wells, *Nat. Materials* **3**, p. 601.

Okawachi, Y. *et al.* (2005). Tunable All-Optical Delays via Brillouin Slow Light in an Optical Fiber, *Phys. Rev. Lett.* **94**, p. 153902.

Ooi, K. (2013). *Plasmonic devices for on-chip optical interconne*, Ph.D. thesis, Nanyang Technological University, Singapore.

Oulton, R. (2012). Surface plasmon lasers: sources of nanoscopic light, *Materials Today* **15**, pp. 26–34.

Oulton, R. *et al.* (2009). Plasmon lasers at deep subwavelength scale, *Nature* **461**, pp. 629–632.

Painter, O., Lee, R. K., Scherer, A., Yariv, A., O'Brien, J. D., Dapkus, P. D., and Kim, I. (1999). Two-dimensional photonicband-gap defect mode laser, *Science* **284**, pp. 1819–1821.

Parazzoli, C. G. *et al.* (2003). Experimental Verification and Simulation of Negative Index of Refraction Using Snell's Law, *Phys. Rev. Lett.* **90**, p. 107401.

Parfenyev, V. M. and Vergeles, S. S. (2014). Quantum theory of a spaser-based nanolaser, *Optics Express* **22**, pp. 13671–13679.

Park, H., Kim, S., Kwon, S., Ju, Y., Yang, J., Baek, J., Kim, S., and Lee, Y. (2004). Electrically driven single-cell photonic crystal laser, *Science* **305**, p. 1444-7.

Park, H.-G., Barrelet, C., Wu, Y., Tian, B., Qian, F., and Lieber, C. (2008). A wavelength-selective photonic-crystal waveguide coupled to a nanowire light source, *Nature Photonics*.

Pendry, J. B. (2000). Negative Refraction Makes a Perfect Lens, *Phys. Rev. Lett.* **85**, p. 3966.

Pendry, J. B. (2008). Time Reversal and Negative Refraction, *Science* **322**, p. 71.

Pendry, J. B., Schurig, D., and Smith, D. R. (2006). Controlling Electromagnetic Fields, *Science* **312**, p. 1780.

Pendry, J. B. *et al.* (1996). Extremely low frequency plasmons in metallic mesostructures, *J. Phys.: Condens. Matter* **76**, p. 4773.

Pendry, J. B. *et al.* (1998). Low frequency plasmons in thin-wire structures, *Phys. Rev. Lett.* **10**, p. 4785.

Pendry, J. B. *et al.* (1999). Magnetism from conductors and enhanced nonlinear phenomena, *IEEE Trans. Microw. Theory Tech.* **47**, p. 2075.

Perdew, J. and Zunger, A. (1981). Self-interaction correction to density-functional approximations for many-electron systems, *Phys. Rev.* **B 23**, pp. 5048–5079.

Pickering, T. W. (2013). *Passive and active stopped-light in plasmonic and metamaterial waveguides*, Ph.D. thesis, Imperial College London.

Pitarke, J. M. *et al.* (2007). Theory of surface plasmons and surface-plasmon polaritons, *Rep. Prog. Phys.* **70**, p. 1.

Popov, A. K. and Shalaev, V. M. (2006). Compensating losses in negative-index metamaterials by optical parametric amplification, *Opt. Lett.* **31**, p. 2169.

Prade, B., Vinet, J. Y., and Mysyrowicz, A. (1991). Guided optical waves in planar heterostructures with negative dielectric constant, *Phys. Rev. Lett.* **B 44**, p. 13556.

Premaratne, M. and Stockman, M. I. (2017). Theory and technology of spasers, *Advances in Optics and Photonics* **9**, pp. 79–128.

Pusch, A., Wuestner, S., Hamm, J., Tsakmakidis, K., and Hess, O. (2012). Coherent amplification and noise in gain-enhanced nanoplasmonic metamaterials: a Maxwell-Bloch Langevin approach, *ACS NANO* **6**, pp. 2420–2431.

Qing, D. K. and Chen, G. (2004). Enhancement of evanescent waves in waveguides using metamaterials of negative permittivity and permeability, *Appl. Phys. Lett.* **84**, p. 669.

Raether, H. (1988). *Surface Plasmons on Smooth and Rough Surfaces and on Gratings* (Springer Tracts in Modern Physics, Berlin).

Ramakrishna, S. A. *et al.* (2002). The asymmetric lossy near perfect lens, *J. Mod. Opt.* **49**, p. 1747.

Ramo, S. (1939). Space charge and field waves in an electron beam, *Physical Review* **56**, pp. 276–287.

Raoufi, D. *et al.* (2007). Surface characterization and microstructure of ITO thin films at different annealing temperatures, *Applied Surface Science* **253**, p. 9085.

Reza, A., Dignam, M. M., and Hughes, S. (2008). Can light be stopped in realistic metamaterials? *Nature* **455**, p. E10.

Rhodes, C. *et al.* (2006). Surface plasmon resonance in conducting metal oxides, *J. Appl. Phys.* **100**, p. 054905.

Rhodes, C. *et al.* (2008). Dependence of plasmon polaritons on the thickness of indium tin oxide thin films, *J. Appl. Phys.* **103**, p. 093108.

Ritchie, R. H. (1957). Plasma Losses by Fast Electrons in Thin Films, *Phys. Rev.* **106**, p. 874.

Rockstuhl, C. *et al.* (2008). Transition from thin-film to bulk properties of metamaterials, *Phys. Rev. B* **77**, p. 035126.

Rosenzveig, T. (2011). *Plasmonic nanowire waveguides and devices at telecom wavelengths*, Ph.D. thesis, University of Iceland.

Rotman, W. (1962). Plasma simulation by artificial dielectrics and parallel-plate media, *IRE Trans. Antennas Propagat.* **10**, p. 82.

Ruppin, R. (2002). Electromagnetic energy density in a dispersive and absorptive material, *Phys. Lett. A* **299**, p. 309.

Sarid, D. (1981). Long-Range Surface-Plasma Waves on Very Thin Metal Films, *Phys. Rev. Lett.* **47**, p. 1927.

Sarid, D. and Challener, W. (2010). *Modern Introduction to Surface Plasmons* (Cambridge University Press, Cambridge).

Savage, K. J., Hawkeye, M. M., Esteban, R., Borisov, A. G., Aizpurua, J., and Baumberg, J. J. (2012). Revealing the quantum regime in tunnelling plasmonics, *Nature* **491**, pp. 574–577.

Schneider, C., Rahimi-Iman, A., Kim, N. Y., Fischer, J., Savenko, I. G., Amthor, M., Lermer, M., Wolf, A., Worschech, L., Kulakovskii, V. D., Shelykh, I. A., Kamp, M., Reitzenstein, S., Forchel, A., Yamamoto, Y., and Höfling, S. (2013). An electrically pumped polariton laser, *Nature* **497**, p. 348-352.

Scholl, J. A., Koh, A. L., and Dionne, J. A. (2012). Quantum plasmon resonances of individual metallic nanoparticles, *Nature* **483**, p. 421.

Schouten, H. F., Kuzmin, N., Dubois, G., Visser, T. D., Gbur, G., Alkemade, P. F. A., Blok, H., 't Hooft, G., Lenstra, D., and Eliel, E. R. (2005). Plasmon-assisted two-slit transmission: Young's experiment revisited, *Phys. Rev. Lett.* **94**, p. 053901.

Schurig, D. *et al.* (2006). Metamaterial Electromagnetic Cloak at Microwave Frequencies, *Science* **314**, p. 977.

Scully, M. O. and Zubairy, M. S. (1997). *Quantum Optics* (Cambridge University Press, Cambridge).

Seidel, J., Grafstrom, S., and Eng, L. (2005). Stimulated Emission of Surface Plasmons at the Interface between a Silver Film and an Optically Pumped Dye Solution, *Phys. Rev. Lett.* **94**, p. 177401.

Shadrivov, I. V., Sukhorukov, A. A., and Kivshar, Y. S. (2003). Guided modes in negative-refractive-index waveguides, *Phys. Rev. E* **67**, p. 057602.

Shadrivov, I. V., Sukhorukov, A. A., and Kivshar, Y. S. (2005). Complete Band Gaps in One-Dimensional Left-Handed Periodic Structures, *Phys. Rev.* **95**, p. 193903.

Shalaev, V. M. (2007). Optical negative-index metamaterials, *Nat. Photonics* **1**, p. 41.

Shalaev, V. M. *et al.* (2005). Negative index of refraction in optical metamaterials, *Opt. Lett.* **30**, p. 3356.

Shamonina, E. *et al.* (2007). Magnetoinductive waves in one, two, and three dimensions, *J. Appl. Phys.* **92**, p. 6252.

Shelby, R. A., Smith, D. R., and Schultz, S. (2001a). Experimental Verification of a Negative Index of Refraction, *Science* **292**, p. 77.

Shelby, R. A. *et al.* (2001b). Microwave transmission through a two-dimensional, isotropic, left-handed metamaterial, *Appl. Phys. Lett.* **78**, p. 489.

Sidiropoulos, T., Roder, R., Geburt, S., Hess, O., Maier, S., Ronning, C., and Oulton, R. (2014). Ultrafast plasmonic nanowire lasers near the surface plasmon frequency, *Nat. Phys.* **10**, pp. 870–6.

Sivan, Y., Xiao, S., Chettiar, U., Kildishev, A., and Shalaev, V. (2009). Frequency-domain simulations of a negative-index material with embedded gain, *Optics Express* **17**, p. 24060.

Smith, D., Schultz, S., Markos, P., and Soukoulis, C. (2008). Determination of effective permittivity and permeability of metamaterials from reflection and transmission coefficients, *Phys. Rev.* ***B*** **65**, p. 195104.

Smith, D. D. *et al.* (2004a). Coupled-resonator-induced transparency, *Phys. Rev.* ***A*** **69**, p. 063804.

Smith, D. R., Pendry, J. B., and Wiltshire, M. K. (2004b). Metamaterials and Negative Refractive Index, *Science* **305**, p. 788.

Smith, D. R., Schurig, D., and Pendry, J. B. (2002). Negative refraction of modulated electromagnetic waves, *Appl. Phys. Lett.* **81**, p. 2713.

Smith, D. R. *et al.* (2000). Composite Medium with Simultaneously Negative Permeability and Permittivity, *Phys. Rev. Lett.* **84**, p. 4184.

Smith, D. R. *et al.* (2003). Limitations on subdiffraction imaging with a negative refractive index slab, *Appl. Phys. Lett.* **82**, p. 1506.

Smith, G. D. (1985). *Numerical Solution of Ordinary and Partial Differential Equations* (Oxford University Press, England).

Smith, R. E. and Houde-Walter, S. (1993). The migration of bound and leaky solutions to the waveguide dispersion relation, *IEEE J. Lightwave Technol.* **11**, p. 1760.

Smith, R. E., Houde-Walter, S., and Forbes, G. W. (1992). Mode determination for planar waveguide using the four-sheeted dispersion relation, *IEEE J. Quantum Electron.* **28**, p. 1520.

Smith, R. L. (1970). The Velocities of Light, *Am. J. Phys.* **38**, p. 978.

Snyder, A. and Love, J. D. (1983). *Optical Waveguide Theory* (Chapman and Hall, New York).

Song, M. (2012). *Surface plasmon propagation in metal nanowires*, Ph.D. thesis, Universite de Bourgogne.

Soukoulis, C. M. *et al.* (2008). The science of negative index materials, *J. Phys.: Condens. Matter* **20**, p. 304217.

Stegeman, G. I., Wallis, R. F., and Maradudin, A. A. (1983). Excitation of surface polaritons by end-fire coupling, *Opt. Lett.* **8**, p. 386.

Stern, M. S. (1988a). Semivectorial polarised finite difference method for optical waveguides with arbitrary index profiles, *IEE Proc. Optoelectron.* **135**, p. 56.

Stern, M. S. (1988b). Semivectorial polarised H-field solutions for dielectric waveguides with arbitrary index profiles, *IEE Proc. Optoelectron.* **135**, p. 333.

Stern, M. S. (1991). Rayleigh quotient solution of semivectorial field problems for optical waveguides with arbitrary index profiles, *IEE Proc. Optoelectron.* **138**, p. 123.

Stockman, M. (2010). The spaser as a nanoscale quantum generator and ultrafast amplifier, *Journal of Optics* **12**, p. 024004.

Stockman, M. (2018). Spasers to speed up CMOS processors. US patent: 10,096,675, .

Stockman, M. (2020). Brief history of spaser from conception to the future, *Advanced Photonics* **2**, p. 054002.

Stockman, M. I. (2004). Nanofocusing of Optical Energy in Tapered Plasmonic Waveguides, *Phys. Rev. Lett.* **93**, p. 137404.

Stockman, M. I. (2008). Spasers explained, *Nature Photonics* **2**, pp. 327–329.

Stockman, M. I. (2013). Spaser, plasmonic amplification, and loss compensation, in A. V. Zayats and S. A. Maier (eds.), *Active Plasmonics and Tuneable Plasmonic Metamaterials* (Wiley, Hoboken, New Jersey).

Stockman, M. I. *et al.* (2018). Roadmap on plasmonics, *Journal of Optics* **20**, p. 043001.

Strauf, S. and Jahnke, F. (2011). Single quantum dot nanolaser, *Laser Photon. Rev.* **5**, p. 60733.

Stuart, R. D. (1961). *An Introduction to Fourier Analysis* (Methuen and Co LTD, New York).

Sudbo, A. S. (1992). Why Are Accurate Computations of Mode Fields in Rectangular Dielectric Waveguides Difficult? *J. Lightwave Technol.* **10**, p. 418.

Sumi, M. (1966). Travelling-wave amplification by drifting carriers in semiconductors, *Appl. Phys. Lett.* **9**, pp. 251–253.

Sydoruk, O. (2014). Amplification and generation of terahertz plasmons in gated two-dimensional channels: Modal analysis, *J. Appl. Phys.* **115**, p. 204507.

Taflove, A. and Hagness, S. C. (2000). *Computational Electrodynamics: The Finite-Difference Time-Domain Method*, 2nd edn. (Artech House, Norwood).

Takahara, J. (2009). Negative dielectric optical waveguides for nano-optical guiding, in S. Bozhevolnyi (ed.), *Plasmonic. Nanoguides and Circuits* (Pan Stanford, Singapore).

Takahara, J., Yamagishi, S., Taki, H., Morimoto, A., and Kobayashi, T. (1997). Guiding of a one-dimensional optical beam with nanometer diameter, *Optics Letters* **22**, pp. 475–477.

Tamir, T. (ed.) (1979). *Integrated Optics, Ch. 2* (Springer-Verlag, New York).

Tandaechanurat, A., Ishida, S., Guimard, D., Nomura, M., Iwamoto, S., and Arakawa, Y. (2011). Lasing oscillation in a three-dimensional photonic crystal nanocavity with a complete bandgap, *Nat. Photon.* **5**, p. 914.

Tatel, G. and Wartak, M. (2020). Amplification of surface plasmons, in K. D. Sattler (ed.), *21st Century Nanoscience - A Handbook. Nanophotonics, Nanoelectronics, and Nanoplasmonics*, Vol. 6 (CRC Press), pp. 16–1–16–14.

Thraskias, C. A., Lallas, E. N., Neumann, N., Schares, L., Offrein, B. J., Henker, R., Plettemeier, D., Ellinger, F., Leuthold, J., and Tomkos, I. (2018). Survey of photonic and plasmonic interconnect technologies for intra-datacenter and high-performance computing communications, *IEEE Communications Surveys and Tutorials* **20**, pp. 2758–2783.

Totsuka, K., Kobayashi, N., and Tomita, M. (2007). Slow Light in Coupled-Resonator-Induced Transparency, *Phys. Rev. Lett.* **98**, p. 213904.

Tsakmakidis, K. L., Boardman, A. D., and Hess, O. (2007). "Trapped rainbow" storage of light in metamaterials, *Nature* **450**, p. 397.

Tsakmakidis, K. L. *et al.* (2005). Systematic modal analysis of 3D dielectric waveguides using conventional and high accuracy nonstandard FDTD algorithms, *IEEE Photon. Technol. Lett.* **17**, p. 2598.

Tsakmakidis, K. L. *et al.* (2006a). Single-mode operation in the slow-light regime using oscillatory waves in generalized left-handed heterostructures, *Appl. Phys. Lett.* **89**, p. 201103.

Tsakmakidis, K. L. *et al.* (2006b). Slow and Fast Light 2006 Technical Digest, paper No. WA5. *Optical Society of America.*

Tsakmakidis, K. L. *et al.* (2006c). Surface plasmon polaritons in generalized slab heterostructures with negative permittivity and permeability, *Phys. Rev.* **B 73**, p. 085104.

Tsakmakidis, K. L. *et al.* (2009). "Trapped rainbow" storage of light in metamaterials, *(submitted).*

Ullrich, C. A. and Yang, Z.-H. (2014). A brief compendium of time-dependent density functional theory, *Braz. J. Phys.* **44**, pp. 154–188.

Vahala, K. J. (2003). Optical microcavities, *Nature* **424**, pp. 839–846.

Valentine, J. *et al.* (2008). Three-dimensional optical metamaterial with a negative refractive index, *Nature* **455**, p. 376.

van Vugt, L., Ruhle, S., and Vanmaekelbergh, D. (2006). Phase correlated nondirectional laser emission from the end facets of a zno nanowire, *Nano Lett.* **6**, pp. 2707–11.

Varas, A., García-González, P., Feist, J., García-Vidal, F., and Rubio, A. (2016). Quantum plasmonics: from jellium models to ab initio calculations, *Nanophotonics* **5**, pp. 409–426.

Veselago, V. G. (1967). Properties of materials having simultaneously negative values of the dielectric and magnetic susceptibilities, *Sov. Phys. Solid State* **8**, p. 2854.

Veselago, V. G. (1968). The Electrodynamics of Substances with Simultaneously Negative Values of ε and μ, *Sov. Phys. Usp.* **10**, p. 509.

Vlasov, Y. A. and McNab, S. J. (2006). Coupling into the slow light mode in slab-type photonic crystal waveguides, *Opt. Lett.* **31**, p. 50.

Vlasov, Y. A. *et al.* (2005). Active control of slow light on a chip with photonic crystal waveguides, *Nature* **438**, p. 65.

Wahab, M. A. (ed.) (2008). *Dynamics and Vibration: An introduction* (Wiley, West Sussex).

Wang, L. J., Kuzmich, A., and Dogariu, A. (2000). Gain-assisted superluminal light propagation, *Nature* **406**, p. 277.

Wang, S., Zhang, J., Wu, F., Tian, W., Dai, J. N., Tian, Y., and Chen, C. Q. (2016). Long-range surface plasmon polaritons for efficient optical coupling in AlGaN/GaN quantum well infrared photodetector, *Plasmonics* **11**, pp. 833–838.

Waterman, P. C. and Pedersen, N. E. (1986). Electromagnetic scattering by periodic arrays of particles, *J. Appl. Phys.* **59**, p. 2609.

Webb, K. J. and Thylen, L. (2008). Perfect-lens-material condition from adjacent absorptive and gain resonances, *Opt. Lett.* **33**, p. 747.

Wegener, M., García-Pomar, J., Soukoulis, C., Meinzer, N., Ruther, M., and Linden, S. (2008). Toy model for plasmonic metamaterial resonances coupled to two-level system gain, *Optics Express* **16**, p. 19785.

West, P. *et al.* (2010). Searching for better plasmonic materials, in *Laser and Photonics Reviews* (Willey), p. 795.

Wiltshire, M. C. *et al.* (2001). Microstructured Magnetic Materials for RF Flux Guides in Magnetic Resonance Imaging, *Science* **291**, p. 849.

Wong, H. M. K. (2011). *Extending Plasmonics in Semiconductors to Higher Operating Frequencies*, Master's thesis, University of Toronto.

Working Group 1, COST-16 (1989). Comparison of different modeling techniques for longitudinaly invariant integrated optical waveguides, *IEE Proc. J.* **136**, p. 273.

Wu, B. I. *et al.* (2003). Guided modes with imaginary transverse wavenumber in a slab waveguide with negative permittivity and permeability, *J. Appl. Phys.* **93**, p. 9386.

Wu, W., Bonakdar, A., and Mohseni, H. (2010). Plasmonic enhanced quantum well infrared photodetector with high detectivity, *Appl. Phys. Lett.* **96**, p. 161107.

Wuestner, S. and Hess, O. (2014). Active optical metamaterials, in E. Wolf (ed.), *Progress in Optics*, Vol. 59 (Elsevier B.V.), pp. 1–88.

Wuestner, S., Pusch, A., Tsakmakidis, K., Hamm, J., and Hess, O. (2010). Overcoming losses with gain in a negative refractive index metamaterial, *Phys. Rev. Lett.* **105**, 12, p. 127401.

Wuestner, S., Pusch, A., Tsakmakidis, K., Hamm, J., and Hess, O. (2011). Gain and plasmon dynamics in active negative-index metamaterials, *Phil. Trans. R. Soc. A* **369**, pp. 3523–3550.

Xiang, C. and Wang, J. (2013). Long-range hybrid plasmonic slot waveguide, *IEEE Photonics Journal* **5**, p. 480031.

Xiao, S. *et al.* (2010a). Loss-free and active optical negative-index metamaterials, *Nature* **466**, p. 735.

Xiao, S. *et al.* (2010b). Loss-free and active optical negative-index metamaterials, *Nature* **466**, p. 735-738.

Xu, C. L. *et al.* (1994). Full-vectorial mode calculations by finite difference method, *IEE Proc. Optoelectron.* **141**, p. 281.

Xu, L., Liu, Y., Yao, F., and Liu, S. (2019). Surface plasmon nanolaser: principle, structure, characteristics and applications, *Appl. Sci.* **9**, p. 861.

Yao, P. *et al.* (2009). Ultrahigh Purcell factors and Lamb shifts in slow-light metamaterial waveguides, *Phys. Rev. B* **80**, p. 195106.

Yariv, A. and Armstrong, D. (1973). Traveling wave oscillations in the optical region: A theoretical examination, *J. Appl. Phys.* **44**, p. 1664.

Yee, K. S. (1966). Numerical Solution of Inital Boundary Value Problems involving Maxwell's Equations in Isotropic Media, *IEEE Trans. Antennas Propagat.* **14**, p. 302.

Yoshie, T., Loncar, M., Okamoto, K., Qiu, Y., Shchekin, O., Chen, H., Deppe, D., and Scherer, A. (2004). Photonic crystal nanocavities with quantum well or quantum dot active material, in A. Abidi, A. Scherer, and S.-Y. Lin (eds.), *Photonic Crystal Materials and Devices II*, Vol. 5360 (Proc. SPIE, Bellingham), p. 16-23.

Zayats, A., Smolyaninov, I., and Maradudin, A. (2005a). Nano-optics of surface plasmon polaritons, *Physics Reports* **408**, pp. 131–314.

Zayats, A. V., Smolyaninov, I. I., and Maradudin, A. A. (2005b). Nano-optics of surface plasmon polaritons, *Phys. Rep.* **408**, p. 131.

Zhang, H., Zhu, J., Zhu, Z., Jin, Y., Li, Q., and Jin, G. (2011). Surface-plasmon-enhanced GaN-LED based on a multilayered M-shaped nano-grating, *Optics Express* **21**, pp. 13492–13501.

Zhang, J., Zhang, L., and Xu, W. (2012). Surface plasmon polaritons: physics and applications, *J. Phys. D: Appl. Phys.* **45**, p. 113001.

Zhang, Q., Li, G., Liu, X., Qian, F., Li, Y., Sum, T., Lieber, C., and Xiong, Q. (2014). A room temperature low-threshold ultraviolet plasmonic nanolaser, *Nat. Commun* **5**, p. 4953.

Zhang, S. *et al.* (2005). Experimental Demonstration of Near-Infrared Negative-Index Metamaterials, *Phys. Rev. Lett.* **95**, p. 137404.

Zhang, Y., Saxena, D., Aagesen, M., and Liu, H. (2019). Towards electrically-driven semiconductor nanowire lasers, *Nanotechnology* **30**, p. 192002.

Zhao, Y., Belov, P., and Hao, Y. (2009). Accurate modelling of left-handed metamaterials using a finite-difference time-domain method with spatial averaging at the boundaries, *J. Opt. A: Pure and Applied Optics* **9**, p. S468.

Zheng, J., Nicovich, P. R., and Dickson, R. M. (2007). Highly fluorescent noble-metal quantum dots, *Annu. Rev. Phys. Chem.* **58**, p. 409-431.

Zhou, J., Koschny, T., and Soukoulis, C. M. (2008). An efficient way to reduce losses of left-handed metamaterials, *Opt. Exp.* **16**, p. 11147.

Zhou, J. *et al.* (2005). Saturation of the Magnetic Response of Split-Ring Resonators at Optical Frequencies, *Phys. Rev. Lett.* **95**, p. 223902.

Zia, R. and Brongersma, M. L. (2007). Surface plasmon-polariton analogue to Young's double-slit experiment, *Nature Nanotechnology* **2**, pp. 426–429.

Zia, R., Chandran, A., and Brongersma, M. L. (2005a). Dielectric waveguide model for guided surface polaritons, *Optics Letters* **30**, pp. 1473–1475.

Zia, R., Schuller, J. A., and Brongersma, M. L. (2006). Near-field characterization of guided polariton propagation and cutoff in surface plasmon waveguides, *Phys. Rev. B* **74**, p. 165415.

Zia, R., Selker, M. D., and Brongersma, M. L. (2005b). Leaky and bound modes of surface plasmon waveguides, *Phys. Rev. B* **71**, p. 165431.

Ziolkowski, R. W. (2003). Pulsed Gaussian beam interactions with double negative metamaterial slabs: errata, *Opt. Exp.* **11**, p. 1596.

Ziolkowski, R. W. and Heyman, E. (2001). Wave propagation in media having negative permittivity and permeability, *Phys. Rev. E* **64**, p. 056625.

Zuloaga, J. (2011). *Quantum Plasmonics: A first-principles investigation of metallic nanostructures and their optical properties*, Ph.D. thesis, Rice University, Houston, Texas.

Zuloaga, J., Prodan, E., and Nordlander, P. (2009). Quantum description of the plasmon resonances of a nanoparticle dimer, *Nano Lett.* **9**, pp. 887–891.

Index

CPSIA information can be obtained
at www.ICGtesting.com
Printed in the USA
BVHW010312290822
645559BV00002B/8